RENEWABLE ENERGY POLICY CONVERGENCE IN THE EU

Global Environmental Governance

Series Editors: John J. Kirton, Munk Centre for Global Affairs, Trinity College, Canada and Miranda Schreurs, Freie Universität Berlin, Germany

Global Environmental Governance addresses the new generation of twenty-first century environmental problems and the challenges they pose for management and governance at the local, national, and global levels. Centred on the relationships among environmental change, economic forces, and political governance, the series explores the role of international institutions and instruments, national and sub-federal governments, private sector firms, scientists, and civil society, and provides a comprehensive body of progressive analyses on one of the world's most contentious international issues.

Recent titles in the series (full listing continued at the back of the book)

Renewable Energy Policy Convergence in the EU

The Evolution of Feed-in Tariffs in Germany, Spain and France

DAVID JACOBS
Freie Universität Berlin, Germany

Routledge
Taylor & Francis Group

LONDON AND NEW YORK

First published 2012 by Ashgate Publishing

Published 2016 by Routledge
2 Park Square, Milton Park, Abingdon, Oxfordshire OX14 4RN
711 Third Avenue, New York, NY 10017, USA

First issued in paperback 2016

Routledge is an imprint of the Taylor & Francis Group, an informa business

British Library Cataloguing in Publication Data
Jacobs, David, 1978 Dec. 11–
    Renewable energy policy convergence in the EU : the evolution of feed-in tariffs in Germany, Spain and France. – (Global environmental governance series)
    1. Electric utilities–Government policy–Germany. 2. Electric utilities–Government policy–Spain.
    3. Electric utilities–Government policy–France. 4. Renewable energy sources–Government policy–Germany. 5. Renewable energy sources–Government policy–Spain. 6. Renewable energy sources–Government policy–France. 7. Renewable energy sources–Economic aspects–Germany. 8. Renewable energy sources–Economic aspects–Spain. 9. Renewable energy sources–Economic aspects–France. 10. Renewable energy sources–Law and legislation–International unification.
    I. Title II. Series
    333.7'9323'094–dc23

Library of Congress Cataloging-in-Publication Data
Jacobs, David, 1978 Dec. 11–
    Renewable energy policy convergence in the EU : the evolution of feed-in tariffs in Germany, Spain and France / by David Jacobs.
      p. cm. – (Global environmental governance)
    Includes bibliographical references and index.
    ISBN 978-1-4094-3909-7 (hbk. : alk. paper) 1. Energy policy–European Union countries. 2. Renewable energy sources–Government policy–European Union countries. 3. Energy development–Environmental aspects–European Union countries. 4. Electric utilities–Rates–European Union countries. I. Title.
    HD9502.E82J33 2011
    333.79'4094–dc23

                                                                                    2012003877

ISBN 13: 978-1-138-27165-4 (pbk)
ISBN 13: 978-1-4094-3909-7 (hbk)

Contents

List of Figures

List of Tables

Acknowledgements

This book is based on a research project conducted at the Freie Universität Berlin, Germany (D 188). I am especially thankful to Assistant Professor Dr Danyel Reiche, PD Dr Lutz Mez and Professor Miranda Schreurs for their helpful guidance. I would also like to thank the Deutsche Bundesstiftung Umwelt for their financial support. I also would like to express my special gratitude to those who reviewed the manuscript, namely (in alphabetical order) Dr Mischa Bechberger, Thomas Chrometzka, Toby Couture, Guillaume Duclos, Dr Michael James, Dr Helge Jörgens, Markus Lohr, Benjamin Miethling, Dr Volker Oschmann, Fabian Pause, Wilson Rickerson, Sibyl Steuwer and Fabian Zuber. One anonymous reviewer also provided useful suggestions for improvement. Finally, I express my appreciation for the support for all those experts in Germany, Spain, France and Belgium who agreed to be interviewed. This project would not have succeeded without their cooperation. (The views expressed in the interviews do not necessarily represent the official policy of any particular company, institution, or agency. Individuals spoke entirely – and only – on their own behalf.)

Acknowledgements

List of Acronyms

ADEME	Agence de l'Environnement et de la Maîtrise de l'Energie (French Environment and Energy Management Agency)
AEE	Asociación Empresarial Eólica (Spanish Wind Energy Association)
AGEE-Stat	Arbeitsgruppe Erneuerbare Energien-Statistik (German Working Group for Renewable Energy Statistics)
ALDE	Alliance of Liberals and Democrats for Europe (European Parliament)
ASIF	Asociación de la Industria Fotovoltaica (Spanish Industry Association for Solar Photovoltaics)
APPA	Asociación de Productores de Energías Renovables (Spanish Renewable Energy Association)
BDEW	Bundesverband der Energie- und Wasserwirtschaft (German Energy and Water Association)
BEE	Bundesverband Erneuerbare Energie (German Renewable Energy Association)
BGBl	Bundesgesetzblatt (German Federal Law Gazette)
BMU	Bundesministerium für Umwelt, Naturschutz und Reaktorsicherheit (German Federal Ministry for the Environment, Nature Conservation and Nuclear Safety)
BMWi	Bundesministerium für Wirtschaft und Technologie (German Federal Ministry of Economics and Technology
BSW	Bundesverband Solarwirtschaft (German Solar Industry Association)
BWE	Bundesverband Wind Energie (German wind Energy Association)
CDU	Christlich Demokratische Union Deutschlands (Christian Democratic Union Germany)
CEER	Council of European Energy Regulators
CHP	Combined Heat and Power
CIRED	Centre de Recherche sur l'Environnement et le Développement (French Research Centre for the Environment and Development)
CLER	Comité de Liaison Energies Renouvelables (French Liaison Comittee for Renewable Energies, NGO)
CNE	Consejo National de Energía (Spanish National Energy Council)
CRE	Commission de Régulation de l'Énergie (French Commission for Energy Regulation)

CSP	Concentrated Solar Power
DBFZ	Deutsches Biomasse Forschungszentrum (German Biomass Research Centre)
DG	Diretorate-General (European Commission)
DG Tren	Directorate-General Energy and Transport (European Commission)
DGEMP	La Direction Générale de l'Énergie et du Climat (Devision of French Ministry MEEDDAT)
DIW	Deutsches Institut für Wirtschaftsforschung (German Institute for Economic Research)
DLR	Deutsches Zentrum für Luft- und Raumfahrt (German Aerospace Centre)
DSO	Distribution System Operator
ECN	Energy Research Centre of the Netherlands
EDF	Électricité de France
EDF-EN	Électricité de France – Énergies Nouvelles
EEG	Erneuerbare-Energien-Gesetz (German Renewable Energy Source Act)
EFET	European Federation of Energy Traders
EnBW	Energie Baden-Würtemberg (German utility)
EnWG	Energiewirtschaftsgesetz (German Energy Industry Act)
EPIA	European Photovoltaic Industry Association
EPP-ED	European People's Party-European Democrats (European Parliament)
EREC	European Renewable Energy Council
EREF	European Renewable Energy Federation
EU	European Union
EU Commission	Commission of the European Union
EU Council	Council of the European Union
EU Parliament	Parliament of the European Union
Eurelectric	Union of Electricity Industy
EWEA	European Wind Energy Association
EWI	Energiewirtschaftliches Institut an der Universität zu Köln (Institut of Energy Industry and Economic Theory)
FDP	Freie Demokratische Partei (German Free Democratic Party)
FFU	Forschungsstelle für Umweltpolitik, Freie Universität Berlin (Environmental Policy Research Centre)
FIT	Feed-in tariff
GO	Guarantee of Origin
HWWA	Hamburgische Welt-Wirtschafts-Archiv (Hamburg Institute of International Economics)
IDEA	Instituto para la Diversifición y Ahorro de Energía (Spanish Institute for Energy Diversification and Savings)
IEA	International Energy Agency

IFIC	International Feed-in Cooperation
IFNE	Ingenieurbüro für Neue Energien (German Consulting Engineers for New Energy)
IRR	Internal rates of return
ISET	Institut für Solare Energieversorgungstechnik (Institute for Solar Energy System Technology)
J.O.	Journal Officiel de la République Française (Official Journal of the Republic of France)
MEEDDAT	Ministère de l'Écologie, de l'Énergie, du Développement durable et de l'Aménagement du territoire (French Ministry of Ecology, Energy, Sustainable Development and Sea)
MEEDDM	Ministère de l'Écologie, de l'Énergie, du Développement durable et de la Mer (French Ministry of Ecology, Energy, Sustainable Development and Sea)
MINEFI	Ministère de l'Économie, des Finances et de l'Industrie (French Ministry of Economy, Finance and Industry)
MITYC	Ministerio de Industria, Turismo y Comercio (Spanish Ministry of Industry, Tourism and Commerce)
NGO	Non-Governmental Organisation
NREL	National Renewable Energy Laboratory (US Department of Energy)
PPI	Programmations Pluriannuelles des Investissements (French Pluriannual Plan for Investment in the Power Sector)
PER	Plan de Energías Renovables para España (Spanish Renewable Energy Plan)
REE	Red Eléctrica de España (Spanish transmission system operador)
RECS	Renewable Energy Certificates System
REN21	Renewable Energy Policy Network for the 21st Century
RES	Renewable Energy Sources
RES-e	Electricity produced from Renewable Energy Sources
RTE	Gestionnaire du Réseau de Transport d'Électricité (French transmission system operator)
SER	Syndicat des Énergies Renouvelables (French Renewable Energy Industry Association)
SPD	Sozialdemokratische Partei Deutschlands (German Social Democratic Party)
SRU	Sachverständingenrat für Umweltfragen (German Advisory Council for the Environment)
TAC	Transfer Accounting Certificate
TGC	Tradable Green Certificates
TREN	Transport and Energy (Directorate General of the European Commission)
TSO	Transmission System Operator

UBA	Umweltbundesamt (German Federal Environmental Agency)
UNESA	Asociación Española de la Industria Eléctrica (Spanish Industry Association for Electricity Producers)
VDEW	Verband der Elektrizitätswirtschaft (German Association of the Electricity Industry, now BDEW)
WFC	World Future Council

Chapter 1

Introduction

This book deals with the coordination of renewable energy policies in the European Union (EU). It analyses the evolution of feed-in tariffs in Germany, Spain and France in order to understand the reasons for policy convergence in this area in the EU. Why are national support instruments becoming more similar over time even though the harmonisation of these national support frameworks at the European level has failed? Which causal mechanisms lead to cross-national policy similarities? And what are the implications for policy coordination options in the EU?

This book was written with both empirical and theoretical research interests in mind. Empirically, the study analyses policies supporting renewable electricity in the EU. Currently, member states of the EU are free to choose the support instrument they prefer to increase the share of renewable electricity. Between 1999 and 2009, two attempts by the European Commission to harmonise these financial support instruments failed. Interestingly, recent assessments of support instruments in the EU have nonetheless revealed substantial similarities between national support instruments despite this failure of harmonisation. This apparent contradiction raised the author's interest. Why and how have national support policies become more alike over time even though 'top down' harmonisation from the European level has been unsuccessful?

This question leads us to an interesting theoretical discourse in the discipline of political science: international policy coordination. An increasing number of scholars are analysing cross-national policy similarities and potential ways of increasing policy convergence within the EU. Researchers in comparative politics have identified a number of causal mechanisms which lead to cross-national policy similarities – even without coercive measures being implemented at a higher political level. In European studies, scholars have elaborated on the impact of European politics at the national level, and also focus on alternative steering modes which work without legally binding measures.

In this book we argue that legally binding, top-down harmonisation is not necessarily required to coordinate national support policies for renewable electricity in the EU. Increasing national similarities can also be achieved through voluntary cooperation between member states and the horizontal diffusion of specific policy design features. Our analysis of renewable energy policy convergence in the EU is intended to shed a new light on theoretical research in comparative politics and European studies on international policy coordination.

This book will therefore be of interest to students and scholars in political science (comparative politics and European studies) and practitioners and experts in the field of renewable energy policies. With respect to the latter, actors involved in the political decision-making process at European level (European Commission, European Parliament, industry associations, non-governmental organisations, and so forth) will certainly find this research project interesting. In addition, national policymakers can acquire information from this project about support conditions for renewable electricity in other member states and thus obtain information of how to further improve national support mechanisms.

1.1. The Theoretical Interest: European Policy Coordination

International policy coordination – 'mutual adjustment of the interests, goals, and actions of collective actors in the international system' (Busch and Jörgens 2009: 21) – has become an increasingly important research focus over the last decades. Political scientists in comparative politics and European studies have tried to answer the question why national policies are becoming increasingly similar over time and what role international organisations and other actors in the international arena play in this respect.

In European studies, scholars originally assumed that European institutions were able to shape national policies and that a common market would exert pressure on national policymakers to implement similar policies (Héritier et al. 1996, Mény et al. 1996). However, Europeanisation studies soon revealed that the impact of the EU on national policy design was limited (Héritier and Knill 2001). Instead of finding homogeneity, researchers detected a 'patchwork' of regulatory policies (Héritier 1996, Knill and Lenschow 1998). In European studies, most scholars have tried to identify the influence of the European level on the national level. Generally it was assumed that the stricter and the more binding the steering from the European level was, the more strictly national governments complied with these international rules (Héritier et al. 1996, Börzel 1999). However, this is not always the case. Analysing the impact of European policies on national administrations, Knill (2001) points out that the less hierarchical and the less binding the European regulation, the more likely were similarities at cross-national level to emerge.

From a more policy advice-oriented perspective, scholars of Europeanisation have also elaborated on alternative and less hierarchical modes of European policy coordination – in case top-down harmonisation based on binding regulations is difficult to achieve. Given the complexity of classical forms of decision making in the EU (Peterson and Bomberg 1999), political scientists and policymakers have tried establish new modes of policymaking.[1] These new modes of policy coordination in the EU include self-regulation of private actors, co-regulation of

---

1 A comprehensive collection of publications on new modes of governance is available from the Integrated Project on New Modes of Governance (NEWGOV) (http://www.eu-newgov.org/).

public and private actors, and voluntary cooperation among public actors. Their primary advantage is in avoiding lengthy and complicated decision-making processes. Besides, new modes of policy steering are expected to meet less resistance from actors in a given policy area as they do not involve strict regulatory requirements. They are guided by three major principles: voluntarism, subsidiarity and the inclusion of a wide range of actors engaged in a policy subsystem (Héritier 2002b).

In comparative politics, in contrast, researchers started out looking for cross-national differences. In fact, in comparative politics policy diversity is a general supposition, hypothetically reflecting different prevailing national ideas, national interests and national institutions. However, these researchers were surprised to find many similarities in policies and programmes among European member states (Busch and Jörgens 2009). Comparative studies have mostly relied on three distinct but interrelated theories to explain increasing cross-national policy similarities and policy coordination, namely, theories of policy convergence, policy diffusion and policy transfer. All three theories purport to explain cross-national policy similarities, even though with slightly different analytical foci. In this book we make use primarily of policy convergence theory (see Chapter 2).

There is little dispute about the definition of policy convergence as 'the tendency of policies to grow more alike, in the form of increased similarity in structures, processes and performances' (Drezner 2001: 53, Kerr 1983: 3). Knill has extended this basic definition in order to operationalise policy convergence theory for empirical research. He defines policy convergence as any increase in the similarity between one or more characteristics of a certain policy (for instance, policy objectives, policy instruments, policy settings) across a given set of political jurisdictions (supranational institutions, states, regions, local authorities) over a given period of time. Policy convergence thus describes the end result of a process of policy change over time towards some common point, regardless of causal process. (Knill 2005).

These causal processes, or causal mechanisms, have been identified by political scientist over the last two decades. The analysis of the causes of policy convergence started with recurring empirical findings. Large-scale research projects that analysed a large number of case studies came up with first indicators of the factors driving cross-national policy convergence (Bennet 1991, DiMaggio and Powell 1991, Dolowitz and Marsh 1996, Drezner 2001, Hoberg 2001). Conclusions were drawn not only from convergence theory but also from related research concepts such as policy diffusion (Holzinger and Knill 2008). The theory of policy convergence has been applied to number of policy areas, including social policy, fiscal policy, trade policy, health policy, banking regulation and environmental policy (Heichel et al. 2005). Especially in the field of environmental policies, researchers have tried to identify causal mechanisms in a more structured way since 2000. State-of-the-art policy convergence theory is currently dealing with five causal mechanisms.

One the one hand, convergence is based on mechanisms mandating certain policies or programmes based on top-down regulation, namely, harmonisation and imposition. On the other hand, the theory has identified a number of causal mechanisms that lead to cross-national policy similarities based on voluntary cooperation. Here, three causal mechanisms have been identified, namely, transnational communication, regulatory competition and independent problem solving (Holzinger and Knill 2005a, Holzinger et al. 2008a). These are of special interest for us, since they might help us explain the convergence of national policies for supporting renewable electricity in the EU even though the attempted harmonisation of support instruments in the past was not successful. The analysis of support mechanisms for renewable electricity in the EU is an ideal case for our research interest. We are confronted with a situation where all major European institutions wish to achieve better coordination of renewable electricity policies in the future. However, top-down harmonisation does not seems to be a feasible political option for steering, since already two harmonisation attempts by the European Commission have failed. Therefore, alternatives to the traditional forms of policy steering in the EU might help to overcome the existing policy dilemma.

1.2. The Theoretical Research Gap

Since this book is based on a comparative analysis, its main objective is to test and potentially extend the theory of policy convergence – the theoretical basis of the study. First, the extension of the existing theoretical framework is intended by applying this theory to a policy area which has so far largely been ignored by convergence researchers: the (renewable) energy sector. The closest that researchers have come to support of renewable electricity is the analysis of policy convergence in environmental policies (for instance, Botcheva and Martin 2001, Hoberg 2001, Kern et al. 2000, Busch and Jörgens 2005a, Holzinger and Knill 2005a, Holzinger et al. 2008a, Busch and Jörgens 2009). In addition, Haase (2008a) has applied convergence theory to the liberalisation of European gas markets.

Second, research projects operating with policy convergence theory are now mostly based on quantitative analyses that include a large number of case studies. In contrast, very few studies include a detailed qualitative analysis of the process leading to transnational policy similarities over time. Qualitative analyses have generally been applied in research projects applying the theory of policy transfer (Heichel and Sommerer 2007). However, a qualitative analysis – such as that undertaken by this research project – is crucial for further elaborating the role of and interaction between the causal mechanisms that have been identified so far. This is possible by applying a micro-level focus, assessing the development of detailed design options of just one specific policy mechanism in various countries over time. In addition, this very detailed analysis of a single policy instrument is crucial for revealing the limits of voluntary policy coordination.

Third, the present research project is one of the first to apply the theory of policy convergence to a policy instrument that seeks to promote technological innovation, contrary to regulatory policies which try to prohibit certain behaviours – as is often the case with environmental policies. It will be interesting to observe whether a theory that was largely developed on the basis of command and control policies will also be applicable to market-based policy instruments that operate with financial incentives.

Even though this book is based on a comparative analysis, and largely relies on policy convergence theory, its results could well be of interest also to scholars in the field of European studies. Jordan underlines that policy convergence theory has the potential to shed new light on active debates in EU studies 'such as those concerning Europeanization, globalization, policy transfer and "new" modes of co-ordination' (Jordan 2005: 994). As long ago as 1991 Bennett predicted that policy convergence would be a focal point of European policy research, as the logic of a single market 'forces convergence in a number of policy areas and political processes' (Bennett 1991: 215).

Jordan (2005) suggests that policy convergence theory could be an important ingredient in the search of new governance mechanisms in the EU based on voluntary cooperation. Although top-down harmonisation is still the dominant governance mode in the EU (Knill and Lenschow 2005), different steering modes might be a political alternative, especially in issue areas where harmonisation seems to be difficult to achieve. If we bear in mind that harmonisation of European support mechanisms is a tool for reaching certain policy objectives and not an objective itself, the desired outcome might well be achieved through alternative, less hierarchical steering options. This way, our project will contribute to research into international policy coordination through non-binding, vertical policy diffusion (Busch and Jörgens 2009).

1.3. The Empirical Interest: Renewable Electricity Policies in the European Union

Empirically, this book deals with support mechanisms for renewable electricity in the EU. We analyse the evolution of feed-in tariffs in three European member states (Germany, Spain and France) in order to understand increasing cross-national policy similarities in the EU – despite the failure of harmonisation. By identifying the causal mechanisms behind international policy convergence we hope to shed new light on policy coordination options in the EU.

The promotion of renewable energy sources is a major pillar of European energy policy (see Chapter 3). The European Parliament and the European Commission started to support the wider use of renewable energy sources in the 1990s (EU Commission 1996, 1997), followed by the first European directive for renewable electricity in 2001 (EU 2001). Simultaneously, the EU started to liberalise electricity markets; the first directive for the liberalisation of energy markets was enacted in 1996 (EU 1996), followed by an acceleration directive in 2003 (EU 2003). Since one of the main objectives of market liberalisation was to establish a level playing field in the European energy sector, the European institutions sought to establish a single harmonised European support mechanism for all member states.

However, the European Commission, the European Parliament and the member states could not agree on a single common support instrument for the entire EU. Between 1999 and 2009 two attempts on the part of the European Commission to (partly) harmonise these financial support instruments failed. Therefore, each country has the power to select the support instrument of its choice. Currently, a number of support instruments are applied in the EU, including feed-in tariffs, quota obligations based on tradable green certificates, and auctioning mechanisms. Nonetheless, both the European Commission and the European Parliament still aim to achieve greater policy coordination in the future. Both institutions are targeting a harmonised European support instrument in the long term (EU Parliament 2007, EU Commission 2008a). However, in the short and medium terms the European institutions seek to foster international policy coordination through cooperation between member states and optimisation based on international best practice (EU Commission 2005a, 2008a, EU Parliament 2007).

So far, the issue of policy coordination of renewable electricity support in the EU has been approached almost exclusively from a top-down perspective. This one-sided research focus largely reflects the political objective of the major European institutions. As indicated above, both the European Parliament and the European Commission aim to harmonise renewable energy support mechanisms in the long term. This ambition has influenced the research agenda in this issue arena. The European institutions have been able to influence the research agenda not only indirectly, by proclaiming certain policy objectives. In addition, the European Commission has co-financed several international studies analysing support instruments in the EU, for instance the Optres project (Ragwitz et al. 2007a), the Realise project (Arentsen et al. 2007a) and the Futures-e project (Resch et al. 2009).

In 1995–2001, when the European Commission was promoting a harmonised European support instrument based on a quota obligation and certificate trading – another support mechanism for renewable electricity – researchers focused on this coordination option (Menges 1998, Höpner 1999, Berry and Jaccard 2000, Morthorst 2000, 2003, Drillisch 2001, Espey 2001b, Gual et al. 2001, Timpe et al. 2001, Voogt et al. 2001, Bräuer 2002, Montero 2002).

After the first harmonisation attempt based on certificate trading failed in 2001, researchers mostly focused on a comparison between different support mechanisms. These comparative analyses were also in line with the policy objectives of the European institutions. In 2001, the European Commission did not fully abandon its intention to harmonise support instruments. On the contrary, harmonisation was only postponed. According to Article 4 of the renewable electricity directive 2001, a future proposal for a harmonised support instrument would be based on an analysis of the effectiveness and efficiency of different support instruments. Therefore, it is not surprising that most research analyses focused on a comparison between support instruments in the EU – namely, feed-in tariffs and tradable green certificate schemes – analysing their effectiveness and efficiency (Espey 2001a, Lauber 2001, 2004, Bird et al. 2002, Bräuer 2002, Bechberger et al. 2003, Menanteau et al. 2003, Finon and Menanteau 2004, Sohre 2004, Bustos 2005, Grotz and Fouquet 2005, Häder 2005, Hvelplund 2005, Reiche and Bechberger 2005, Springmann 2005, Blok 2006, Dinica 2006, Harmelink et al. 2006, Held 2006, Mitchell et al. 2006, Finon and Perez 2007, Lewis and Wiser 2007, Lipp 2007, Midttun and Gautesen 2007, Butler and Neuhoff 2008, Bürer and Wüstenhagen 2009, Campoccia et al. 2009). Most researchers approached support instruments from an economic perspective. Only a small number of studies also took into account non-economic factors such as stability, participation, transparency and ownership (for instance, Mendonça et al. 2009a).

In some cases researchers have also evaluated best practice of certain support instruments (for instance, Haas et al. 2004, Huber et al. 2004, Sawin 2004, Morthorst et al. 2005, Mendonça 2007, Resch et al. 2007, Grace et al. 2008, Jacobs 2008, Klein et al. 2008, Mendonça et al. 2009b) and suggested different steps towards full harmonisation, such as the implementation of certain minimum design criteria in all European countries (Ragwitz et al. 2006a), the clustering of policy instruments with similar features (Morthorst et al. 2005) and others (van Sambeek 2002, Darflot 2004, Connor and Mitchell 2004, Jansen et al. 2005).

In its 2005 and 2008 assessments of the performance of support instruments in European member states, the European Commission concluded that 'well-adapted feed in tariff regimes are generally the most efficient and effective support schemes for promoting renewable electricity' (EU Commission 2008a: 3). Therefore, some researchers also started to assess the possibility of establishing a Europe-wide, harmonised feed-in tariff (Muñoz et al. 2007, Kratzat and Langniß 2007, Müller and Bitsch 2007, BEE 2008a, Hinrichs-Rahlwes 2008). However, other researchers have continued to assess the economic benefits of a harmonised, quota-based support instrument (Jansen and Uyterlinde 2004, Wissen et al. 2006, Fürsch et al. 2010). Very few researchers have pointed out the potential disadvantages (for instance, del Río 2005, Toke 2005).

In order to offer a more comprehensive viewpoint on the harmonisation debate, in the recent years several research projects have adopted a more actor-centred research focus. Instead of merely focusing on the potential economic advantages of harmonisation, two larger-scale studies have analysed the opinions of actors in certain European member states with respect to different support instruments and the issue of harmonisation (Arentsen et al. 2007a, b, Resch et al. 2009). Several political scientists have adopted this actor-centred focus in analyses of different coalitions of actors in favour of either feed-in tariffs or alternative support mechanisms (Brand-Schock 2010, Jacobsson et al. 2009, Nilsson et al. 2009, Bechberger 2009, Hirschl 2008).

1.4. The Empirical Research Gap

In several ways, this book sheds new light on the debate over how to coordinate renewable electricity support mechanisms in the EU. First, the project analyses alternative modes of policy coordination that are based not on legally binding, top-down harmonisation but on voluntary cooperation that works through horizontal policy diffusion. Comparing the development of national support mechanisms in three countries allows us to analyse international interdependencies and consequently the reasons for increasing policy similarities in Germany, Spain and France. This is one of the first research projects to analyse the evolution of national support mechanisms over time. Until now, studies have generally depicted a snapshot of national support instruments at a given moment of time. In contrast to this rather static analysis of previous research, our dynamic analysis over a long period of time will enable us to depict the evolution of national support policies.

Second, instead of comparing various support mechanisms, this project focuses on a single support mechanism, namely, feed-in tariffs. Our qualitative research approach and the micro-level perspective of the project – the analysis of all feed-in tariff design options in Germany, Spain and France – allows us to take detailed account of the evolution of national policy. On the one hand, this is important for identifying cross-national policy convergence and potentially the spread of international best practice. On the other hand, it enables us to identify the barriers to the spread of certain design features.

1.5. Research Question

Empirically, the main objective of this book is to assess the development of national feed-in tariffs and to establish whether and why they have become more similar over time even without legally binding harmonisation. We wish to identify the causal mechanisms leading to cross-national policy similarities and eventually to assess the implications for policy steering options in the EU.

At start, a more descriptive analysis of feed-in tariff development over time is needed in order to track the development of national feed-in tariffs. Even though some research projects and reports have pointed to the convergence of European support policies (Busch 2003, EU Commission 2008a), no rigorous assessment has so far been executed. By comparing the individual feed-in tariff design options that have developed over time in three EU member states, we will try to answer our first research question:

Q1: Are feed-in tariffs becoming more similar over time?
Once we have shown that feed-in tariffs have indeed become more similar over time, we will try to explain this cross-national policy convergence by examining the reasons for national policy modifications and international interdependencies. This will help us to answer our second research question:

Q2: Why are feed-in tariffs becoming more similar over time?
Answering these empirical research questions should enable us to contribute to the theoretical debate on international policy coordination in political science. We aim to acquire an in-depth understanding of the causal mechanisms leading to cross-national policy similarities in order to contribute to the extension of policy convergence theory. In addition, the research project is intended to reveal the functioning of less hierarchical, voluntary modes of policy steering in the EU, thus contributing to related research in European studies.

1.6. A Note on Terminology

In order to avoid confusion, a precise definition of terms related to policy steering in the EU is crucial. The terminology of this book is closely related to the terminology used by European institutions in the debate about support mechanisms in the EU (see, for instance, EU 2001, EU Parliament 2007, EU Commission 2005a, 2008a). When we speak of 'policy coordination' we refer to all forms of policy steering, including both mandatory and voluntary mechanisms and measures. When we want to refer to voluntary policy coordination, that is, mechanisms that work without legally binding top-down regulations, we use the term 'cooperation'. In this case, EU member states implement policies unilaterally, unconditionally, and based on horizontal diffusion. When we want to refer only to binding and coercive steering based on top-down regulation, we use the term 'harmonisation'. In this case, obligatory and binding rules are established at EU level, thus mandating policies in the member states. It should be noted that when other researchers in political science use these terms they sometimes refer to different concepts (see, for instance, Busch and Jörgens 2009, Knill and Lenschow 2005).

1.7. Preview of Chapters

This book consists of a total of nine chapters. The exploration of whether and why feed-in tariffs in EU countries are becoming more similar over time is structured as follows. Chapter 2 sets out the theoretical framework, explaining the state-of-the-art policy convergence theory. In this chapter we present the mechanisms that researchers have identified as explaining cross-national policy similarities. We focus on the three causal mechanisms that are based on voluntary cooperation and function without coercive measures, namely, transnational communication, regulatory competition and independent problem solving.

Chapter 3 is of a more descriptive nature and has the purpose of familiarising the reader with the general empirical framework of support for renewable energy in the EU and its member states. The chapter focuses on the two European directives for renewable energy from 2001 and 2009 and the reasons for the failure of harmonisation within the European political decision-making process. It also introduces national support mechanisms in the EU.

In the following four chapters we compare the evolution of different feed-in tariff design options in Germany, Spain and France over time. In Chapter 4 we analyse the evolution of design options for basic feed-in tariffs. These design options define the fundamental rules under which renewable electricity producers can sell electricity to the grid and which suffice to establish a basic but comprehensive feed-in tariff scheme in a given territory. In Chapter 5 design options for tariff differentiation and adjustment in the three countries are compared. The primarily purpose of these design options is the reduction of windfall profits or overcompensation for renewable electricity generators. In Chapter 6 we contrast the design options for improved system and market integration in the three countries under investigation. Those design options are generally implemented to improve the 'quality' of electricity supply, that is, electricity supply that better matches electricity demand and enhances grid stability. In Chapter 7 tariff level developments, that is the levels of tariff payment, are compared. Because of the complexity of the tariff payment regimes in the EU member states, we focus on only one technology, solar photovoltaics. In each of Chapters 4–7, the evaluation of design options over time is followed by a preliminary analysis of the effective causal mechanisms of policy similarities.

In Chapter 8, the preliminary results of the previous comparisons are structured and depicted in more detail. In Chapter 8 we basically test our hypotheses by analysing the reasons for cross-national policy similarities based on voluntary and unilateral policy adjustments. In short, we test how far transnational communication, regulatory competition and independent problem solving have resulted in increasing similarities of feed-in tariffs in Germany, Spain and France over time. In the final Chapter 9, we summarise the major findings of this research project, from both theoretical and empirical perspectives.

Chapter 2

The Theoretical Framework:
Policy Convergence and Related Theories

This book deals with the coordination of renewable energy policies in EU member states. We analyse the evolution of feed-in tariffs in Germany, Spain and France in order to understand the reasons for increasing policy similarities in the EU. Why are national support instruments becoming more similar over time, even though attempts at EU harmonisation of these support frameworks has failed?

The theory of policy convergence has identified a number of causal mechanisms that lead to cross-national policy similarities, some of them even without coercive measures. These mechanisms are threefold: transnational communication, regulatory competition and independent problem solving. We analyse how far these mechanisms have been responsible for policy similarities of feed-in tariffs in different member states of the EU. This way, we analyse how far these mechanisms can contribute to decentralised policy coordination in the EU. So far the issue of coordinating renewable electricity support mechanisms has been discussed from a top-down perspective based on strict hierarchical steering in a multi-level governance system. The approach adopted by this research project reverses the angle and looks at horizontal policy learning. This approach might offer new answers to the question of how to coordinate to renewable energy support in the EU.

2.1. Policy Convergence

Policy convergence has been described as 'one of, if not the, central questions addressed by comparative policy studies' (Goggin et al. 1990: 20). The first publication on cross-national policy convergence dates back to the 1960s.[1] But this research area has become increasingly popular especially since the 1990s, particularly in the field of environmental policy research (see Heichel et al. 2005). Cross-national policy convergence has been analysed from both a comparative

---

1   From a social science perspective, convergence theory is much older. Weinberg (1969) pointed out that classical sociological theorist, including Tocqueville, Marx and Weber, operated with the assumption that cross-national similarities in political and economic systems reflect the impact of international industrialisation.

politics perspective and an international relations perspective.[2] Depending on the discipline's research focus, policy convergence has been approached from contrary starting points and with different methodologies (see also Holzinger and Knill 2008, Busch and Jörgens 2009).

Early studies in the field of comparative politics generally focused on national factors in policy choices and variations, such as the institutional framework, actor constellations and regulatory styles. Therefore, these studies assumed and found diversity in cross-national policy comparison (see, for example, Lundqvist 1974). Later it became apparent that national policies were growing more and more alike. Jänicke was among the first to realise that global interdependencies led to policy learning, convergence and diffusion, which largely determine a country's capacity to tackle environmental problems (Jänicke 1992, 1996a).[3] In the following period, convergence of environmental policies was systematically tested on the basis of large-N studies. In the field of international studies, researchers found a theoretical ground for the assumption of cross-national convergence in an international environment. Most prominently, researchers have hypothesised about an international 'race to the bottom' with respect to social standards and environmental protection (see also the next section). Even though this hypothesis has been confirmed by empirical evidence in the case of international tax rate development (Genschel 2002, Winner 2005, Genschel and Ganghof 2008), it has been challenged by various other empirical studies in the field of environmental policies (Vogel 1995, 1997, Jänicke et al. 1997, Drezner 2001, Jänicke 2005, Holzinger et al. 2008a).[4]

2.1.1. Causes of Cross-National Policy Convergence

The problem of demonstrating causality is not unique to policy convergence research – it complicates all forms of comparative political analysis (Peters 1998). However, policy convergence theory has made considerable progress in recent years in identifying causal mechanisms of cross-national convergence. Mechanisms are recurring processes which create links between causes and effects

---

2   Several authors have published review articles on policy convergence theory, including Bennett (1991), Drezner (2001), and Heichel and Sommerer (2007). These papers offer a deeper analysis of the evolution of policy convergence theory.

3   Later, Jänicke and Weidner (1997) extended their research on environmental policy convergence by systematically comparing policies in 30 industrialised and developing countries.

4   Despite the apparent economic and political pull factors towards harmonisation and homogenisation, researchers in the field of Europeanisation studies have detected a 'patchwork' of institutions, instruments and policy styles, 'differential' impacts on national legislation (Héritier 1996, Héritier and Knill 2001), and only limited convergence. According to Jordan and Liefferink, the content, structure and style of environmental policies in the EU have only marginally converged. Their analysis shows that the 'national decision-making procedures, coordination mechanisms and political traditions ... have remained essentially the same in spite of EU member-ship' (Jordan and Liefferink 2005: 111).

(Mayntz 2002a). In other words, they can be defined as 'the causal pathway, process or intermediate variable by which a causal factor of theoretical interest is thought to affect an outcome' (Gerring 2008: 163, quoted in Busch and Jörgens 2009: 38).

The current state-of-the-art policy convergence theory – the theoretical basis for this research project – distinguishes two types of causal mechanism though which governments coordinate policies internationally.[5] First, there are mechanisms which oblige national legislators to implement or modify certain policies based on coercive measures and top-down regulation. Second, there are causal mechanisms which rely on unilateral and voluntary policy adjustments, thus leading to cross-national policy similarities based on horizontal exchange (Drezner 2001, Hoberg 2001, Holzinger and Knill 2004, 2005a, Holzinger et al. 2008a, Busch and Jörgens 2009).[6] Altogether, five different causal mechanisms have been identified (Figure 2.1) – imposition and harmonisation (based on coercive measures and

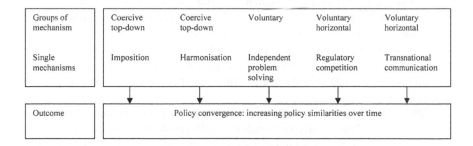

Figure 2.1   Causal mechanisms of policy convergence

Source: Author, adapted and modified from Holzinger et al. (2008)

---

5   Busch and Jörgens (2009) have differentiated three different groups, namely cooperation, coercion and diffusion. In this research project only two groups are differentiated (coercion and cooperation) because this reflects the language used by the European Commission in the debate on European support instruments.

6   This section largely draws on the definitions and concepts provided by Holzinger and Knill (2008), one of the most comprehensive recent studies on policy convergence theory. Since policy convergence theory is still rather young, researchers are not all operating with the same set of causal mechanisms when explaining policy convergence. As well as the causal factors applied in this research project, other researchers have identified such mechanisms as the influence of different policy styles (Vogel 1987) and cultural similarities (Lenschow et al. 2005). Holzinger et al. (2008) have summarised these additional causes of policy convergence with the term 'facilitating factors'. These more general facilitating factors often act as intervening variables, but are not taken into account by the present research project due to its micro-level focus and the limited number of cases. As well, three rather homogeneous countries have been chosen for the case studies. Therefore, these facilitating factors (for instance, cultural and socio-economic similarities) can be set aside.

top-down regulation) and independent problem solving, regulatory competition, and transnational communication (based on voluntary cooperation and horizontal interdependencies) (Levi-Faur 2005).

Imposition occurs when, based on asymmetries in politics or economic power, some countries or international actors force other countries to implement a certain policy (Bennett 1991, Dolowitz and Marsh 1996). Countries that try to impose a certain policy or programme try to export their own values and principles. Either the legislators of the country where a certain policy is implemented because of imposition will avoid confrontation with the politically or economically stronger imposing country, or the imposing country offers certain benefits once the weaker country has adopted this policy. Voluntary elements which influence the national decision-making process are practically non-existent (Busch and Jörgens 2009).[7]

Harmonisation refers to the implementation of policies at the national level as mandated by international or supranational law. Governments are obliged to comply with certain international rules which they have agreed to in multilateral negotiations, and therefore implement similar policies. This obligation to comply with international rules is generally connected to their membership of international organisations (Holzinger and Knill 2008). In this case, national governments hand over certain competencies to a higher political level in order to solve a common problem (Drezner 2001). In the case of harmonisation, the degree of legal specification largely determines the convergence effects (Holzinger and Knill 2005a).

With independent problem solving, different national governments implement similar policies independently as a result of parallel problem pressure. In this case, policy convergence occurs voluntarily and almost coincidentally. Holzinger and Knill (2008: 49) point out that this phenomenon has also been discussed under different names, including 'technocratic' or 'technological determinism' (Bennett 1988: 417, Rose 1991: 9), 'clustering' (Simmons and Elkins 2004) and 'parallel domestic pressure' (Hoberg 2001: 127). Some researchers argue that independent problem solving has to be taken into account in convergence studies in order to 'avoid the pitfall of inferring from transnational similarity of public policy that transnational explanations must be at work' (Bennett 1991: 231). Other researchers, however, have considered independent problem solving a 'null hypotheses', since no causal relations regarding the process of the implementation of similar policies in different countries are explained (Simmons and Elkins 2004: 172).

Regulatory competition emerges from increasing economic interdependencies at regional level or worldwide level. Regulatory competition is thought to lead to policy convergence because countries are subject to competitive pressure and therefore adjust their policies or programmes in similar ways (Garrett et al. 2008, Holzinger et al. 2008a). Indeed, regulatory competition requires economic integration. Convergence theory suggests that regulatory competition is based on

---

7   Busch and Jörgens (2009) use the more encompassing term 'coercion' to describe this and related causal mechanisms.

the economic integration process within the EU but also in other regions of the world, resulting from the abolition of trade barriers and the free movement of goods, capital, services and workforces. Regulatory burdens that might restrict the competitiveness of national industries are removed since economic actors can threaten to shift their economic activities to places with more favourable conditions (Holzinger and Knill 2008). On the other hand, national legislators try to establish attractive framework conditions in order to attract investors from other countries. This type of regulatory competition is usually termed 'economic competition'.[8]

The state-of-the-art theory of regulatory competition does not provide a clear answer concerning the direction of convergence, that is, convergence 'at the top' or 'at the bottom' (Botcheva and Martin 2001). In the case of production processes, early theoretical assessments often assumed convergence towards the least common denominator (the 'bottom'), since strict national regulations were expected to lead to higher production costs and eventually to competitive disadvantages for the national industry in international markets. National industry can threaten to shift production facilities to countries where such regulations do not exist (Drezner 2001, Hoberg 2001, Simmons and Elkins 2004, Knill and Lenschow 2005).[9] This convergence to the least common denominator has also been called 'the race-to-the-bottom hypothesis'.

However, this race-to-the-bottom hypothesis has been refuted by many empirical studies. In the case of product standards, the direction of convergence is less predictable. If policymakers in countries with high standards of regulation manage to enforce stricter standards internationally, for example through the establishment of trade barriers, there will likely be convergence at the top (Holzinger and Knill 2008). Similarly, Jänicke (2005) pointed to the existence of 'innovative races' in the field of environmental policies. This race-to-the-top hypothesis has also been confirmed by other researchers (Vogel 1995, Kern et al. 2000). Holzinger and Sommerer (2008) pointed out that in the case of trade-irrelevant standards there will mostly likely be a variety of standards with no specific direction of convergence.

Besides this focus on economic competition resulting from economic linkages, researchers have identified the influence of competition in the political sphere. The term 'political competition' was introduced by theorists of Europeanisation (see Héritier et al. 1996). By implementing national policy innovations in a certain sector, individual member states are trying to actively shape the European

---

8   The concept of regulatory competition is related to economic theories of systems competition and regulatory competition (Oates and Schwab 1988,Long and Siebert 1991). Contrary to the normative focus of researchers in the field of economics, political scientists tried to assess whether regulatory competition actually works and whether it will lead to convergence at the top or at the bottom (Holzinger and Sommerer 2008).

9   Lazer (2001) states that in this case jurisdictions are in a classical Prisoner's Dilemma and therefore national governments will implement less-demanding regulations in order to be able to compete with low-regulating countries.

regulation in accordance with their own policy styles and regulatory traditions, thus minimising the political and economic costs of policy adjustment. For instance, if a country is pushing for the implementation of a certain policy instrument for renewable energy support at European level and this policy instrument already exists at national level, no or only minimal changes in the national policy framework will become necessary. The primary objective of member states in arguing for the implementation of their national policy at European level is 'the desire to preserve national problem-solving traditions and institutions – hence minimising the cost of legal adjustment to European legislation' (Héritier et al. 1996: 2). If a critical mass of countries has implemented a certain policy or programme, previously reluctant countries will no longer be able to ignore them, and eventually the promotion of such policies at international level might force them to implement similar programmes, even if they collide the with national regulatory traditions (Busch et al. 2005, Busch and Jörgens 2009).

Transnational communication covers a variety of mechanisms leading to increasing cross-national similarities, including lesson drawing from policies in other countries, joint problem solving, emulation and international policy promotion. All of them are based on theories of international cooperation, and their common ground is that they are 'purely based on communication among countries' (Holzinger and Knill 2008: 43). It should be noted that communication between actors is of course also of importance in the case of harmonisation and regulatory competition. In transnational communication, however, communication among actors is the main driver of cross-national policy convergence.

In the case of lesson drawing, governments use the experience of certain policies or programmes elsewhere and consciously solve the domestic problem accordingly. The concept was introduced by Rose, who defined it as a voluntary process whereby one government learns from another government's programme and draws an action-oriented conclusion (Rose 1991). Lessons can be drawn from policies or programmes in different places (for example, cities, regions, countries) or at different times. This is true for 'positive lessons' where transfer takes place as well as for 'negative lessons' where no transfer takes place since the policy did not bring about the desired effects. Therefore, lesson drawing does not necessarily involve a change of behaviour (Rose 1991: 7). Lesson drawing can take place by mere copying, but it might also lead to hybrid systems or domestically modified designs. In contrast to lesson drawing, where communication takes place on a bilateral level in one direction, transnational problem solving assumes a joint international effort to solve a common problem and the subsequent implementation at national level. Key actors in transnational problem solving are transnational elite networks or epistemic communities. They usually share a set of normative and principled beliefs, causal beliefs about means and common standards of accruing and validating knowledge (Haas 1992).[10] Emulation is based on the desire for conformity. Instead of extensive research on various solutions to a problem, policies or programmes are generally copied from other countries (Holzinger and Knill 2008). There are several reasons for this process. First, the mere fact that a large number of political entities have already applied a certain policy or programme will persuade policymakers to do the same. In other words, policymakers will think, 'better to join it than to fight it' (Levi-Faur 2002: 34). Second, emulation can increase the social legitimacy of one organisation in a wider social and institutional environment (DiMaggio and Powell 1991: 70).[11] In order to identify emulation of policies, there should be a clear point of reference (for instance, a state with an innovative policy), a proved awareness of this policy, and a similarity in the goals, content and instruments of this policy or programme (Bennett 1991: 223). Third, within an international system actors wish 'not to be left behind' (Tews 2002a, Holzinger and Knill 2008: 46). According to DiMaggio and Powell (1991: 69), this process occurs especially in cases where there is still little knowledge about the policy innovation and its potential effects.[12] Fourth, rational motivations such as time pressure (Bennett 1991: 223) or the cost of acquiring information (Tews 2002a: 180) might facilitate policy emulation. Finally, policies or programmes are promoted by international institutions and thus exert pressure on countries to become active. Here, in contrast to the aforementioned 'communicational' drivers of policy convergence, international institutions take an active role. Kern (2000:144) concludes that cross-national policy transfer is often facilitated through the promotion of a certain policy or programme by international institutions, which usually make use of non-binding international agreements, overall goals or standards, reviewing and identifying 'best practice' (Humphreys 2002: 54, Tews 2002a: 174). This process can be enhanced by either governmental or societal actors – for example, the United Nations (UN), the Organisation for Economic Cooperation and Development (OECD), the EU or international non-governmental organisations like Greenpeace or World Wide Fund for Nature

---

10   However, international institutions often provide a platform for the formation of epistemic communities. Transnational problem solving can take place in those communities without the need for coercive measures at international level (Holzinger and Knill 2008). Therefore, Kern (2000) argues that international organisations play an important role in providing a framework for transnational policy learning, even without establishing binding international rules.

11   Bennett points out that emulation is generally hard to identify, just as 'plagiarism in student essays is exceedingly difficult to prove'(Bennett 1991: 223). Therefore, the identification of policy emulation often requires some personal judgement.

12   DiMaggio and Powell (1991) use the term 'mimetic isomorphism' rather than 'emulation'.

(WWF) (Kern 2000). These actors can be mediators of policy diffusion in a multi-level framework (Kern et al. 2000).[13]

2.2. Supplementary Theories: Policy Diffusion and Transfer

The section above described five causal mechanisms leading to cross-national policy convergence. In order to fully understand potential process of policy convergence, the reader of this study needs to be introduced to two supplementary theories – namely, theories of policy diffusion and policy transfer. All of these theoretical concepts try to explain why national are becoming more and more similar over time from different perspectives.

As shown above, the question of how to reach decentralised policy coordination in international systems is becoming increasingly important in political science research. Based on the empirical observation of policies becoming increasingly similar in an international context, three different but related theoretical concepts have been developed: policy diffusion, transfer and convergence (Holzinger et al. 2007b).[14] Despite their common approach of analysing cross-national similarities, the three theories have different analytical foci. While policy convergence focuses on the outcome of cross-national policy development, policy transfer and policy diffusion centre on the process leading to such similarities. In this book, policy diffusion and transfer theory are applied in order to answer the question of why feed-in tariffs are becoming more similar over time – in line with our research questions formulated in section 1.5. These theories will be crucial for the qualitative analysis of the policy learning process leading to cross-national policy similarities in European member states. Tracing this process is crucial to identifying causality. Yet, these theories will not be used to formulate further hypotheses.

However, the strict theoretical line between 'process' (policy diffusion and policy transfer theories) and 'effects' (policy convergence theory) is often blurred in empirical research projects as soon as we want to know the reasons for cross-national policy similarities. In order to identify causality, the process leading to policy convergence needs to be analysed. If this research project were to find out

---

13   In many cases, national forerunners also promote their own innovative policies via international organisations. If one country is successful in depicting the advantages of a certain policy measure or programme, this policy might eventually be promoted by international institutions (Héritier et al. 1996). This aspect is discussed above under 'regulatory competition'.

14   Another closely related research field is isomorphism, which differs from policy convergence only by focusing solely on organisational structures. DiMaggio and Powell (1991) define isomorphism as a process of homogenisation that 'forces one unit in a population to resemble other units that face the same set of environmental conditions'. Since a structural focus is not intended in this research project, this theory is not considered further.

only whether feed-in tariffs are converging (Q1), it would be possible to take policy convergence as the sole theoretical framework. However, we also want to learn about the reasons for policy convergence (Q2). Accordingly, it is not enough to focus merely on the effects of cross-national policy developments. In order to understand the causal factors of policy convergence, the policy process has to be analysed. This explains why we will have to 'back up' the theory of policy convergence with other, related theories, namely theories of policy transfer and policy diffusion.

Policy transfer research is considered to be a sub-field of comparative politics (see Rose 1991, Dolowitz and Marsh 1996) and can be defined as 'the process by which knowledge about policies, administrative arrangements, institutions and ideas in one political system (past or present) is used in the development of policies, administrative arrangements, institutions and ideas in another political system' (Dolowitz and Marsh 2000: 5).[15] This transfer process is based on voluntary cooperation. Generally, policy transfer studies seek to explain the 'underlying causes and contents of singular processes or bilateral policy exchange' (Holzinger et al. 2008a: 12). In contrast to policy convergence, therefore, the analytical focus of policy transfer is on the process, not on the results. It is possible that policy transfer leads to policy convergence; but this is not necessarily the case, as major changes of specific policies, administrative arrangements, institutions and ideas can occur within the process of transfer (Evans and Davies 1999, Kern et al. 2000, Holzinger et al. 2008a). For this research project, policy transfer theory will help to explain a bilateral, single process of policy transfer in the development of the national feed-in tariff.

A first, broad definition describes policy diffusion as the spread of policies, programmes and ideas amongst different political systems, or 'any pattern of successive adoptions of a policy innovation' (Eyestone 1977: 441). However, most scholars have adopted a more restrictive definition of diffusion which depicts it as 'the process by which an innovation is communicated though certain channels over time' (Rogers 2003: 5). Thus, the study of policy diffusion also focuses on the process rather than on the outcome (Elkins and Simmons 2005). In contrast to Eyestone's broader definition, Rogers's definition assumes a direct connection between the implementation of one and the same policy or programme in different political systems. Accordingly, policy diffusion takes place when a policy decision by one government influences similar decisions taken by other governments (Simmons and Elkins 2004). This concept of uncoordinated interdependence suggests that policymakers are not legally bound to implement a certain policy or programme but that they consider the actions taken by other governments (Elkins and Simmons 2005).

Policy diffusion scholars, however, disagree on how policies are spread and communicated in an international environment (Braun and Gilardi 2006,

---

15   In the process of policy transfer, major changes can occur to specific policies, administrative arrangements, institutions and ideas (Kern et al. 2000).

Braun et al. 2007). One group of researchers considers all possible modes of 'communication' to be part of the policy diffusion theory, including voluntary adoption, legally binding harmonisation, and imposition through external actors (see, for example, Radaelli 2005). Another group argues that the term 'policy diffusion' should be restricted to the voluntary adoption of ideas, programmes and policies (see Kern 2000, Jörgens 2004, Busch and Jörgens 2005a, Elkins and Simmons 2005, Tews 2005a, Holzinger and Knill 2008, Busch and Jörgens 2009). Accordingly, Tews defines diffusion as 'the spreading of innovations due to communication instead of hierarchy or collective decision-making within international institutions' (Tews 2005b: 65), thus excluding from this definition the obligatory or coercive spread of policies. Therefore, policy diffusion has many features in common with the voluntary causal mechanisms of policy convergence. It occurs with the 'absence of formal and contractual obligation' (Busch and Jörgens 2009: 32). This study relies on this more restrictive definition of policy diffusion. Accordingly, policy diffusion will serve as a theoretical background explaining the voluntary process leading to cross-national policy similarities.

In sum, policy convergence, policy diffusion and policy transfer are three related theoretical concepts. All three concepts are based on the empirical observation that in the case of many issue areas national policies are becoming increasingly similar. The concepts all try to explain why programmes, policies and ideas resemble each other in different political systems. Despite this common point of departure, the three theoretical concepts can be differentiated. As shown above, policy convergence describes the outcome of cross-national policy development. Consequently, this theoretical concept is required to answer the research question 'are feed-in tariffs becoming more similar over time?' (Q1). The analytical focus of policy diffusion and policy transfer is the process.[16] Therefore, these theories are required to answer the research question 'why are feed-in tariffs becoming more similar over time?' (Q2).[17] Policy transfer analysis focuses on a singular process of policy transfer, usually a bilateral one. Policy diffusion, by contrast, assumes a macro-level perspective and focuses on general patterns of policy adoption (Holzinger et al. 2008a). The interaction of policy convergence theory with policy diffusion and transfer theory is depicted in Figure 2.2.

---

16   Due to the similar focuses of policy diffusion and policy transfer, some authors frequently equate the two theoretical concepts (Kern 2000,Tews 2002a).

17   Policy diffusion has sometimes been identified as an independent causal mechanisms leading to policy convergence (see, for example, Busch and Jörgens 2005c,2009,Tews 2005b). However, this research project relies on the five aforementioned causal factors (imposition, harmonisation, transnational communication, regulatory competition and independent problem solving). Policy diffusion and policy transfer are theoretical concepts explaining the process that may (or may not) lead to policy convergence. However, they are not considered to be independent, additional causal mechanisms. Radaelli (2005) has clarified the interdependencies of policy convergence and policy diffusion. He points out that there might be cases where diffusion does not lead to policy convergence.

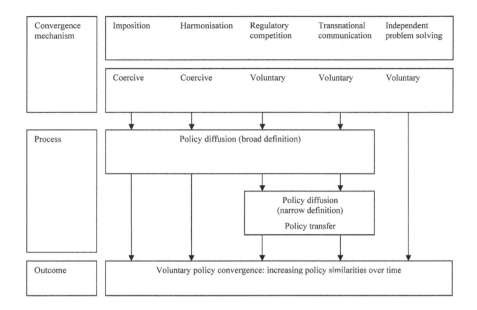

Figure 2.2   Interaction of policy convergence, diffusion and transfer

Source: Author, adapted and modified from Holzinger and Knill (2008)

2.3. Hypotheses

The partial inclusion of supplementary theories, namely policy diffusion and policy transfer, serves as a vehicle to establish a theoretical background for the qualitative analysis of feed-in tariff development over time, tracing the process which leads to cross-national policy similarities. This qualitative approach allows us to better understand the causal mechanisms leading to increasing policy similarities. Nonetheless, policy convergence theory remains the primary theoretical approach of this research project. It identifies a number of causal factors explaining cross-national policy convergence, that is, the increasing similarity of the policies or programmes of different countries over time.

As shown above, state-of-the-art policy convergence theory operates with five causal mechanisms: harmonisation, imposition, independent problem solving, regulatory competition and transnational communication. Of the five, two are based on coercive, top-down policy coordination and only three (regulatory competition, transnational communication and independent problem solving) are based on voluntary and horizontal policy learning. As shown in Chapter 3, the harmonisation of support instruments for renewable electricity has failed, and so coercive measures cannot explain the existing cross-national policy similarities.

Therefore, in this research project we focus on the three remaining convergence mechanisms which are based on voluntary cooperation.

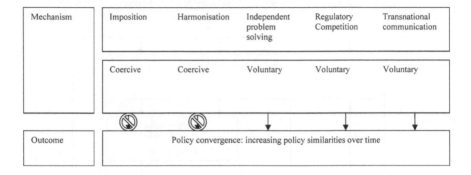

Figure 2.3   Causal factors for policy convergence

Source: Author, adapted and modified from Holzinger et al. (2008)

In order to assess state-of-the-art policy convergence theory, the following three hypotheses will be tested. Feed-in tariffs in Germany, France and Spain are becoming more similar over time because of

- transnational communication (H1)
- regulatory competition (H2)
- independent problem solving (H3)

2.4. Research Design

Most policy convergence studies analyse a large number of cases and therefore apply a quantitative research approach (see, for instance, Holzinger et al. 2008b, Haase 2008).[18] In contrast to these large-N studies, this research project is based on a micro-level, qualitative analysis of only a limited number of cases (three

_____

18   In large-N studies, quantitative assessments of policy convergence have been used to guarantee comparability of completely different policies or even policy types and to come up with general conclusions about convergence or divergence tendencies. Holzinger et al. (2008b), for example, compare the development of environmental policies as diverse as 'sulphur content in gas oil', 'efficient use of waste in industry', and 'glass reuse/recycling targets'. Clearly, in this case it makes sense to develop certain numerical indicators that allow for quantitative comparability in order to determine whether national environmental policies have become more similar over time. However, the objective of this

countries). This research project analyses the development of all the design features of just one policy instrument over time, thus adopting a micro-level perspective. As well, we are interested in the process and not only the results of policy convergence, and therefore a qualitative analysis is crucial – which, in turn, forces us to limit the number of cases that can be analysed. We need to track the process of policies becoming more similar over time, which absolutely requires a qualitative research approach – including the analysis of a large number of actors and institutions which are crucial for facilitating policy learning. Data collection is based on archival research, textual analysis, field research and expert interviews.

---

research project is not the comparability of the policies and general statements about policy convergence in the field of energy policies.

Chapter 3

The Empirical Framework:
EU Support Instruments for Renewable
Electricity and the Failure of Harmonisation

This chapter is largely descriptive. It gives an overview of the framework for renewable energy support in the EU, describing the support mechanisms applied in the EU and analysing why the attempted European harmonisation of these national mechanisms failed. First, this chapter familiarises the reader with the different support instruments for renewable electricity. Even though feed-in tariffs (FITs) are applied by most countries, it is important to understand how other instruments basically function in order to interpret actors' positions when they argue either in favour or against one or another instrument. We introduce the reader to the most widely used support instruments for renewable electricity, namely, tendering schemes (section 3.1.1), quota-based mechanisms (section 3.1.2), feed-in tariffs (section 3.1.3) and fiscal incentives (section 3.1.4). In the subsequent section we describe the two unsuccessful attempts to harmonise these support instrument in the years preceding the adaptation of the two renewable energy directives in 2001 and 2009 (see section 3.2 and 3.3). The arguments for and against harmonisation will be summarised in section 3.4. Finally, section 3.5 will outline the detailed analytical approach to the comparison of feed-in tariffs in Germany, Spain and France that follows in the Chapter 4–7.

3.1. Support Mechanisms for Renewable Electricity in the European Union

A variety of support instruments are applied in EU member states. The first national support policies were already implemented in the 1970s and 1980s, long before the intention to implement a single European support scheme was first tabled at European level in the mid-1990s. Support policies have been classified both from a theoretical point of view (for instance, Héritier 1987, Braun and Giraud, Tews and Jänicke 2005, Mickwitz et al. 2007) and with respect to renewable electricity support (for instance, Bechberger et al. 2003, Meyer 2003, IEA 2004, Johansson and Turkenburg 2004, Resch et al. 2005, Morthorst et al. 2005, Haas et al. 2008, Mendonça et al. 2009b). The European Commission classifies support mechanisms in two different categories: price-based market instruments and quantity-based market instruments. The latter consist of quota obligations and tendering mechanisms, while price-based support schemes include feed-in tariffs and fiscal incentives (EU Commission 2005a, 2008a).

3.1.1. Tender Schemes

Tender or auctioning mechanisms are quantity-based support instrument where the legislator issues a call for tender, that is, auctions a certain amount of renewable electricity capacity. The financial support can be based on either the total equipment cost (support per kW installed) or the power generation cost per electricity unit (support per kWh produced). The bidder that requires the lowest level of necessary financial support generally wins the tender. In theory, auctioning capacity should guarantee power production at the lowest possible cost. Tender procedures are usually technology-specific. For instance, the legislator can ask power producers to make bids for 100 MW of wind energy and 20 MW of solar photovoltaics.

Tender schemes have been used in a number of EU member states, including the UK (until 1999) and Ireland (until 2005). As of 2010, this support instrument is still applied in Denmark for offshore wind energy. France also made use of this support mechanism from 1996 to 2000. Today, calls for tender are still issued in exceptional circumstances in France[1] (Resch et al. 2005, Ragwitz et al. 2006).

3.1.2. Quota Obligations

Under a quota obligation the legislator requires electricity suppliers to provide a certain share of electricity from renewable energy sources. The supplier can either produce electricity itself or buy it from other green electricity producers. In order to increase the flexibility of the system, in many countries the supplier is also allowed to make up the required share by trading certificates, which serve as proof of compliance with legal obligation. Therefore, these mechanisms are also called tradable green certificate (TGC) schemes in the EU.[2]

---

1   The French feed-in tariff mechanism operates with only technology-specific programme-size caps (see section 4.2.3), that is, tariff payment is guaranteed only until a certain amount of capacity has been installed in the country. Therefore, the legislator sets out technology-specific targets within multi-year investment plans for the French electricity system (Programmation pluriannuelle des investissements, PPI). If the technology-specific targets specified within the PPI are not reached, the government has the option of auctioning the remaining capacity (J.O. 2000a; 2003a). In France, the tender scheme is also used to gain information about the generation costs for certain technologies. The French government issued a call for tender and later fixed the feed-in tariff rates in line with the project proposals of the bidders. This 'two-part FIT' approach has also been proposed by Lesser and Su for the Californian feed-in tariff scheme (Lesser and Su 2008).

2   In the US and other parts of the world TCG schemes are often called renewable portfolio standards (RPS), as supply companies are obliged to provide a certain share of the electricity portfolio from renewable energy sources. RPS mechanisms sometimes operate without certificate trading.

The electricity supplier can obtain certificates either by itself producing renewable electricity or by buying certificates on the national certificate market. If suppliers fall short of reaching the required share they must pay a penalty. Under this type of support mechanism renewable electricity producers have two income sources. First, they sell their electricity on the spot market for electricity at the given (hourly) price. Second, they can sell their certificates on the national green certificate market. In theory, the certificate sales compensate for 'greenness' of the electricity, that is, the positive attributes of renewable electricity compared with conventionally produced 'grey electricity'.

As of 2010, six countries in the EU applied this type of support mechanism. Quota obligations were first implemented in the UK, Italy and Belgium in 2002. Sweden followed in 2003 and Poland in 2004. Romania was the last EU country to implement this support instrument, in 2007 (EU Commission 2005a, 2008a, Resch et al. 2005).

3.1.3. Feed-in Tariffs

Feed-in tariffs consist of at least three design options: a purchase obligation, a predefined (fixed) tariff level and a long duration of tariff payment. First, the purchase obligation forces the nearest grid operator to buy all renewable electricity – independent of electricity demand. Second, the renewable electricity producer is guaranteed a certain amount of money per unit of electricity produced. Third, this payment is guaranteed over a long period of time (usually 15 to 20 years), which increases investment security and ideally allows for cost recovery.

The great majority of EU member states use feed-in tariffs for renewable electricity support. As of 2010, feed-in tariffs were established in 23 out of the 27 EU member states.[3] The first European country to implement a feed-in tariff was Portugal in 1988, followed by the Netherlands (1989), Germany (1990), Italy (1992), Denmark (1993), Spain and Greece (1994), Latvia (1998), France (2000), and Austria (2002). Since 2000, most new member states of the EU have also adopted this support instrument. In 2001, the Czech Republic introduced feed-in tariff legislation, followed by Slovenia, Hungary, Lithuania (2002), and Cyprus, Estonia, and the Slovak Republic (2005). In the following years, this group of countries was joined by Ireland (2006), Bulgaria and Italy (2007). In 2010, feed-in tariffs were implemented in the UK, Finland, and Malta (Busch 2003, Bechberger and Reiche 2007, Coenraads et al. 2008a, 2008b, Mendonça et al. 2009b, REN21 2009, 2010).

---

3   By early 2010, this policy mechanism had been implemented at the national level in at least 50 countries (REN21 2010).

3.1.4. Fiscal Incentives

Fiscal incentives include rebates, tax incentives and soft loans. Rebates are used to 'buy down' the initial investment costs for renewable power equipment. In this case, the legislator pays a certain share of the total cost of each installed unit of electricity generation capacity. Tax incentives can be offered in form of tax reductions or tax credits. In addition, legislators can offer soft loans, that is, loans offering reduced interest rates (Mendonça et al. 2009b).[4]

In the EU, fiscal incentives were the primary support mechanisms for renewable energies in the 1970s, the 1980s and at the start of the 1990s (Resch et al. 2005). Today, fiscal incentives are generally applied in addition to the support instruments described above. Malta and Finland used fiscal incentives as their primary support instrument until 2010. Malta had already switched to feed-in tariffs in the same year and Finland was planning to implement a feed-in tariff for wind energy in 2011 (EU Commission 2008a, Gipe 2010b, c).

3.2. Directive 2001/77/EC: The First Harmonisation Attempt

In order to understand fully the debate on the harmonisation of support mechanisms for renewable electricity, we first need to take a closer look at the energy policy framework in the EU. It has to be noted that the EU did not have any direct competence in the energy sector until 2009. Although two of the three founding treaties of the European Economic Community were energy-related,[5] the European treaties of Rome, Maastricht, Amsterdam and Nice did not explicitly include any European competences in the field of energy. Therefore, the European institutions were able to develop a European energy policy only on the basis of other, more general competences. In this respect, the EU's role in establishing an internal European market and protecting the environment were – and still are – crucial (Matlary 1997, Brand and Witthohn 2002, Hirschl 2009). Only under the Lisbon Treaty, ratified in 2009, did the EU gain 'shared competences' in the field of energy (EU 2007).[6]

---

4   Some countries also support renewable electricity producers by allowing for accelerated depreciation of renewable energy projects (Beck and Martinot 2004).

5   The European Coal and Steel Community (ECSC) and the European Atomic Energy Community (Euratom) (Schreyer et al. 2008).

6   The competences are included in Article 194 of the 'The Treaty on the Functioning of the European Union'. The text explicitly refers to the functioning of the energy market, security of supply, the promotion of energy efficiency and renewable energies, and the interconnection of energy networks (EU 2008).

Although the first policy initiatives for the support of renewable energy sources were taken in the 1980s,[7] the debate on harmonising support instruments for renewable electricity was largely triggered by the liberalisation of European energy markets in the 1990s. The Single European Act of 1986 (EU 1986) laid the foundation for the creation of an internal market in the EU, which became effective in 1992. The concept of creating an internal market also applied to the energy sector (EU Commission 1985, 1988, Baur and Blask 2002, Oschmann 2002). Directive 96/92/EC concerning common rules for the internal market in electricity (EU 1996) was a major step towards a common European power market. Notwithstanding the lack of direct competences in the energy sector, the European Commission exerted indirect influence because it was now responsible for guaranteeing the functioning of the internal energy market.

First indications of a harmonised European support framework can be found in the White Paper of 1995 on an energy policy for the EU. The document stated that 'certain forms of energy like renewables may need to be supported initially through specific programmes or subsidies in order for them to find a place on the market' (EU Commission 1995: 18). Nonetheless, support for renewable energies should be given in a way that is least harmful for competition. Consequently, 'further policy development may need to be considered' and the organisation of national markets should be monitored closely with the aim of identifying 'possible needs for further harmonisation' (EU Commission 1995: 19). In the following years, the European Commission and the European Parliament were looking for the adequate support schemes for future harmonisation (Busch 2003).

In its 1996 Green Paper *Energy for the Future*, the European Commission criticised the fragmented and rapidly changing nature of national support mechanisms. The Commission argued for the implementation of 'a stable and Community wide framework for renewable energy sources' (EU Commission 1996: 27) in order to achieve faster market penetration of renewable energy technologies. In the same paper the Commission expressed a clear preference for quota-based support instruments by stressing the need to phase out and replace regulatory policies by 'more market oriented measures', making explicit reference to the functioning of quota-based support instruments (EU Commission 1996: 34).[8]

In 1997 the European Parliament also published its opinion on the future support framework for renewable electricity sources in the EU. But it pointed in a different direction. In a report from 1997, rapporteur Rothe from the Party of European Socialists called for the establishment of a European feed-in tariff

---

7 The European Parliament passed several resolutions to support renewable energies and the European Council included the promotion of renewable energy sources among its energy objectives (EU Council 1986, Hirschl 2008).

8 In its White Paper from 1997, the language was more diplomatic. The Commission stated that it would monitor 'closely the different schemes proposed or introduced by the member states in order to propose a Directive which will provide a harmonised framework' (EU Commission 1997: 15).

scheme. Even though the term 'feed-in tariff' was not explicitly mentioned, the report argued for a purchase obligation of green electricity and a 'minimum payment by the utilities ... which would at least cover all of the current costs of the producer' (EU Parliament 1997: 6).[9]

Despite this obvious disagreement on the 'best' support instrument among the European institutions, the European Commission continued to push for an EU-wide, quota-based support instrument, pressuring several member states to adopt this type of support instrument by promising a first-mover advantage once this support instrument was introduced Europe-wide (Lauber 2007, Agnolucci 2008). At the same time, the diffusion of feed-in tariff schemes slowed down because many legislators expected a change in the support system and feared legal insecurity (Busch 2003, Bechberger and Reiche 2007). The European Commission openly criticised the German feed-in regulation, which granted high levels of remuneration for wind energy (Hustedt 1998, Advocate General Jacobs 2000), and the German feed-in law of 1990 was referred to the European Court (*PreussenElektra vs Schlesag*) for its assumed incompatibility with the rules of the internal energy market.

The European Commission's 1999 working paper 'Electricity from renewable energy sources and the internal electricity market' increased the pressure on feed-in tariff mechanisms. The paper found that, given the shift towards electricity market competition, a move away from a 'fixed price tariff approach towards one based on trade and competition is at some stage inevitable' (EU Commission 1999: 17). In addition, the paper highlighted the distorting effects of a multitude of national support mechanisms on the functioning of a common European electricity market, and criticised the limited competition between different renewable electricity technologies under feed-in tariffs. The new energy commissioner, Loyala de Palacio, in 1999 made one more attempt to push for a harmonised European support framework based on a quota obligation (Lauber 2007). However, this political initiative failed, too, following criticism from the European Parliament, industry associations and environmental organisations (for instance, Greenpeace et al. 1999, BMU 2001).

Therefore, the next proposals for a European renewable electricity directive envisaged no policy harmonisation in the short term (EU Commission 2000a, b). Commissioner Palacio tried to take the concerns of member states and industry associations into account, allowing different support instruments to work in parallel – at least for a certain number of years (Neue Energie 2000). After an interim period of five year, the European Commission planned to table a proposal for a harmonised support mechanism which had 'to be compatible with the principles of the internal electricity market' (EU Commission 2000a: 18). Lauber argued that this formulation was another indication of the intended future implementation of a

---

9   However, the concept of a harmonised European feed-in system was rejected by the European conservative parties (EED-EP) and the liberal group (ALDE) (Neue Energie 1998).

Europe-wide, quota-based mechanism, since being 'compatible with the principle of the internal electricity market' was usually the code for this type of support mechanism (Lauber 2007: 18).

In March 2001 the European Commission had to cope with yet another indirect defeat as the European Court rejected the aforementioned complaint against the German feed-in tariff mechanisms of 1990. As a consequence, this judgement also undermined the Commission's case against the German feed-in tariff of 2000 and the Commission's intention to make explicit reference to state aid provisions within the planned directive (Lauber 2007). Finally, the European directive of 2001 accepted heterogeneity of support instruments in the short term. The European directive fixed indicative renewable electricity targets for the European member states; but – in line with the principle of subsidiarity – national legislators were free to choose any support instrument to achieve these targets. However, no later than 2005 the European Commission was expected to table 'a well-documented report on experience gained with the application and coexistence of the different mechanisms' which might be 'accompanied by a proposal for a Community framework with regard to support schemes' (EU 2001: 36).[10] Interestingly, the term 'harmonisation' was no longer explicitly used.

In 2001 and 2006, the political debate on European policy coordination for renewable electricity was largely influenced by the assessment reports of the European Commission. Under Article 3 of directive 2001/77/EC, the European Commission was required to assess the progress of European member states in reaching their indicative targets for 2010. Under Article 4 of the directive, the European Commission should assess the experience gained from the application and coexistence of the different mechanisms and possibly make suggestions for a harmonised support mechanism based on their effectiveness, efficiency, simplicity, and compatibility with the principles of the internal electricity market.

The first assessment report on target achievement in European member states was tabled in 2004 (EU Commission 2004a, b). The European Commission's analysis showed that only four member states were 'on track' to reach their targets, three of which (Germany, Spain and Denmark) were operating feed-in tariffs. In the 2006 assessment (EU Commission 2006), only three countries were perfectly 'on track' for target achievement – again, two of which (Germany and Denmark) operate feed-in tariffs.

The 2005 assessment of support instruments was first tabled in December 2005 (EU Commission 2005a, b). The report stated that in most countries the support level had been insufficient to attract investment in renewable electricity generation plants. Interestingly, and contrary to theoretical economic assumptions about the effectiveness and efficiency of support instruments (cf. EU Commission 1999), the report found that feed-in tariffs had been more successful in promoting renewable

---

10   In order to grant a certain degree of investment security for the industry, the harmonised support scheme could enter into force only seven years later, that is, in 2012 at the earliest.

energy sources. Even though quota-based systems provided significantly higher support levels, they had been less successful in triggering an increase in newly installed capacity. This was especially true in the case of wind energy.

Despite the better performance of feed-in tariffs, the European Commission's assessment report did not propose harmonisation based on this support instrument. It argued that more experience had to be gained from different support schemes and that competition among them 'could be seen as healthy at least over a transitional period' (EU Commission 2005a: 16). Therefore, the Commission wanted to foster coordination via cooperation and optimisation in the short term. In the long term, however, the report of the European Commission considered the harmonisation support mechanisms to be 'an important component of the drive towards completion of the internal electricity market' (EU Commission 2005b: 7). Finally, the report stated that the next assessment of support mechanisms should be tabled 'no later than December 2007' (EU Commission 2005a: 18). Eventually, the second assessment report was issued together with the first proposal of the new directive for renewable energy sources in January 2008.

3.3. Directive 2009/28/EC: The Second Harmonisation Attempt

The preparation for the new directive on renewable energies – the 2001 directive included targets only up to 2010 – started as early as 2007. On 10 January 2007 the European Commission presented 'an integrated energy and climate change package', the so-called Energy Package. The package consisted of 12 different reports, including a road map for renewable energy sources (EU Commission 2007b, c). Among others, in this package the European Commission called for longer-term, mandatory renewable energy targets, including for the electricity sector, the heating and cooling sector and the transport sector. Only two months later, in March 2007, the European heads of states agreed on a binding renewable energy target of 20 percent of the total final energy consumption by 2020 (EU Council 2007).[11] The European Commission was asked to prepare a new directive in order to 'put in place an integrated policy on energy combining actions at European and member states' level' (EU Council 2007: 14) – implicitly suggesting the need for an at least partially harmonised support mechanism.

The European Parliament welcomed the new political initiative and issued a related report in 2007. The so-called Thomson Report stated that before 'the ultimate objective' of harmonising support schemes for renewable energies should be envisaged, an internal European electricity market with a level playing field for electricity from all energy sources would have to be fully implemented (EU Parliament 2007). The report also stated that a harmonised support mechanism should 'assure technological diversity' and draw 'on the best practices in the

---

11   According to the Futures-e project, a target of 20 percent final renewable energy consumption by 2020 translated into a 35 percent target for RES-e (Resch et al. 2008).

Member States'. (EU Parliament 2007: 9). Both statements point in the direction of a harmonised feed-in tariff, since quota-based mechanisms generally offered technology-neutral support, and feed-in tariffs had proven to be most successful.

The first draft of the new European directive for renewable energy sources was tabled after a long delay in January 2008 (EU Commission 2008b). The delay largely arose from the difficulty of allocating the total target of 20 percent renewable energy consumption by 2020 among all 27 member states in a fair and transparent way.[12] In contrast to the first directive, the targets were no longer of a voluntary nature, and therefore the allocation of national targets was much more controversial.[13] In order to make the achievement of national targets easier for member states, the European Commission planned to implement certain flexibility mechanisms (Johnston et al. 2008). In 2007 and 2008, the debate centred mostly on the implementation of certificate trading based on guarantees of origin (GOs). This became the hook for the newly emerging harmonisation debate.

Even though the European Commission did not officially speak of the harmonisation of support mechanisms for renewable energies – officials of the Commission argued that they would not prescribe one single support instrument – the effects of Europe-wide certificate trading based on GOs would have been similar to those of implementing a harmonised, technology-neutral quota obligation (Ragwitz et al. 2007b, 2009, Toke 2008, Jacobsson et al. 2009). Therefore, the first proposal of the European Commission can be described as a second attempt at harmonisation.[14]

The implementation of flexibility mechanisms was considered to be necessary since the national targets were not exclusively set according to the renewable energy potential in the different countries. Instead, the national economic strength of a country was also taken into account when national targets were allocated to European member states. Originally, GOs had been implemented for electricity disclosure, that is, proof that a certain unit of electricity was produced via renewable energy sources (EU 2001). Now, GOs were intended to be used for the virtual trade of renewable electricity among EU member states.[15] If a member state

---

12   It was in this regard that the term 'burden sharing', as used in international GHG emission negotiations, was established. Later, proponents of renewable energies used the term 'benefit sharing' to stress the potential advantages of achieving leadership in renewable energy development, as opposed to the idea of putting a 'burden' on member states. More neutrally, the term 'target sharing' was introduced.

13   The European Spring Council 2007 had already set out the basic rules for any 'target sharing agreement' by stating that the 'different national circumstances, starting points and potentials' have to be taken into account (EU Council 2007: 25).

14   The impact assessment of the draft directive confirms this interpretation. It states that 'transferable GO regime requires national support schemes to open up to non national GOs. This would result in price convergence, regulatory reforms and eventual harmonisation of support schemes' (EU Commission 2008d: 105).

15   In October 2008, a producer and a trader of renewable energy filed a complaint with the European Commission about territorial restrictions under national support mechanisms.

were unable to produce a unit of renewable electricity within the country, it could buy a certificate from another member state in order to prove compliance with the national renewable energy target.

An early (leaked)[16] internal draft of the new renewable energy directive included non-restricted GO trading between private actors (renewable electricity producers) and governments (EU Commission 2007d). However, this leaked proposal attracted much criticism, mostly from countries operating with feed-in tariffs, as their national support mechanisms would have been undermined by such an approach (Clos and Gabriel 2008). In additional, researchers claimed that unrestricted GO trade (including trade among European renewable energy companies) would increase the costs of renewable energy support (Klessmann et al. 2007, Ragwitz et al. 2007, Fouquet and Johansson 2008). Therefore, the final draft limited the scope of GO trading. Member states were allowed to trade GOs only when they had achieved their interim targets in 2012, 2014, 2016 or 2018. Private parties had permission to trade GOs if their power plants were commissioned after the directive came into force. As well, the European Commission implemented an opt-out clause for member states that did not want to participate in certificate trading. However, this opt out was criticised because of legal uncertainties (Fouquet and Johansson 2008, Johnston et al. 2008, Müller and Bitsch 2008).

As offenders against free trade in energy, the complaint explicitly named the German, UK, Polish and Swedish support mechanisms, thus including countries with tradable green certificate (TGC) schemes and feed-in tariff schemes (EFET 2008).

16   The European Commission was internally divided over the design of GO trading for target achievement. Therefore, Nilsson et al. (2009) supposed that the documents were leaked intentionally. DG Environment emphasised the need for a support mechanism that was compatible with the European emissions trading scheme. A certificate trading scheme for renewable energy support was believed to be more compatible with the emission trading scheme. DG Competition favoured a trading mechanism based on its primary policy objective, that is, the creation of an internal market and market efficiency (Nilsson et al. 2009). Only the DG Tren was opposed to certificate trading because it believed that volatile certificate prices would undermine investment security. Besides, DG Tren acknowledged factors beyond mere macroeconomic analysis, including the promotion of local economies and public acceptance of renewable electricity support (Nilsson et al. 2009). This conflict within the European Commission also becomes apparent when one compares different documents from the European Commission. While the impact assessment of the new draft directive states that the cheapest option for achieving the 20 percent target would be a harmonised TGC scheme (EU Commission 2008c), the report on the assessment of support schemes concludes that harmonisation would be premature (EU Commission 2008a). These assessments of future renewable energy support options differ largely because different DGs have been responsible for drafting the documents. While the assessment report of European support mechanism was carried out by DG Tren, the draft directive and the attached impact assessment were a compromise between all branches within the European Commission.

The first assessment by the European Parliament's rapporteur Claude Turmes was very critical of certificate trading based on GOs. The draft report of May 2008 – which was later approved by the industry committee of the European Parliament – states that '[t]he triple function given in Article 8 to the GOs – disclosure, support accounting/trade and target accounting – generates legal difficulties and undermines national support schemes by creating a mandatory good for which trade cannot be limited' (EU Parliament 2008: 84). The report explicitly states that 'the concept favoured by the large power producers (e.g. EURELECTRIC) and the traders of electricity (EFET) to bring legal certainty by creating an EU wide renewables certificate market is not the way forward', as such a scheme would generate large windfall profits due to non-technology-specific support (EU Parliament 2008: 84). According to Rapporteur Turmes, GOs should be limited to their original purpose of electricity disclosure and not be used in order to establish an EU-wide certificate trading scheme for target compliance (EU Parliament 2008: 84). Instead of relying on GO trade, the rapporteur suggested different mechanisms for flexible target achievement.[17] First, the report suggests increasing flexibility via the statistical transfer, so called Transfer Accounting Certificates (TAC). Second, flexibility can be increased via bilateral cooperation based on target sharing or the voluntary implementation of a common support instrument. Third, the report suggests putting more emphasis on energy efficiency measures as the 'biggest and cheapest "flexibility" instrument' (EU Parliament 2008: 85).

Certificate trading based on GOs was also assessed critically by the Council of the European Union. In the first months of 2008, the Council was only calling for modifications of the proposed certificate trading scheme (EU Council 2008a). In the summer of 2008, the debate in the Council of the European Union shifted towards alternatives to GO trade. The UK was the first country to internally propose an alternative flexibility mechanism (UK Government 2008a, b). According to this (leaked) internal proposal, the UK administration favoured a trading scheme 'which does not rely on certificates' (UK Government 2008b). Instead, member states should be given flexibility for target achievement by 'statistical transfer between Member States or project-based agreements between two Member State governments for an operator to build a renewable installation in one Member State, and for the renewable energy generated by this project to count towards another Member State's share of the target' (UK Government 2008b). In this respect, the proposal was very much in line with the Turmes Report from the European Parliament. The proposal had a substantial impact on the debate in the Council of the European Union, since the UK was formerly one of the strongest proponents of certificate-based flexibility mechanisms. The shift towards alternative mechanisms, which was not proposed even by sceptical countries like Germany or Spain at the start of the debate, considerably weakened

---

17 These flexibility mechanisms have been largely inspired by the already existing regulations under the Kyoto protocol (Luhmann 2009).

the position of the European Commission and the few countries that still supported the implementation of GO trade for target compliance.

A few weeks later, at the start of June 2008, a group of three countries – Germany, Poland and the UK – further elaborated the initial UK proposal. The joint proposal included a number of flexibility mechanisms for target achievement, including statistical transfer between member states, joint projects between member states, and joint target compliance. The Council of the European Union's proposal for alternative flexibility mechanisms was the ground for the final version of the EU Directive 2009/28/EC. The option of GO trade for target compliance was explicitly excluded. The joint proposal made it clear that GOs should be used only for their original purpose of disclosure (EU Council 2008b).

As both the Council of the European Union and the European Parliament rejected the idea of certificate trading for target compliance, the Commission's proposal for GO trade was not included in the final version of the directive. The European Commission's second (hidden) attempt at harmonisation had failed. On 9 December 2008, less than one year after the proposal of the European Commission was tabled, Parliament and Council agreed on the final wording of the renewable energy directive. The directive was adopted by the December Council a few days later.[18]

3.4. Arguments For and Against Harmonisation

In the two previous sections we have traced two unsuccessful attempts to harmonise support instruments in the EU. Our analysis has revealed that the harmonisation debate was rather fragmented. A large number of actors argued either for or against the use of certain support instruments and for or against an EU-wide, harmonised system. We summarise the arguments in this section. Even though in the previous section we largely focused on governmental actors, the arguments have also been picked up by industry associations and environmental groups.

There is no common understanding of the objectives of harmonisation. Proponents of harmonisation mostly stressed the need for compatibility with the internal market rules, which would increase the effectiveness of European support and reduce the total cost of support. First, proponents of harmonisation argue that the establishment of a common European energy market requires the creation of a level European playing field in the (renewable) energy sector. National support instruments are expected to create market distortions and therefore an EU-wide solution is needed (for instance, EU Commission 1996, 1998, 1999, EU Parliament 1997).

---

18   This comparatively fast procedure, from the first proposal to the factual agreement between Commission, Parliament and Council, was due to the applied 'first reading agreement'.

Although European utilities, their industry associations and other actors frequently use this argument when pushing for harmonisation based on a quota obligation and certificate trading (Eurelectric 1999, 2007, ETSO 2003, Eurelectric and RECS 2004, Bleuel, M. and Hillebrand, B. 2005, Business Europe 2007, Eurelectric et al. 2007, EFET 2007, Power in Europe 2007, RECs International 2008), the large incumbent industry also pursues an additional 'hidden agenda'. Many European utilities have argued for the implementation of quota-based mechanisms, at both national and European levels (Hirschl 2008), since under quota-based support mechanisms primarily large actors will invest in new power generation capacity due to the system's inherent financing risk (a fluctuating electricity and market price). Therefore, utilities would have to face less competition and could make higher profits. In addition, a quota obligation implicitly sets a cap on support for renewable electricity. A prohibition on open-ended support enables utilities to protect their investment in conventional power-generation capacity (Fouquet 2008, Ragwitz et al. 2007, Jacobsson et al. 2009, Mendonça et al. 2009b).[19]

Second, harmonisation has often been expected to accelerate market penetration of renewable energy sources. In the 1990s and the early 2000s, only a few member states offered sufficient support for renewable electricity. Therefore, a harmonised support instrument was intended to force laggard countries to take action, too. This argument was advanced by both proponents of quota-based support instruments (for instance, EU Commission 1996, 1998, Drillisch 2001) and proponents of a harmonised European feed-in tariff (for instance BEE 2008a, Hinrichs-Rahlwes 2008).

Third, it is assumed that a harmonised support framework (based on quota obligations) would create competition among renewable energy sources and therefore reduce costs for the final consumer (for instance, EU Commission 1998, 1999, 2008c, Business Europe 2007). In addition, it is believed that an EU-wide support framework would make better use of the available resources. As renewable energy sources are mostly stationary and cannot be moved (with the exception of biomass), they ought to be used (and supported) in the regions with the best resource potential (for instance, EU Commission 2005a, EU Parliament 2007). In other words, wind power plants should be built at the windiest spots along the west European coastline while photovoltaics should primarily be implemented in the sunnier southerly countries (for instance, Fürsch et al. 2010, McKinsey 2010).

Opponents of harmonisation never have really formulated a clear alternative method of policy coordination; Most of the time the opponent have only reacted to the claims made by proponents of harmonisation.[20] They have usually responded

---

19　Similarly, one could argue that the hidden agenda of the European institutions' push for harmonisation is to gain more power and competences in the energy sector.

20　However, some opponents of harmonisation based on quota obligations have elaborated on the possibility of establishing a harmonised feed-in tariff scheme (Muñoz et al. 2006, Kratzat and Langniss 2007, Hinrichs-Rahlwes 2008).

to the internal market argument by pointing out that a level playing field currently does not exist in the conventional energy sector either. Therefore, critics of harmonisation frequently argue that it was premature to harmonise support mechanisms for renewable electricity (for instance, EWEA 2001, 2004, EREC 2004, 2005a, 2005b, EPIA et al. 2007, EU Parliament 2007). In fact, several reports from the European institutions have confirmed that the process of market liberalisation has not yet been concluded (for instance, EU Commission 2004c, 2005c, 2006b, 2007a, Newbery 2002). In addition, it has been argued that the share of renewable energy sources is marginal, and so interferences with the principles of an internal market have been insignificant (for instance, Advocate General Jacobs 2000, EWEA 2004, Kjaer 2004). However, some researchers have pointed out that this argument will lose its teeth as member states are expected to provide a significant share of energy demand from renewable energy sources (Sensfuss et al. 2007).

Moreover, whereas proponents of harmonisation argue that a Europe-wide support instrument could lead to cost reductions, critics argue that – depending on the type and design of the harmonised support mechanism – support renewable energy might also present Europe with higher costs in the event of harmonisation. This argument was mostly advanced during the debate on restricted or non-restricted GO trade in 2007 and 2008 (Klessmann et al. 2007, Ragwitz et al. 2007, Fouquet and Johansson 2008). Critics of harmonisation have also feared that the regulatory change would have immense disruptive effects on the development of a European renewable energy industry (for instance, EWEA 2005, Arentsen et al. 2007, Greenpeace 2008).

In addition, it has frequently been argued that the experience gained from the existing support mechanisms is insufficient to indicate the 'best' support mechanism for the entire continent. Interestingly, this argument has been primarily advanced by European institutions, mostly after they realised that harmonisation based on certificate trading and quota obligations did not enjoy the desired political support (EU Commission 2000b, 2005a, EU 2001).

Finally, criticism of harmonisation can be fully understood only when one keeps in mind that both harmonisation attempts have been based on quota obligations and/or certificate trading. Therefore, the debate on harmonisation largely reflects the confrontation between proponents of quota-based support instruments and proponents of feed-in tariffs. Supporters of feed-in tariffs have tried to protect these successful national support instruments against a European harmonisation based on other, less successful support mechanisms (for instance, Bustos 2005, Fouquet et al. 2005, Greenpeace 2008).

The confrontation between proponents of feed-in tariffs and proponents of quota-based mechanisms also reflects the choice between technology-specific and technology-neutral support. As shown above, the European Commission had frequently argued for a Europe-wide, quota-based support instrument because it was believed that this type of support would be more cost-effective as it focused on the least-cost renewable energy technologies. However, proponents of feed-in tariffs argue that that these short-term cost reductions will endanger the development of technologies that will be needed in the long term (Meyer 2003, Lauber 2004, Sandén and Azar 2005).

In fact, technology-specific support is one of the major advantages of feed-in tariffs over other support schemes, as it allows a wide range of technologies to be supported, the most cost-effective and also less mature ones (Mendonça et al. 2009b). Quota-based support is generally non-technology-specific and therefore only the least costly renewable energy technologies will be supported. This explains why harmonisation (based on quota obligations or certificate trading) has been so strongly contested by representatives of less mature technologies (EPIA 2005, 2007, 2008), whereas representatives of more mature technologies, such as wind energy, have been slightly less critical (C. Kjaer, personal communication, 11 January 2011). As indicated above, quota obligations also favour large actors, such as utilities. Therefore, it is not surprising that attempts at harmonisation have been fiercely criticised by representatives of small (renewable) power producers and by actors that are in favour of a less centralised power system (for instance, EPIA et al 2007, Greenpeace 2008, F. Thies, personal communication, 10 April 2008).

3.5. The Analysis of Feed-in Tariffs in Germany, Spain and France

Following the analysis of the European framework for renewable energy support, this section briefly introduces the analytical focus of this research project: feed-in tariffs in Germany, Spain and France. In order to analyse the development of this support mechanisms in the three countries and to detect the reasons for cross-national policy similarities, we adopt a predefined analytical approach. Even though feed-in tariffs have been implemented in the majority of EU member states, no one feed-in tariff mechanism is exactly like any other. Since around 1990, feed-in tariffs have become increasingly complex, and evince a large number of design features. Therefore, their specific design can vary significantly from one country to another. Feed-in tariff design options have been analysed by a number of researchers since 2005 (Klein et al. 2006, 2008, 2008, Mendonça et al. 2009b, Grace et al. 2008, DB Climate Change Advisors 2009, Couture et al. 2010).

For the purpose of this research project, we separate feed-in tariff design options into three groups: first, design options for basic feed-in tariffs (Chapter 4), defining the fundamental rules under which renewable electricity producers can sell electricity to the grid and which suffice to establish a basic but comprehensive feed-in tariff scheme in a given territory; second, design options for tariff differentiation and adjustment (Chapter 5), reducing windfall profits or overcompensation for renewable electricity generators; and third, design options for system and market integration (Chapter 6), improving the 'quality' of electricity supply, that is, electricity supply that better matches electricity demand and enhances grid stability. In addition, we analyse the general evolution of tariff levels of selected technologies (Chapter 7).[21] The single design options that are included under each category (see Table 3.1) are presented in more detail at the start of each of the following four chapters (4–7). Based on these design options, the development of this support mechanism in three countries over time will be analysed. Reflecting the chronological order in which feed-in tariffs were first implemented, our analysis in the following chapters always starts with Germany (1990), continues with Spain (1994), and closes with France (2000).

Table 3.1   Categorisation of feed-in tariff design options

CHAPTER 4	CHAPTER 5	CHAPTER 6	CHAPTER 7
Design options for basic feed-in tariffs	Design options for tariff differentiation and adjustment	Design options for system and market integration	Tariff levels evolution
Eligible technologies	Technology-specific tariffs	Alternative sales options (market sales, premium FIT, direct sales)	The case of solar photovoltaics
Capacity caps and targets	Size-specific tariffs		
Tariff calculation methodology	Location-specific tariffs	Tariff payment for improved system integration (grid services, demand-oriented payment)	
Duration of support Financing mechanism	Tariff degression		
Cost-sharing methodology for grid connection		Regulations for controlling power output (forecast obligation, remote-controlled output)	

Source: Author

Germany was one of the first European countries to implement a feed-in tariff scheme; and it is still using this support instrument today. No other country in the world has made continuous use of this support mechanism for such a long period of time. From an international perspective, Germany can be considered to be a pioneering country in the development and deployment of renewable energies (Bechberger and Reiche 2004, Reiche 2004).

---

21   Besides, special design options might have to be considered in the case of emerging economies or developing countries (Mendonça et al. 2009b, Jacobs and Kiene 2009, Solano-Peralta et al. 2009). These, however, are not relevant to this research project.

In Germany, the feed-in tariff mechanism has the legal status of a law.[22] This support mechanism was first implemented in 1990 (BGBl 1990), followed by five amendments in 2000 (BGBl 2000), 2004 (BGBl 2004) 2009 (BGBl 2008) and 2012 (BMU 2011).[23] These pieces of legislation serve as the major points of reference for the analysis of the evolution of the feed-in tariff in Germany. Minor modifications were incorporated in 1994 (BGBl 1994), 1998 (BGBl 1998), 2001 (BGBl 2001), 2003 (BGBl 2003), 2005 (BGBl 2005), 2006 (BGBl 2006), 2009 (BGBl 2009a, 2009b) and 2010 (BGBl 2010). These are taken into consideration where applicable.

The promotion of renewable energy has been a priority for the Spanish energy policy for more than two decades. Even though the first important legislative texts for the promotion of certain renewable energy sources were issued in the 1980s (BOE 1981a, 1981b), Spain is normally considered to have been operating a feed-in tariff mechanism since 1994 (see Ragwitz et al. 2007a, REN21 2009). The Spanish feed-in tariff mechanism does not have the legal status of a law. It is largely based on decrees which are issued by the responsible ministry. Only the several basic feed-in tariff design features, such as guaranteed grid access and the range of tariff levels, are defined within the national electricity laws of 1997 (BOE 1997b, del Río and Gual 2007, del Río 2008, Bechberger 2009).

After the first implementation of a feed-in tariff in 1994 (BOE 1994a), the Spanish support scheme saw a number of major amendments, in 1998 (1998a), 2004 (BOE 2004a) 2007 (BOE 2007c, 2007d) and 2010 (BOE 2010b, 2010c, 2010d). These pieces of legislation are also the major point of reference for our analysis of feed-in tariff design options. Minor modifications of the Spanish feed-in tariff mechanisms were implemented in 2002 (BOE 2002a), 2006 (BOE 2006a, 2006c), 2008 (BOE 2008a), 2009 (BOE 2009) and 2010 (BOE 2010a). These are taken into consideration where applicable.

In France, the legal foundation for the national feed-in tariff mechanism was laid in 2000 with the electricity law on the modernisation and evolution of public services in the electricity sector (J.O. 2000a). The electricity law of 2000, like its Spanish equivalent, established only the very basic framework conditions of the feed-in tariff mechanisms, such as the obligation to purchase renewable electricity, the definition of the eligible technologies, and grid access rules.

In the following years, more specific regulations were issued individually for each technology through legally lower-ranking *arrêtés*. These are of importance

---

22 Note that this also explains the greater availability of information regarding the different amendments of the policy. The responsible German ministry frequently has to report to the parliament; and the parliamentary minutes allow for a much more elaborate actor analysis (see Hirschl 2008, Jacobsson and Lauber 2006) than in the case of countries where the respective legislation is issued in the form of ministerial orders or decrees, such as Spain and France.

23 At the time of printing, the law has already been passed by the upper and lower house of the German parliament, however not yet published in the Bundesgesetzblatt.

for our comparative analysis in Chapters 4–7. From 2001 to 2011, these pieces of legislation were issued in three waves. The first feed-in tariffs were issued in 2001 and 2002, including regulations for wind energy (J.O. 2001b), hydro power (J.O. 2001d), biogas (J.O. 2001e), solar energy (J.O. 2002a), biogas (methanisation) and biomass (J.O. 2002b, 2002c), and geothermal (J.O. 2002d). In June 2005 the French Ministry announced revisions of feed-in tariffs for renewable electricity (Rigaud 2006), leading to new legislation for wind power (J.O. 2006a), biogas (J.O. 2006c), geothermal (J.O. 2006e), solar power (J.O. 2006f), and hydro power (J.O. 2007a). In 2009, 2010 and 2011 – the third wave of feed-in tariff regulation in France – new tariff regimes were drafted for organic waste (J.O. 2009b), solar energy (J.O. 2010a, 2010d, 2011), and geothermal (J.O. 2010b).

Chapter 4

The Evolution of Design Options for Basic Feed-in Tariffs

The detailed analysis of basic feed-in tariff design features seeks to identify national policy similarities in order to answer our research questions as presented in section 1.5: Are feed-in tariff becoming more similar, and if so, why? The extended time frame of the analysis (more than two decades) and the micro-level analysis – the detailed analysis of many design features of just one policy instrument – will enable us to identify causal factors for policy similarities. When identifying reasons for policy similarities, we will focus on our three hypotheses: transnational communication, regulatory competition and independent problem solving. Preliminary results are presented after the analysis of each design option. The reasons for policy similarities will be summarised in Chapter 8.

Basic feed-in tariff design options define the fundamental rules under which renewable electricity producers can sell electricity to the grid. These design features suffice to establish a basic but comprehensive feed-in tariff scheme in a given territory. A feed-in tariff should have at least two design features, namely, a purchase obligation and a fixed tariff payment per unit of renewable electricity guaranteed over a long period of time (REN21 2010, 2011, Jacobs 2010).

First, the purchase obligation obliges the nearest grid operator to buy all renewable electricity, regardless of electricity demand. This design feature is essential to guarantee that tariff payment will actually take place as soon as power production is initiated. The purchase obligation thus plays an important role in increasing investment security. Exemptions from this rule are usually allowed only in cases of potential grid instability.[1] Second, the renewable power producer is guaranteed a certain amount of money per unit of electricity that is produced. This tariff payment is generally fixed and often depends on the technology, the plant size or even the location of the power plant in question (see Chapter 5). Besides, the payment is usually guaranteed over a long period of time (usually 15 to 20 years), which increases investment security and allows for full cost amortisation.

---

1   It should also be noted that the purchase obligation is no longer included under a premium feed-in tariff option. Some European countries guarantee green power purchasing only to the extent of electricity network losses (Klein et al. 2008: 23). However, all other countries operating a feed-in tariff mechanism, including Germany, Spain and France, guarantee the unrestricted purchase of renewable electricity power. Therefore, this design option will not be analyzed in more detail.

As long payment terms have not been a feature of early feed-in tariff mechanisms, this design option will be analyzed in more detail in section 4.4.

Besides the aforementioned essential design features, a basic feed-in tariff usually consists of further design options. It consists of a definition of eligible technologies (see section 4.1), clarifies whether capacity caps will be implemented, thus limiting the scope of the support instrument (see section 4.2), establishes a clear and transparent tariff calculation methodology (see section 4.3), regulates the financing of the entire support mechanism (see section 4.5) and defines a cost-sharing approach for grid connection (see section 4.6).

4.1. Eligible Technologies

In order to establish the producers that will be eligible for tariff payments under the feed-in tariff scheme, the legislator has to determine the eligible technologies and the territory to which the legislation applies. With regard to territory, today all European support mechanisms apply only within national boundaries.[2] For maritime installations (offshore wind power and ocean energy power plants), boundaries can be drawn either by the UN definition of Territorial Waters (12 nautical miles offshore), or by the Exclusive Economic Zone (200 nautical miles offshore).

Normally, legislators support a large portfolio of technologies, as technology differentiation is one of the major advantages of feed-in tariff schemes over other support mechanisms.[3] However, some countries (usually those which operate with a combination of different support mechanisms) support very few technologies or even only one under the national feed-in tariff.[4] At the outset, the potential of different technologies might have to be assessed in order to determine the eligibility of different technologies. For this purpose, it might become necessary to draft national wind and solar maps, along with other resource maps (Mendonça et al. 2009b).

---

2   In 2001 the Netherlands became the first – and is so far the only – European country to open up the national support mechanism to certified imports from other European countries. However, the practice was abandoned in 2003 because of its distorting effects and its high costs (Rooijen and Wees 2006).

3   Besides, by supporting both fluctuating technologies (wind energy and solar energy) and technologies that can produce renewable power steadily (like biomass, solar thermal, geothermal, and hydroelectric), the legislator can lay the foundations for backing fluctuating renewable energy sources with other renewable energy technologies, thus decreasing the dependency on conventionally produced electricity (Mendonça et al. 2009b).

4   Italy, for instance, applied a tradable green certificate system for all renewable energy technologies. As solar photovoltaic usually does not benefit from these non-technology-specific support mechanisms, the Italian legislator established an additional feed-in tariff mechanisms. Under the legislation of 2006, only solar photovoltaic (PV) power producers are eligible for tariff payment.

For some renewable electricity technologies, further specification and definition might be necessary. For certain renewable energy technologies there are no commonly accepted definitions. In particular, the generic term 'biomass' encompasses a wide variety of primary materials, such as forestry products, animal waste, energy crops, and sometimes municipal waste. Besides, feed-in tariff legislation has to stipulate whether impure biomass and waste resources will be eligible under the support mechanism. In most cases, the non-degradable fraction of waste is not eligible. Accordingly, the EU Directive 2001/77/EC has defined biomass 'the biodegradable fraction of products, waste and residues from agriculture (including vegetal and animal substances), forestry and related industries, as well as the biodegradable fraction of industrial and municipal waste'(EU 2001).[5] Moreover, the cost of different transformation processes of biomass products, such as co-combustion and gasification, might have to be reflected in the tariff design.

4.1.1. Germany

The German feed-in tariff is open to producers on German territory.[6] In the 1990s, the most common renewable energy technologies were eligible for tariff payment. The law explicitly refers to hydro and wind power, solar energy, landfill gas, sewage treatment gas, and biomass products. However, the number of technologies eligible under the StrEG was subject to controversy. While the Ministry for Economy wanted to include only a limited number of technologies, the Ministry for the Environment and the Ministry of Agriculture, together with several renewable energy industry associations, pushed for the inclusion of biomass (Kords 1993).

Under the EEG 2000, more technologies were included, namely, geothermal power, mine gas and offshore wind power. The legislator thus tried to stimulate the promotion of less mature technologies with important long-term potential (C. Pfeiffer, personal communication, 4 September 2008). In the case of solar energy, the EEG 2000 incorporated the term 'solar radiation', referring to both solar photovoltaic and solar thermal installations. In 2004 a more precise definition of hydro power was adopted, also including less mature technologies such as wave power, tidal power, salt gradient and flow energy. The 2009 and 2012 amendment did not include any additional technologies.

---

5   Further specifications might also be necessary for other renewable energy technologies. In the case of solar photovoltaics, for instance, feed-in tariff schemes sometimes differentiate between free-standing photovoltaic installation and integrated photovoltaic installations.

6   However, some efforts have been made to open up national FIT schemes so that they can incorporate renewable electricity produced in other countries. For instance, the German parliamentarian Hans-Josef Fell suggested extending tariff payment under the German support mechanism to electricity produced in northern Africa and then imported via high-voltage direct current transmission lines to Europe, including Germany (Fell 2009b).

Neither the first feed-in tariff scheme of 1990 nor the amendment of 2000 included a precise definition of biomass products for electricity generation. For clarification, the German Ministry for the Environment (BMU) issued the so-called Biomass Ordinance (*Biomasseverordnung*) of 21 June 2001 (BGBl 2001). The biomass ordinance specifieswhich substances and technical processes for electricity generation from biomass are within the scope of the feed-in tariff legislation. The BMU decided that that biomass originating in vegetables or animals (but not animal carcasses) were eligible, as were secondary sources of energy like biogas or alcohol (biomethanol or bioethanol) derived from it. As well, the biomass ordinance included biological waste and waste wood (that is, used wood or industrial waste wood so long as it was not strongly contaminated). The ministerial order excluded peat, mixed municipal solid wastes, paper, cardboard, pasteboard, sewage sludge, textiles, sewage and landfill gas.An amendment to the ordinance of 2005 primarily concerned regulations related to the eligibility of animal by-products (BBGl 2005).

4.1.2. Spain

The Spanish legislator started operating with price-based support instruments for hydro power producers in the early 1980s (BOE 1981a, b). Accordingly, the title of the first feed-in tariff mechanism of 1994 explicitly refers to the support of hydro power, but also mentions 'other renewable energy technologies' without specifying them. Article 2 of the decree of 1994 refers to 'solar, wind power, tidal power, geothermal and similar technologies', 'municipal and industrial waste, biomass and similar fuel types' and hydro power (BOE 1994a). The royal decree of 1998 was more precise and, for the first time, established clear technological categories that are still observed in Spanish renewable energy legislation today. Besides solar, wind and hydro power, the RD 2818/1998 differentiated between 'geothermal power, wave energy, tidal power and hot dry rock' (the last of these is a special technique for harnessing geothermal power). In 2002 the generic term 'solar energy' was replaced by the more precise terms 'solar photovoltaics' and 'solar thermal power' (BOE 2002a). In 2004 several technologies which were still at a technological stage of research and development were included. Group b3, which formerly included geothermal power, wave energy, tidal power and hot dry rock, now also included 'ocean thermal energy conversion' and 'tidal stream energy' (BOE 2004a). In 2007 a special category was created for offshore wind energy (BOE 2007c).

The definition of eligible biomass resources was frequently changed in Spain. The 1998 legislation also included a more precise definition of biomass, differentiating 'primary' biomass sources – basically energy plants grown in less than one year – and 'secondary' biomass sources – including waste products from primary usage of biomass sources such as agricultural waste and biogas. In 2004 three categories of biomass products were established: (a) energy crops, agricultural waste and forestry by-products; (b) electricity generation via anaerobic digestion (biofuels and biogas); and (c) biomass from industrial-sized installations from the forestry and agricultural industries.[7] The categories were revised once again in 2007 to include energy crops (group b.6.1), waste products from agricultural activities (group b.6.2), waste products from forestry activities, biogas from landfills (group b.7.1), biogas produced in digesters via biodegradable industrial waste and others (group b.7.2), and biogas from manure via combustion of liquid biofuels (BOE 2007c).Over the years, the categories have thus become increasingly precise and elaborate.

4.1.3. France

The electricity law of 2000 does not precisely identify eligible renewable energy technologies. It simply states that renewable energy sources and co-generation power plants will be eligible for tariff payment (J.O. 2000a). After this law on the modernisation and evolution of public services in the electricity sector was approved, the French legislator issued a decree stipulating the technologies eligible for tariff payment, including hydro power (from lakes, running water and the ocean), wind power, solar energy, biomass, biogas and geothermal power (J.O. 2000b). In the following years, the responsible ministry issued a number of decrees, each providing for tariff payments for these technologies. In 2001 and 2002, the first feed-in tariffs decrees were issued for wind power, hydro power and solar photovoltaics. The first tariff for geothermal power was established in 2002 (J.O. 2002d) and amended in 2006 (J.O. 2006e). In 2007 the revised tariff for hydro power established a new category of less mature hydro power technologies, including wave power, tidal power and hydrokinetic power plants (J.O. 2007a).

Over the years, a number of decrees were issued for different biomass and biogas categories, including 'methanisation' (biogas from household and municipal waste) (J.O. 2002b) and biogas from landfills (J.O. 2001e).[8] These two categories were combined and amended in 2006 (J.O. 2006c). In 2001 another category was created for electricity generation based on household and municipal waste. In contrast to the aforementioned categories, this category excluded power generation based on biogas. A final category for electricity generation in combustion plants via biomass from plants and animal waste (animal meal) was established in 2002 (J.O. 2002c) and amended in 2009 (J.O. 2009b).

4.1.4. Preliminary Conclusion

In all three countries, the number of technologies eligible under the national feed-in tariff mechanisms has increased over time. This is partly because of the

---

7   The precise biomass sources eligible for tariff payment were specified in annex II of RD 436/2004 (BOE 2004a).

8   A decree defining the tariff level of co-generation power plants was also issued in 2001.

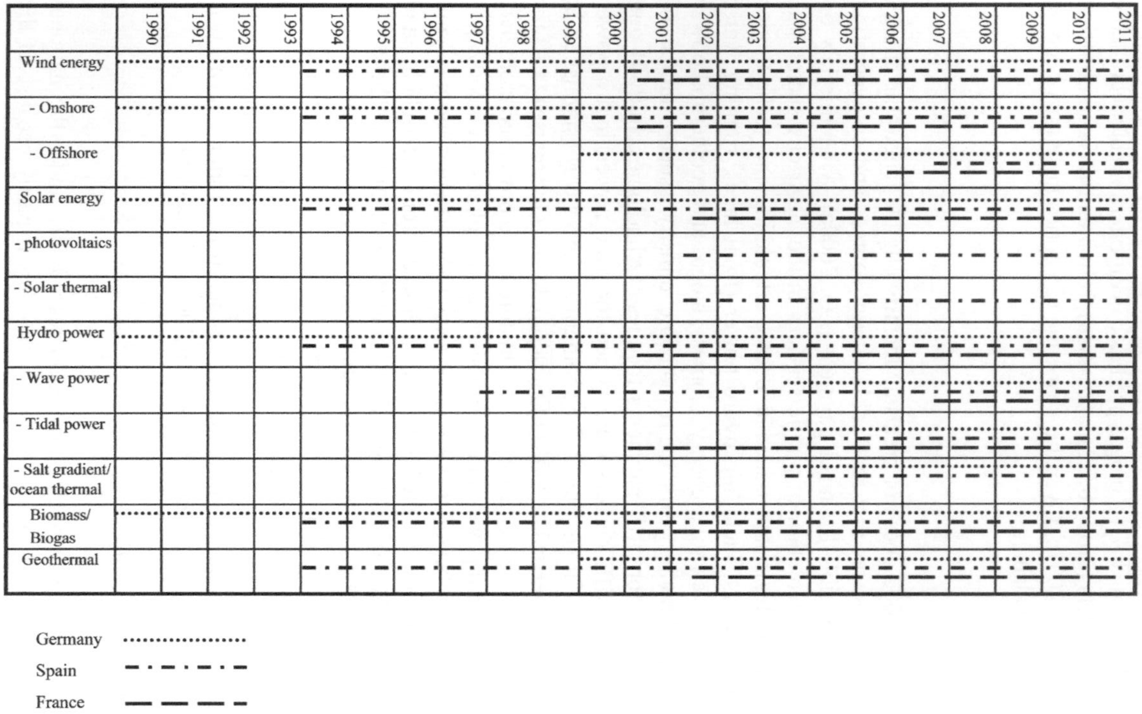

Figure 4.1 Inclusion of eligible technologies in Germany, Spain and France (1990–2011)

Note: Various biomass and biogas categories have not been included in this figure because they are difficult to compare.

Source: Author

Germany ·············

Spain – · – · –

France – – – –

inclusion of new renewable energy technologies (such as geothermal and biogas) and partly because of more precise definitions of generic terms. In the case of hydro power in Germany, for instance, the generic term 'hydro' was replaced by a more concise definition, namely, 'wave power, tidal power, salt gradient and flow energy' (BGBl 2008). In Spain the general term 'solar energy' was replaced by 'solar photovoltaics' and 'solar thermal' (BOE 2004a). As Figure 4.1 shows, convergence can be identified in the more mature and most-utilised renewable energy technologies, like wind power, solar power and hydro power. The degree of convergence is minor in less mature technologies and subcategories of more generic terms, for instance flow energy and wave energy as a subcategory of hydro power. As resource availability for all technologies is rather evenly spread, it had little effect on the choice of eligible technologies in Germany, Spain and France.

The inclusion of more and more technologies under the feed-in tariff schemes in Germany, Spain and France primarily depended on two factors: the ambitiousness of renewable electricity targets – an indication for the influence from the European level – and the general technological development – an indication for independent problem solving due to technological development. Transnational communication and regulatory competition apparently had no influence on the development of this design option. Regarding circumstances of independent policy solutions, the analysis above shows that renewable electricity technologies were included in feed-in tariff legislation once they reached a certain level of maturity. At an earlier stage, research and development funding is generally used in order to initiate demonstration projects. Once the next stage (towards market integration) is targeted, technologies such as geothermal, tidal power, offshore wind energy, and hot-dry rock became eligible for tariff payment in order to increase the deployment rate.[9] National legislators decided to include an increasing number of technologies independently from each other. This clearly shows that national policy makers have independent adjusted their support mechanisms due to international technological advances.

Besides, ambitious renewable energy targets force national legislators to include a larger number of technologies. National targets have often been set in the light of EU renewable energy directives. The renewable energy directives of 2001 and 2009 included an indicative target (2001) and a binding target (2009) for each member state (see section 4.2). For instance, in Germany the inclusion of offshore wind energy (in combination with increased tariff payment) has become necessary to achieve national long-term targets and compliance with the European 2020 target. Similarly, knowledge about the national potential for geothermal-based power generation persuaded the German legislator to include this technology (V. Oschmann, personal communication, 1 September 2008). In Spain, where the growth of the renewable electricity sector is primarily based

---

9   Less mature technologies, such as geothermal power, were included in Germany in 2000, because the legislator wanted to also promote less mature technologies (C. Pfeiffer, personal communication, 4 September 2008).

on wind energy projects, governmental actors have expressed their intention to increase their support for other renewable energy technologies in order to reach the national target. In order to achieve the target of 20 percent of final renewable energy consumption by 2020, Spain will have to increase the share of renewable electricity to 40 percent or more. Preparing the national Renewable Energy Action Plan, a representative of the Institute for Diversification and Saving of Energy (IDEA) stated that consolidated technologies such as wind power and biomass will have to be further developed. As well, new technologies such as geothermal and ocean technologies will have to be promoted further (Lucas 2009). Generally speaking, the ambitious national targets for 2020, as mandated by the new European Directive of 2009, will force national legislators to further increase the number of eligible technologies or at least to increase support for technologies which have a relatively small share today. This reveals an influence from the European level on national policy decisions.

4.2. Capacity Caps and Targets

Legislators can implement capacity caps, thus limiting the eligibility of power generation under the feed-in tariff mechanism. These caps can be related either to the overall renewable energy capacity installed in a given country (so-called program-size caps) or to the size of renewable energy power plants (so-called plant-size caps). In the case of program-size caps, tariff payment for new installations can be stopped once renewable electricity units reach a certain overall installed capacity (for instance, 5 GW) or renewable electricity makes up a certain share of the overall power demand (for instance ten percent). In some cases, program-size caps can also apply to selected renewable electricity technologies. A legislator might decide, for instance, that the tariff payment for wind energy will be guaranteed only until the accumulated capacity of all wind power plants in the country reaches 5 GW. Plant-size specific caps can be related to all or just a few specific technologies.

If a legislator makes use of program-size caps, it is very likely that plant-size caps will also be established. This is especially true for countries with very small feed-in tariff programs, where, for instance, only 200 MW of installed wind energy capacity will be eligible for tariff payment. In this case, the legislator will probably also limit the size of each wind park because otherwise only one producer (or only very few of them) will profit from the national support mechanism.[10] Capping the eligible renewable electricity capacity is sometimes interrelated with the establishment of national targets. In this case, the short- or mid-term targets established by the national legislators – often because of international commitments

---

10   This has been the case under the Kenyan feed-in tariff, where the legislator implemented both program-size and plant-size caps for all technologies (Mendonça et al. 2009b).

– have two, partially contradictory, objectives. On the one hand, targets create investment security because they signal to the industry that the policymaker wants renewable electricity to reach a certain share at a given moment of time. On the other hand, the interlinkage of targets and caps might create investment insecurity because there is the risk that tariff payment will be stopped once a certain share for renewable electricity has been reached.

4.2.1. Germany

The first German feed-in tariff of 1990 included a program-size cap. The so-called hardship clause limited the electricity that the regional monopolist was obliged to purchase to five percent of their overall power supply (BBGl 1990). If this limit was exceeded, the upstream grid operator had to bear the additional cost. Additionally, the local grid was exempt from the obligation to purchase renewable electricity, if this substantially increased the electricity price for the final consumer. This system-size cap was further specified and extended by a 1998 amendment in order to 'protect' transmission system operators. As mentioned before, regional grid operators were permitted to transfer the costs of green power promotion to the transmission grid operators once the five percent cap was exceeded. Therefore, another five percent cap was introduced at this transmission level, under which the tariff payment would not be granted for new installation if the overall share at the transmission system level surpassed five percent (BBGl 1998). Therefore, the regulation was termed the 'dual-five-percent capacity cap' (Lauber and Mez 2007). By the end of the 1990s, the five percent limit had already been reached in certain areas. To solve the problem, the new Social Democrat–Green Party governing coalition enacted an amendment in 2000 that introduced a new financing mechanism and a burden-sharing mechanism. From then on, the costs related to the feed-in tariff mechanism were equally distributed amongst all electricity consumers (see section 4.5). Consequently, single utilities or grid operators that had to purchase larger amounts of renewable electricity were no longer disadvantaged. As an indirect consequence, the aforementioned system-size cap (hardship clause) was abolished.

The German feed-in tariff also included program-size caps for certain technologies. The 2000 legislation stipulated that, in the case of solar photovoltaics, tariff payment would be guaranteed only for the first 350 MW installed. This provision was implemented in order to control total costs and therefore to ensure better compliance with the EU rules on state aid (Bechberger and Reiche 2004). Besides, many decision makers opposed the inclusion of solar photovoltaics in the German feed-in tariff because of the high costs involved, but thee critics were eventually persuaded by the argument that this capacity cap would control the costs. After the legislation was passed in 2000 based on this political compromise, the technology-specific program-size cap was modified in the following years. In June 2002 the limit was increased to 1,000 MW, shortly before the 350 MW cap was reached. One year later, the extraordinary amendment of solar photovoltaic tariffs (see section 7.1.1) led to the complete cancellation of the program-size cap for this technology (BBGl 2003).[11]

The feed-in tariff mechanism in Germany also includes plant-size caps for certain technologies. The first piece of legislation of 1990 imposed a capacity limit for a large number of technologies. For hydro power, landfill gas, sewage treatment gas and biomass, the tariff payment was granted only for small-scale installation up to 5 MW (BBGl 1990). In 2000 the 5 MW capacity limit was extended to solar photovoltaics and mine gas. In the case of solar photovoltaic, installations had to be building integrated or roof mounted. For free-standing solar photovoltaic systems an even stricter limit of 100 kW applied. The German legislator wanted to prevent the deployment of large-scale free-standing solar photovoltaic plants, partly in order to avoid soil sealing and partly because the maximum installed capacity of 350 MW (program-size cap) should promote the building of a large number of small-scale installations (Bechberger 2000). Geothermal power and wind power plants were the only technologies without plant-size caps. In the case of geothermal, this was probably because the industry was not expected to grow quickly in the near future. The capacity limit for biomass installation was increased from 5 MW to 20 MW (BBGl 2000).

In 2004 plant-size specific caps have been extended to a number of technologies. In the case of hydro power, the upper limit was raised from 5 MW to 150 MW. However, larger-scale power plants had to fulfil a number of criteria, such as efficiency gains from the renovation of already existing hydro power plants. Besides, the tariff payment for very large hydro power plants was significantly lower. The inclusion of larger-scale hydro power can be interpreted as a political concession to the German utilities that in those years were largely opposed to the feed-in tariff mechanism. At that time, the utility EnBW planned to build a 100 MW hydro power plant near the city of Rheinfelden. Consequently, EnBW was lobbying for the eligibility of larger-scale hydro power plants under the national feed-in tariff scheme. Some sharp tongues drily dubbed the amendment of 2004 'Lex Rheinfelden'.

The 2004 amendment also removed the plant-size cap for solar photovoltaic installations, including the 100kW cap for free-standing modules. However, free-standing power plants also have to comply with certain criteria in order to become

---

11   The only limitation that was set by the newly formulated Article 8 was a timely limitation for free-standing solar photovoltaic plants. Accordingly, the grid operator was not obliged to guarantee tariff payment for installations that were connected after 2015. This deadline was confirmed by the 2004 amendment. It is sometimes argued that the amendment of 2009 reintroduced the cap for solar photovoltaics. As described in detail in section 5.4.1, the 2009 amendment introduced the system of flexible degression, meaning that the annual degression rate is linked to market growth. Certain actors in Germany have for political reasons termed this design feature a 'breathing cap' or a 'flexible cap' by. However, above-target market growth will only reduce tariff payment and will in no way cap the installed capacity. Therefore, the term 'cap' is not correct.

eligible for tariff payment. The amendment of 2009 still includes a 5 MW plant-size cap for landfill gas, sewage treatment gas and mine gas. The maximum size of biomass plants remunerated under the EEG is 20 MW. However, larger-scale biomass plants producers can claim tariff payment, but only for 20 MW of the total installed capacity. Prior to the 2009 amendment the paper industry had called for the eligibility of all biomass plants, regardless of size (Deutscher Bundestag 2008c). No changes were implemented with the 2012 amendment.

4.2.2. Spain

Spain has a long tradition in planning renewable energy deployment via targets and caps. The first national plan for renewable energy sources (Plan de Energías Renovables en España, PER) was tabled in 1986 for the period until 1992.[12] Only three years later, a new renewable energy plan was announced for the period 1989–95, since the target for biomass and small hydro had already been reached in 1988.[13] A national energy plan for the next decade was tabled in 1991, which provided for the promotion of renewable energy sources until the year 2000.[14] At the start of the new millennium, another plan for the promotion of renewable energy sources was issued by the Spanish government, the (Plan de Fomento de las Energías Renovables 2000–10, PFER). Later, the respective technology-specific targets were amended twice: first within the electricity and gas plan for 2002–10, and second within the previewed amendment of PER in 2005 (Bechberger 2009).[15] The aforementioned targets of the national (renewable) energy plans were absorbed into the national electricity law of 1997 and the national feed-in tariff legislation of 1998, 2004 and 2007. Th3e targets often had the function of capping the share of renewable electricity.

The first program cap was established by the feed-in tariff legislation from 1998. The 12 percent target, as defined by the electricity law of 1997 based on the renewable energy plan 1991–2000, was included in the national feed-in tariff

---

12   At that time, the promotion of renewable energy sources was a political objective because of the related economic and technological benefits – contrary to today's focus on its environmental advantages (Nebreda Pérez 2007). The plan was part of the national energy plan (PEN) of 1983. The first renewable energy plan already included quantitative targets for the overall share of renewable electricity and additional technology-specific targets for biomass, solar photovoltaics, small hydro, wind energy and geothermal power.

13   The second national plan for renewable energy sources primarily focused on the most cost-efficient renewable electricity technologies, including municipal waste, biomass and small hydro. The target was increased only marginally to four percent of primary energy consumption in 1995.

14   Accordingly, the government planned to increase the share of renewable electricity (excluding large hydro) from 1.2 percent in 1990 to 3.2 percent in 2000.

15   In order to track the installed capacity over time, the first feed-in tariff of 1994 established a registry, which all power plants that were part of the so-called Special Regime had to sign.

legislation and became a program cap. Interestingly, the primary purpose of capping tariff payment was not to control costs but rather to stabilize the national electricity system (BOE 1998a). Because the feed-in tariff legislation of 1994 and 1998 did not explicitly state whether tariff payment would stop as soon as the national targets had been achieved, the legal situation was anything but clear and therefore caused investment insecurity in the industry (ASIF 2002). The feed-in tariff legislation of 2004 and 2007 was more straightforward in this respect, and explicitly indicated that tariff payment would cease once a certain capacity limit was reached.

However, contrary to the legislation of 1998, the legislation of 2004 and 2007 did not fix general program-size caps. Instead, technology-specific program-size caps were established. Tariff payment was guaranteed only until a certain installed capacity limit was reached for each technology. The technology-specific targets were defined in the national renewable energy plan of 2000. For some technologies, the targets were slightly increased by national electricity and gas plans for the period 2002–11. The 2004 legislation established a 150 MW limit for solar photovoltaics, a 200 MW limit for solar thermal, a 13,000 MW limit for wind power, a 2,400 MW limit for small hydro power and a 3,200 MW limit for biomass. In view of their marginal market shares, no limits were imposed on geothermal and wave power.

This system was further elaborated in 2007. The RD 661/2007 included technology-specific targets that were fixed by the updated renewable energy plan PFER for the period 2005–10. At the same time, the 2010 targets served as a cap for the installed capacity of each technology. This limit was 371 MW for solar photovoltaics, 500 MW for solar thermal, 20,155 MW for wind, 2,400 MW for small hydro and 1,567 for biomass and biogas. As with to the legislation of 2004, no capacity limits were implemented for less mature technologies such as geothermal and wave power (BOE 2007c). In order to track the growth of each technology over time, the National Energy Commission (CNE) publishes data on installed capacity on its website. The Royal Decree specified that, once a technology reached a share of 85 percent of the defined target, tariff payment for new installations would be guaranteed for only one more year (Jacobs 2008). This transitional period was expected to provide a certain degree of investment security. Renewable electricity projects that were already in the planning stage would still have the chance to be connected in the one-year period.

Despite the rigidity of the system of technology-specific program-size caps, it should be noted that the legislator sometimes extended the national mid-term objectives (shortly) before they were reached. This is especially obvious in the case of wind energy. The national renewable energy plan of 2000 imposed a target of about 9,000 MW installed capacity. Once it became apparent that the wind sector was able to grow much faster, the 2010 target was extended to 13,000 MW in 2002 and to more than 20,000 MW by the updated renewable energy plan of 2005.[16] In

---

16　The electricity and gas plan for 2008–16 even envisages an installed wind capacity of more than 30 GW by 2016.

contrast, the solar photovoltaic industry was less successful in extending the 2010 target mid-term. After the extraordinary growth of installed solar photovoltaic capacity in 2007 and 2008, the national industry tried to persuade the legislator to establish a new and more generous target for 2010 within the extraordinary amendment of 2008. However, this initiative was not very successful. Whereas the solar photovoltaic industry had argued for the complete removal of the program-size cap and its replacement with a more flexible degression rate (ASIF 2007, 2008) (see section 5.4.2), the government finally decided to implement an annual cap of 400 MW for the following years (BOE 2008a).[17] The different treatment of wind power and solar photovoltaics can be explained not only by the national legislator's focus on the least costly renewable energy technology but also by the stronger lobbying capacity of the Spanish wind power industry. It also has to be noted that the installed capacity of solar photovoltaic installations was more than eight times greater than the target for 2010. Instead of the envisaged 400 MW, total installed capacity already reached more than 3,300 MW in 2008 (see section 7.1.2). Once it became apparent that other technologies, namely solar thermal power and wind energy, were also exceeding the targets for 2010, the legislator also issued annual capacity caps for those technologies (BOE 2009).[18]

The dramatic solar photovoltaic development of 2007 and 2008 was also indirectly responsible for a legislative change in 2009. Solar photovoltaics substantially overshot the 2010 target primarily because of a design error in of the 2007 feed-in tariff legislation. The one-year transitional period set out by royal decree 661/2007 was designed in light of planning durations for wind energy projects (T. Días, personal communication, 25 February 2008). For this technology, since planning and permitting procedures and the following installation of equipment exceed a one-year period, no new project would be started once the 85 percent threshold of the technology-specific program-size cap had been reached. In the case of solar photovoltaics, however, this regulation had contrary effects. Once the 85 percent threshold was reached in September 2007, hundreds of new projects were started given that tariff payment was guaranteed for only one more year. From September 2007 to September 2008, more than 2 GW of new solar photovoltaic capacity were installed and connected to the grid. The legislator

---

17   For a more detailed analysis of the capacity caps issued by the latest Spanish legislation for solar photovoltaics, see section 7.1.2.

18   The provisions of the law 6/2009 have also been implemented in order to reach a more sustainable market growth for technologies such as wind energy. The Spanish legislator realised that in each year the national feed-in tariff scheme was amended, the installed wind capacity increased significantly since producers still wanted to profit from the existing and in most cases more attractive regulation. In 2004, the installed wind capacity reached 2,300 MW, in contrast to 1,300 MW in 2003 and 1,500 MW in 2005. Similarly, the newly installed wind capacity in 2007 reached 3,500 MW, in contrast to 1,600 MW in 2006 and 2008. The pre-registration mechanism of the law shall guarantee to an installed wind capacity of 1,700 MW annually until 2012 (Wind Directions 2010).

feared that similar developments could occur also with other renewable energy technologies. This fear was not without foundation, as the industry expected the installed capacity of concentrated solar plants to reach 3,000 MW by 2010, six times more than the targeted installed capacity (Energías Renovables 2009). In the case of wind power, too, it was expected that the 2010 targets would be reached prematurely (Bechberger 2009). Therefore, the law 6/2009 abolished the one-year transitional period and stipulated that, once the technology-specific targets were reached, the government had the right to adjust tariff payment and to implement annual capacity caps for the technologies involved (BOE 2009).[19]

New technology-specific caps have been issued in June 2010 within the national renewable energy action plan PANER. In line with the requirements of the European directive on renewable energy sources, the Spanish legislator issued new technology-specific targets for 2020. Wind energy is expected to reach a total installed capacity of 38,000 MW, hydro power 22,362 MW, geothermal 300 MW, solar photovoltaics 8,367 MW, CSP 5,079 and biomass 1,587 MW (MITYC and IDAE 2010).

Besides the aforementioned program-size caps, the Spanish legislation provides for plant-size caps, generally for all technologies and more specifically for certain technologies. The first general plant-size cap applied to all installation eligible under the so-called special regime – renewable power producers, co-generation plants and waste. Under the 1994 legislation, they were not allowed to be larger than 100 MW (BOE 1994a). The focus on small-scale power plants is historically based. When the special regime was first implemented in the early 1980s, its primary objective was to support electricity generation in isolated areas (auto-generation) and small hydro power plants (BOE 1981b). The electricity law of 1997 imposed an even stricter limit of 50 MW on all other renewable electricity technologies. All producers with larger power plants were excluded from the special regime and were consequently obliged to make offers on the wholesale electricity market (BOE 1997b, BOE 1998a). As the electricity law of 1997 has never been replaced, the 50 MW cap is officially still in place. However, the legislation of 2004 and 2007 included special provisions for power plants up to 100 MW.[20] Besides, it should be noted that in reality this capacity limit does not hinder the construction of large-scale power plants, as producers can simply divide large power plants into several smaller units not exceeding 50 MW.

---

19   Not surprisingly, the even stricter enforcement of capacity caps was implemented in a period of economic recession in the summer of 2009. The law 6/2009 explicitly states that the recent developments, that is, the exponential growth in the photovoltaics sector, might risk the sustainability of the overall system in the short term. According to the legislator, the risk arises both from the high costs for the final consumer and from technical problems related to the integration of a large share of renewable electricity (BOE 2009).

20   The feed-in tariff of 1998 already included a small premium payment for installations larger than 50 MW (1 Spanish peseta per kilowatt-hour produced) (BOE 1998a).

The only significant shortcoming of this regulation is the increasing costs of the administrative procedure, as the necessary permits have to be organised for each unit of the power plant.

An even stricter cap was imposed on certain technologies, including hydro power and solar photovoltaics. In 1994 tariff payment was granted only for hydro power plants with a capacity of less than 10 MW. This limit was increased up to the general cap of 50 MW in 1998. The 1998, 2004 and 2007 legislation did not implement any further plant site-related capacity caps, even for solar photovoltaic installations[21] (plant-size caps were first implemented for solar photovoltaic installations in 2008). The new royal decree 1578/2008 limited the size of all roof-mounted systems to 2 MW and the size of all free-standing modules to 10 MW. The limitations on solar photovoltaics have become necessary because of the implementation of a trimestrial, technology-specific cap. The new decree for solar photovoltaics of 2010 also limited the annual operating hours. Accordingly, the tariff payment is only guaranteed for 1250 full-load hours (BOE 2010d).

4.2.3. France

In France, program-size caps had already been debated before the first feed-in tariff came into effect. Already the preparatory report for the French feed-in tariff scheme, the Cochet report, suggested limiting the annually installed capacity. In this way Cochet – a strong proponent of renewable electricity promotion in France – hoped to counter the objection that feed-in tariff schemes generated high costs for final consumers. The Cochet report also envisaged technology-specific caps (Cochet 2000).

Eventually, the French feed-in tariff mechanism operated with only technology-specific program-size caps. The technology-specific caps were related to national targets, as determined by the multi-year investment plans for the French electricity system (Programmation pluriannuelle des investissements, PPI).[22] The periodical publication of PPIs was introduced by the 2000 law on the modernisation and development of electricity as a public service – the legal basis of the French feed-in tariff scheme. If the technology-specific targets specified within the PPI are not reached, the government has the option of auctioning the remaining capacity (J.O. 2000a, J.O. 2003a). If the accumulated installed capacity exceeds the technology-specific targets, the government has the option of ending the feed-in tariff payment for new installations or even of prohibiting the installation of new capacity altogether (J.O. 2000a). However, tariff payment does not stop automatically: the legislator can decide on a case-by-case basis.

---

21   However, in 1998 installations with a capacity exceeding 5 kW received the tariff payment was substantially lower. Therefore, the Spanish market only consisted of small-scale systems (ASIF 2002).

22   The PPI is designed to plan investments into electricity generation capacity for the next decade.

So far, three multi-year investment plans have been issued: in 2002 (MINEFI 2002, J.O. 2003a), 2006 (MINEFI 2006, J.O. 2006b) and 2009 (MEEDDAT 2009a). In each case, the responsible ministry has been prepared a report for the parliament. Based on the data and suggestions, a decree is issued by the parliament fixing the mid-term objectives for each technology. The first (2002) plan included targets for most renewable energy technologies for the year 2007, including biogas (up to 100 MW), biomass (up to 400 MW), biodegradable waste (up to 200 MW), wind energy (up to 6,000 MW), geothermal (up to 60 MW), hydro power and ocean energy (up to 1,000 MW), and solar photovoltaics (up to 50 MW). The multi-year investment plan of 2006 included more precise and more ambitious targets. The overall share of renewable electricity that would be needed to reach the European 2010 target was estimated to be a little bit greater because of increasing electricity demand. The decree of 2006 established a 200 MW target for biodegradable waste, 13,500 MW for wind energy, 90 MW for geothermal, 500 MW for hydro power and 160 MW for solar photovoltaic systems (J.O. 2006b). At the time of writing, the binding legal text for the PPI has not yet been approved by the French parliament. However, in summer 2009 the responsible ministry had already submitted its report to the parliament based on the recommendations of the *Grenelle de l'Environnement*. This draft of the PPI 2009 fixed ambitious targets for a large number of renewable electricity technologies, namely, hydro power (3,000 MW), wind energy (25,000 MW, including 19,000 MW onshore and 6,000 MW offshore), solar electricity (5,400 MW, including solar photovoltaics and concentrated solar power) and biomass (3,200 MW). These targets were confirmed by the French national renewable energy action plan of 2010 (MEEDDAT 2010d). The PPI 2009 did not explicitly fix specific targets for ocean energy sources and geothermal power. However, it dedicated a special chapter to investment planning in the overseas departments. These technologies are expected to contribute to reaching the overall objective of 30–50 percent renewable energies in primary energy consumption (MEEDDAT 2009a). Generally speaking, it should be noted that the targets expressed in the PPIs are ambitious, and most of them even exceed the expectations of the respective industry associations (M. Lettry, personal communication, 9 October 2008).

Additional program-size caps existed in the case of wind energy. The feed-in tariff legislation of 2001 included a 'soft' cap of 1,500 MW. Once the installed capacity of wind energy exceeded this limit, tariff payment would be reduced significantly (J.O. 2001b). This cap for wind energy embodies a political compromise between proponents and opponents of wind power (M. Lettry and G. Duclos, personal communication, 27 July 2007). This program-size limit was removed by the amendment of 2006. However, this technology-specific system cap has not limited the growth of the French wind energy sector and has had no effect on the growth of the wind power market in France (M. Lettry, personal communication, 9 October 2008). In the case of solar photovoltaics, a program-size cap was implemented in 2011. Due to the rapid market growth in 2010 and a waiting line of 3 GW solar photovoltaic capacity, the French legislator decided to limit the annual growth to 500 MW. Small-scale residential system with less than 100 kW installed capacity can account for 100 MW per year, larger scale residential systems for 100 MW, large-scale roof-mounted systems for an addition 300 MW (J.O. 2011).

In addition to the program-size cap, the French legislator has implemented a plant-size specific capacity cap for all renewable energy technologies. In France, only producers with an installed capacity of less than 12 MW are eligible for tariff payment. This general limit was laid down by the electricity law of 2000 and the Decree 2000-1196 (J.O. 2000b). The development of larger power plants was intended to be incentivised by the parallel use of a tender scheme. This plant-size cap is historically based. The 12-MW-limit for renewable electricity producers stems from the Nationalisation Law of 1946, the former French electricity law (J.O. 1946). Up until 1946, the French electricity market was in the hand of thousands of producers. The law of 1946 nationalised all power production units except those with a capacity of no more than 8 MW. Thereafter, the French electricity market was characterised by large-scale, state-owned power plants and a few small-scale, privately owned generation units. The law of 1946 also incorporated other aspects of the current French feed-in tariff scheme: it obliged the newly established national utility EDF to purchase all electricity from independent power producers under a fixed price regime – one basic feature of today's feed-in tariff mechanism.

In addition to the program size cap, the French legislator has implemented a plant-size specific capacity cap for all renewable energy technologies. In France, only producers with an installed capacity of less than 12 MW are eligible for tariff payment. This general limit was laid down by the electricity law from 2000 and the respective Decree 2000-1196 (J.O. 2000b). The development of larger power plants was intended to be incentivised by the parallel use of a tender scheme. This plant-size can is historically routed. The 12-MW-limit for renewable electricity producers stems from the Nationalisation Law of 1946, the former French electricity law (J.O. 1946). Up until 1946, the French electricity market was in the hand of thousands of producers. The law of 1946 nationalised all power production units – except for units with a capacity of no more than 8 MW. Since these days, the French electricity market was characterised by large-scale, state-owned power plants and (a few) small-scale, privately owned generation units. The law of 1946 also incorporated other aspects of the current French feed-in tariff scheme: it obliged the newly established national incumbent EDF to purchase all electricity from independent power producers under a fixed price regime – one basic feature of today's feed-in tariff mechanism.

Shortly after the feed-in tariff mechanism was established in 2001, it became clear that it was severely hindering the development of wind power. According to a study commissioned by the French Renewable Energy association, SER, the 12 MW cap increased the cost of projects since larger wind power plants had to be realised in slices of 12 MW. This increased administrative and grid connection costs (Boston Consulting 2004). Eventually, the new electricity law of 2005 removed the 12 MW limit for wind energy producers (J.O. 2005b).Under a political compromise, wind power plants were no longer subject to size restrictions as long as they were located in specially designated wind power development zones. Interestingly, opponents of wind energy played a crucial role in removing the 12 MW capacity cap. A number of parliamentarians critical of wind power argued that, in place of a large number of small-scale wind power parks throughout the country, special regions should be assigned for fewer, large-scale generation sites. The lower house of parliament tabled and voted for an amendment stipulating that only wind power plants of more than 30 MW capacity would be eligible for tariff payment (Dodd 2005).[23] Since industry associations and non-governmental organisations were opposed to this strict exclusion, in the end the upper house voted for a case-by-case approach, which was implemented by establishing special wind-power development zones (R. Claustre, personal communication, 22 October 2008).[24]

In summer 2010, the 12 MW cap was removed for other renewable electricity technologies. The first *Grenelle de l'Environnement* stated that in future the development of several technologies might be hindered by the 12 MW capacity cap – including PV, geothermal (especially in the DOM), and ocean energy. The report therefore suggests that the capacity limit of 12 MW for these technologies be reconsidered (Dupuis et al. 2008). Consequently, the law related to the second *Grenelle de l'Environnement* abolished the 12 MW cap for marine energy technologies, geothermal power and concentrated solar power (J.O. 2010c).[25]

In the case of solar photovoltaics, the first feed-in tariff of 2002 granted tariff payment only for installations with a capacity inferior to 5 kW for private households, 1,000 kW for professional buildings and 150 kW for apartment building (J.O. 2002a). This limit was removed in 2004 (J.O. 2004a). The latest tariff regime for building integrated solar photovoltaic systems limits tariff payment to installations with a capacity of up to 250 kW (J.O. 2010a).[26] Solar photovoltaic electricity generation is also limited with respect to annual operating hours.

---

23   The latest report of the French parliament on wind energy confirms this intention. It proposes that even within wind power development zones wind parks should not be smaller than 15–20 MW (Assemblée Nationale 2010: 115).

24   In fact, the requirement to produce within a special development zone considerably worsened the framework for small-scale wind power projects. The administrative procedure is so complicated and expensive that the requirement is no longer attractive for small-scale wind power parks (M. Lettry, personal communication, 9 October 2008).

25   Internally, at some stage the participants of the *Grenelle* intended to increase the capacity limit to 50 MW for all technologies (J.-L. Bal, personal communication, 24 October 2008).

26   The responsible ministry imposed this limitation shortly after the new decree was issued in December 2009, arguing that the measure was necessary to control the costs of the promotion of renewable electricity and the electricity price for the final consumer. According to the ministry, the announcement of new tariffs for solar photovoltaics created a financial bubble which could have triggered investments of more than €50 billion over 20 years. This might have led to electricity price increases of ten percent. As large-scale,

This innovative approach was implemented primarily in order to avoid fraud. The French legislator was afraid that power producers would manipulate their photovoltaic plants – for instance, by co-powering them with batteries – in order to increase power output illegally. Instead of controlling each power generation unit, which would have been complicated, especially in overseas departments, the legislator established a cap fixing the maximum number of annual operating hours. For installations on the French mainland the limit was set at 1500 kWh/kWp while producers in the overseas departments were remunerated up to 1800 kWh/kWp (J.O. 2004a).[27] In 2010, this cap was extended to 2,200 kWh/kWp, and 2,600 kWh/kWp in the overseas departments, in the case of sun-tracking solar systems (J.O. 2010a), as it has become clear that this regulation disincentivised the use of such systems (E. Wael, personal communication, 14 October 2008).

4.2.4. Preliminary Conclusion

Capacity caps (both program-size caps and plant-size caps) exist or have existed in all three countries studied. However, as a general tendency capacity caps have been diminished or removed over time. Germany has removed all program-size caps and eliminated or at least considerably eased plant-size limitations. In France, existing program-size caps were always sufficiently ambitious that they did not hinder the growth of the renewable energy sector. Besides, plant-size caps were eased for a large number of technologies. In Spain, technology-specific targets and caps have frequently been eased and the general plant-size cap of 50 MW is largely theoretical.

Despite this general tendency, differences still exist between the designs of caps in the three countries. Germany is the only country where program-size caps have been fully removed. In 2000 the dual-five-percent capacity cap was abolished and in 2004 the last technology-specific program-size cap for solar energy was removed. In this respect, Germany is unique, not only among the three countries in question but also worldwide. In no other country in the world is the feed-in tariff payment not limited by any sort of program-size cap.[28] Program-size caps still exist today in France and Spain; in both cases they are technology-specific program-size caps which are related to national renewable energy targets.

The interdependency of targets and caps differs as between Germany, Spain and France. Spain has established the most rigorous scheme, under which tariff payment will be automatically stopped once the technology-specific targets

---

speculative investment primarily focuses on larger-scale projects, the French legislator has tried to stop speculation by establishing this plant size limitation (MEEDDAT 2010a).

27   Once the limit is reached, each additional unit of electricity is remunerated with only 5 €cent/kWh. This reduced payment is intended to reflect the electricity market price (Assemblée Nationale 2009: 71).

28   Although some other countries are without general program-size caps, they nonetheless have technology-specific program-size caps for solar photovoltaics.

are reached. Even the one-year transitional period, which was established by the legislation of 2007, was abolished in 2009. The Spanish government wants to exercise strict control over market growth in the renewable energy sector, primarily for cost-control reasons. The French approach is less strict. In the event that the technology-specific targets are reached, the legislator has the legal right to end tariff payment but is not obliged to do so; the administration can decide on a case-by-case approach. Besides, the French targets are rather ambitious; up to now, mid-term targets have never been reached. In Germany the targets stipulated by the feed-in tariff legislation of 2004 (a 12.5 percent share by 2010) and 2009 (30 percent by 2020) cannot cap the deployment of renewable energy sources. In contrast to those of Spain and France, the German targets are defined as minimum targets (for instance, 'at least' 30 percent by 2020). Therefore, tariff payment has to be guaranteed even if the target is achieved prematurely. For instance, the 12.5 percent target for 2010 was reached in 2007. Nonetheless, the support instrument remained in place for additional capacity. Technology-specific targets do not exist in Germany. The only indication of the expected development of each technology is given in the annual lead studies (Nitsch 2008, Wenzel and Nitsch 2010). However, these indications have no effect on tariff payment modalities.

Despite the differences as between Germany, Spain and France, one similarity is that technology-specific program-size caps have often been removed shortly before they were reached. This was the case with solar photovoltaics in Germany (2002), wind energy in France (2006) and wind energy in Spain (2002, 2005). Even though targets were often formulated during initial political negotiations, proponents of renewable electricity were often able to remove them at a later stage. This also reflects a common political strategy of political proponents of feed-in tariffs. In order to win the support for the mechanism from critics of renewable electricity at an early stage, political concessions are made by capping the installed capacity of some or of technologies. This way, the objection that renewable electricity promotion leads to high costs for final consumers can be countered. This approach could be observed in the case of solar photovoltaics in Germany (2000) and in the case of wind energy in France (2001). In both cases, the initial technology-specific program cap was removed or extended before it was reached. In other words, the existence of program-size caps can also be attributed to the political influence of actors which do not support the deployment of renewable energy sources.[29]

---

29 An analysis of the solar photovoltaic market in Germany and Spain also reveals an interesting interdependence between technology-specific program caps and tariff levels. When the Spanish solar photovoltaic tariff was implemented in 2008, the government gave the solar photovoltaic industry a choice between a relatively high tariff (32 €cent/kWh) for a limited amount of capacity and a significantly lower tariff (22 €cent/kWh) without any capacity restriction (Photon 2010). Eventually, the Spanish industry opted for a relatively high tariff and, in turn, accepted the program-size cap. The German solar photovoltaic industry was confronted with a similar choice in 2009. The legislator proposed either to

National government have independently decided to remove capacity caps. Transnational communication and regulatory competition did not have an impact on the policy makers' decisions. The removal of capacity caps in all three countries can be largely attributed to technological advances in the renewable electricity industry. First of all, the cost of renewable electricity generation from all technologies has significantly fallen since the 1990s. This has also increased legislators' leeway for promoting renewable electricity and thus has also led to the extension or removal of program-size caps. Renewable electricity technologies are no longer niche market technologies but have the potential of replacing conventional power generation units on a large scale. Second, renewable electricity power plants have become bigger. This is most obvious in the case of wind energy, where the average turbine capacity was 600 kW in 1995 and more than 2 MW in 2010. Similar increases in the size of single plants can be observed in all other technologies. This technological development has refuted the antiquated assumption that renewable energy technologies are by definition small in scale. Therefore, plant-size caps have also been eased or removed over time.

National government have independently decided to remove capacity caps. Transnational communication and regulatory competition did not have an impact on the policy makers' decisions. The removal of capacity in all three countries can be largely attributed to technological advances in the renewable electricity industry. First of all, electricity generation from renewable energy technologies has become cheaper over time. The cost for renewable electricity generation from all technologies was significantly reduced over the past two decades. This has also increased legislators' leeway for promoting renewable electricity and thus has also led to the extension or removal of program-size caps. Renewable electricity technologies are no longer niche market technologies but have the potential of replacing conventional power generation units on a large scale. Second, renewable electricity power plants have become bigger. This is most obvious in the case of wind energy, where the average turbine capacity was 600 kW in 1995 and is more than 2 MW in 2010. Similar increases in the size of single plants can be observed with all other technologies. This technological development has refuted the antiquated assumption that renewable energy technologies are by definition small in scale. Therefore, plant-size caps have also been extended or removed over time.

To some extent, the European framework has forced national legislators to remove or increase program-size caps. As described in Chapter 3, the first European directive of 2001 included indicative national targets for 2010, and the second European directive even included mandatory targets for 2020. These European targets were usually transposed into national legislation in the form of national caps and targets. In Germany, the feed-in tariff law of 2004 stated that the national

---

reduce the tariff by about 15 percent or to grant higher tariffs for limited market growth. As the most important objective of the German solar photovoltaic industry was to avoid a program-size cap, important tariff reductions were finally implemented in 2010.

target of 12.5 percent was to be reached in order to comply with the European target. The 2009 amendment states that in order to reach the 2020 European target (18 percent of final energy consumption from renewable energy sources), the share of renewable electricity shall be of at least 30 percent. In an earlier version of the law, a minimum target of 25–30 percent was envisaged. However, it soon became apparent that, in order to reach the ambitious European renewable energy target, a national electricity share of at least 30 percent would be necessary. In Spain the European renewable electricity target forced the national legislator several times to increase the target renewable capacity, as defined in national renewable energy plans, because the European targets for 2010 and 2020 are consumption target. As electricity demand in Spain has been increasing rapidly in recent decades, even with a rapid increase in the installed renewable capacity electricity's relative share has been stagnating or even decreasing (Jacobs 2008). Spain is now able to comply with the European directive of 2001 but only because of the economic recession and the decreasing demand for electricity.[30] Ambitious national targets will have to be formulated for reaching the 2020 target, especially in the case of wind energy.[31] In France national targets as expressed in the multi-year investment plans were set according to the European targets for 2010 and 2020. Besides, the European 2010 target for France was transposed into the national electricity law of 2005. In fact, the European framework helped proponents of renewable electricity to exert pressure on the national legislator in the teeth of strong national opposition. This strategy is especially important in the case of wind energy, which not only faces opposition from a large number of parliamentarians but also has the greatest mid-term potential for achieving the European targets (Sauter 2006).

4.3. Tariff Calculation Methodology

When designing feed-in tariffs, one of the most difficult tasks for policymakers is to set the tariff level. A tariff which is too low will not trigger any investment, while a tariff that is too high might lead to windfall profits and overcompensation of renewable electricity producers. In order to fix the tariff level, the legislator

---

30 In January and February 2009 the share of renewable electricity was 30 percent, and thus within the range of the European target as set out in the Directive 2001/77/EC. The favourable climatic conditions – strong winds and precipitation – in combination with an economic downturn because of the global economic crisis allowed the European target to be achieved (Méndez 2009).

31 The responsible ministry, MITYC, estimated that in order to reach this target 42 percent of Spanish electricity demand would have to be covered by renewable energy sources in 2020 (Iberdrola 2008). Preparing the Renewable Energy Action, the Spanish government expects that, in order to reach the 20 percent target by 2020, 40 percent of the electricity will have to be provided by renewable energy technologies (Lucas 2009). To achieve this objective, at least 35,000–40,000 MW of wind energy capacity and 3,000 MW of solar thermal capacity will have to be installed (Iberdrola 2008).

generally applies a predefined tariff calculation methodology for all renewable energy technologies. This way, transparency and comparability for power producers from different technologies can be guaranteed (Mendonça et al. 2009b). Feed-in tariffs around the world have basically applied two distinct tariff calculation methodologies, the 'value-based' and the 'cost-based' approach.

The value-based approach takes either the 'value to the utility' or the 'value to the society' as the benchmark when fixing tariffs. By 'value to the utility', legislators usually mean the concept of avoided costs for power generation. This value-based approach operates on the assumption that a unit of electricity is a homogeneous 'good' and that one kilowatt-hour of renewable electricity has the same value as a kilowatt-hour of conventionally produced power. Under this approach, monopolists are obliged to purchase electricity at prices based on the utilities' avoided costs, that is, costs that would have been incurred if the electricity had been produced by other, existing conventional generation units (Hirsh 1999).[32] 'Value to the society' implies the concept of avoided external costs for society. The benchmark is still the cost of power generation from conventional energy sources. However, a top-up is usually paid for the 'greenness' of the renewable electricity. This increase in remuneration reflects the avoided external costs to society, such as climate change or damage to health and buildings from air pollution (Sijm 2002).[33] As this approach to tariff calculation is based on a large set of assumptions, the aforementioned terms have to be clearly defined or even quantified.[34]

In contrast, the cost-based approach directly takes the generation costs of renewable energy technologies into account and fixes tariffs accordingly. Contrary to the tariff calculation methodology discussed above, there is no direct link to the costs of power generation from conventional technology sources. Under this approach, the policy maker sets the tariff level based on the technology-specific generation costs, building in a certain profitability margin for generators, usually

---

32   This approach was first utilised in the United States as part of the Purpa Act of 1978. As the actual definition of avoided costs was left up to the individual states, this led to widely different interpretations across the US (Hirsh 1999, Lipp 2007).

33   Value-based approaches can also combine the concepts of 'value to the utility' and 'value to the society'. For instance, the Portuguese feed-in tariff scheme takes into account the avoided capital investment, the avoided electricity production costs (fuel, operation and maintenance), the avoided costs of network losses and the avoided environmental costs (IEA 2008c, Klein et al. 2008).

34   In line with the notion of avoided costs, some jurisdictions have chosen to fix the tariff level as a certain percentage of the retail electricity price (Espey 2001a; Sijm 2002). This approach is predominantly adopted in oligopolistic or monopolistic energy markets where the electricity price for final consumers is also generally regulated. Like the avoided-cost approach, this approach relates the payment to renewable electricity producers indirectly to the generation cost of conventionally produced power. A wind-power producer, for instance, receives a tariff equivalent to 90 percent of the electricity price for final consumers (Mendonça et al. 2009b).

a return on investments of between five and ten percent.[35] In this case, a number of factors affecting generation cost have to be taken into account, including plant investment costs (together with material and capital costs), grid-related and administrative costs (grid-connection costs, costs of licensing procedures, and so on), operation and maintenance costs, and (in the case of biomass and biogas) fuel costs. Based on these parameters, the policymaker or research institute can calculate the nominal electricity production costs for each technology. Based on the average operating hours of a standard plant and the duration of tariff payment, it is possible to fix the nominal remuneration level per unit of electricity (Mendonça et al. 2009b). In sum, the benchmark for the value-based tariff setting approach is the generation costs of conventionally produced electricity. The point of reference for fixing tariffs under the cost-based approach is the specific generation costs of renewable electricity technologies.

4.3.1. Germany

From 1990 to 2000, feed-in tariffs in Germany were calculated in line with the value-based approach. Tariffs were related to the electricity price for final consumers and therefore indirectly linked to the cost of conventionally produced electricity. Depending on the technology, tariff levels varied between 65 percent and 90 percent of the retail electricity price (BGBl 1990).[36] The avoided external costs of conventional power generation were taken into account when setting the feed-in tariffs. According to the chief negotiator of the conservative party CDU Egelsberger, at that time the external cost for coal-powered electricity generation was about 3 to 5 €cent/kWh (Deutscher Bundestag 1990, Lauber and Mez 2007). Not coincidentally, the tariffs for renewable energy sources were in the range of the cost of conventionally produced electricity plus the above mentioned avoided external costs.

Despite the use of this value-based approach, it should be noted that taking the avoided social costs of conventional power sources into account already meant a big improvement for renewable electricity producers. Before the first feed-in tariff law was passed in 1990, green power producers received only avoided costs based on conventional power generation. In some cases tariffs were even below

---

35   In some cases a higher rate is established to allow levels of profitability comparable to the expected profit from conventional electricity generation. In the UK the legislator even envisaged different rates of return depending on the group of actors being targeted. According to the analysis of a consultancy, the 'hurdle rate' of return for domestic consumers of renewable energy is three percent, while commercial building owners have a hurdle rate of six percent. However, the consultancy argues that professional developers and utilities require higher internal rates of return (exceeding eight percent) in order to make investments (Brown et al. 2009).

36   For practical reasons, this payment was based on the national statistical data of the penultimate year.

the generation costs of coal-based power plants (Bechberger 2000).[37] The feed-in tariff of 1990 can be understood as a first move towards a cost-based calculation approach. At least for some technologies, factoring in the external costs of conventional power sources led to tariff levels which were close to the actual generation costs of the most cost-effective renewable energy sources. It should be noted, however, that this occurred rather coincidentally and not as a result of a thorough analysis of renewable electricity generation costs.

Linking tariff payments for renewable energy producers to the electricity price for final consumers became problematic when electricity prices started to fluctuate and later fell considerably. With the process of market liberalisation initiated at the end of the 1990s, electricity prices were no longer regulated by the legislator. Therefore, linking tariff payment to the retail electricity price not only undermined investment security for renewable electricity producers. In the case of rising prices for final consumers, the *VDEW* and the European Commission feared that renewable power producers could earn from windfall profits (Hischl 2008).

In addition, the new government coalition wanted to extend the scope of the national renewable energy support scheme, as primarily small hydropower and wind energy profited from the existing legislation. In contrast to the payment regime under the StrEG, the new law of 2000 established fixed tariff rates that were no longer linked to the electricity price for final consumers. Moreover, the new tariff calculation approach was based on the generation costs of each technology.[38] The tariff levels were set according to a detailed analysis of the costs of renewable energy projects (Fell 2009a, BGBl 2000). According to the Explanatory Memorandum, Germany's 2000 feed-in tariff were set according to the following approach:

> The compensation rates specified in the RES Act have been determined by means of scientific studies, subject to the provision that the rates identified should make it possible for an installation – when managed efficiently – to be operated cost-effectively, based on the use of state-of-the-art technology and depending on the renewable energy sources naturally available in a given geographical environment (BGBl 2000).

---

37   According to Kords, at that time the remuneration for renewable electricity producers varied between one and 5 €cent/kWh, depending on the electricity price and the renewable energy technology. Therefore, renewable electricity producers avoided feeding electricity into the grid and instead used it for own consumption (Kords 1993).

38   Nonetheless, the new law still referred to the notion of avoided costs in order to justify renewable energy support (Lauber and Mez 2007). More precisely, the new law justified tariff payment for renewable energy sources by the polluter-pays principle, the remaining subsidies for conventional power generation and the intention to establish mass markets in order to reduce generation costs for renewable energy sources (BGB 2000).

This method of 'cost-covering remuneration'[39] was first implemented at the local level in the case of solar photovoltaic (PV) panels in 1993. The cities of Freising, Hammelburg and Aachen established a feed-in tariff scheme for solar photovoltaics which allowed for full cost recovery.[40] In the coming years, these early local feed-in tariff schemes became the model for many other communities. In 1999, the Red–Green coalition decided to implement this scheme at national level and applied the approach of cost-covering remuneration to all renewable energy technologies, guaranteeing an internal rate of return of about seven percent (Fell 2009a).[41]

When the EEG 2000 was drafted no transparent methodology for tariff calculation was available. During the preliminary research for the law, several research institutes conducted studies which were commissioned by the responsible ministry. However, these studies did not focus on the remuneration for each technology. Instead, they looked at more general questions regarding the design of support schemes for renewable electricity (Bechberger 2000). In 1999 the Ministry of the Economy (BMWi) drafted two proposals for amending the German feed-in tariff scheme. However, when it became clear that these proposals did not include suggestions from leading parliamentarians, the feed-in tariff law was eventually drafted by individual members of parliament. Therefore, the tariff level evaluation was largely based on intensive consultation with the industry and not primarily on a transparent tariff-calculation methodology.

A transparent methodology was first developed for the first progress report in 2002. The EEG 2000 stated that, every two years after the entry into force of the Renewable Energy Source Act, a report should be submitted by the responsible ministries, analysing the 'progress achieved in terms of the market introduction and the cost development of power generation installations' (BGBl 2000). If necessary, the progress report should include suggestions for adjustment of tariffs and reduction rates (degression). The first progress report was published in 2002 (Deutscher Bundestag 2002) and updated in 2003 (BMU 2003) as the analytical basis for the 2004 amendment.

The second progress report, published in 2007 (Büsgen and Dürrschmidt 2007), defined the German approach to setting tariffs as 'practice-oriented reference systems' to track the costs of renewable energy systems over time (BMU 2008a). For this report, the number of commissioned studies increased significantly.

---

39   The German legislator used the term *kostendeckende Vergütung*.

40   Even though the German feed-in tariff is often cited as the first to offer cost-based remuneration for solar photovoltaics, the first local cost-covering feed-in tariff for solar photovoltaics was implemented in Burgdorf, a small city in Switzerland, in 1989 (Podewils 2010a).

41   With some technologies, renewable electricity producers could not exclusively rely on tariff payment in order to operate power plants profitably. For instance, solar photovoltaic systems profited from both tariff payment and the investment subsidies granted by the 100,000-roof program.

As shown above, the process for setting the tariffs has been institutionalised over the years, and now includes studies from independent research institutes and proposals from the responsible ministry, followed by potential modification in the general parliamentary legislative process. Whatever the existing methodology for tariff calculation, it should be understood that the actual tariff level might differ from the scientific calculation contained within the progress report. Because the German feed-in tariff legislation is a statute – and not just a decree or a ministerial order, as in other countries – the German parliament can modify the initial tariff proposal of the responsible ministry. Non-governmental actors, such as environmental and consumer protection groups and industry associations, have the chance to lobby for their interests once the scientific progress report has been tabled and government and parliament become active. Their opinions are also taken into account in public hearings.[42] However, the proposals set out in the progress report usually remain unchanged. Any modifications are generally marginal, within the range of one cent per kilowatt-hour.[43]

4.3.2. Spain

The early Spanish feed-in tariff scheme of 1994 based tariff calculation on the avoided cost for conventional power generation.[44] The RD 2366/1994 explicitly stated that the tariffs for power feed into the grid should 'take the avoided costs of the electricity sector into account, based on the concept of power generation, transport and distribution' (BOE 1994a). More precisely, renewable power plants with an installed capacity exceeding 25 MW received the equivalent of the 'variable avoided costs' of conventional power generation. The tariff for smaller scale power plants was calculated based on long-term avoided costs, also taking the retail electricity price in account (BOE 1994a). The value-based approach is also reflected in the fact that no technology-specific tariff was granted by legislation in 1994 (see section 5.1.2).

---

42   In case of exceptional tariff level modifications, as discussed for solar photovoltaics in 2010, the responsible ministry engages in a stakeholder exchange with the interest groups mentioned in order to hear their opinions.

43   The progress report of 2007 proposed tariffs for hydropower, biomass, biogas, geothermal, wind and solar. While the suggestions for biogas, biomass and geothermal were accepted without any changes, the tariffs for wind, solar and hydro were modified within the political decision-making process. Of a total of 36 tariff proposals, 20 remained unchanged and 16 were modified marginally. The maximum deviation between the tariff proposal of the progress report and the actual tariff payment as guaranteed by the new law was of 1.48 €cent/kWh (in the case of large-scale, roof-mounted solar photovoltaic systems) (BMU 2008a).

44   The approach for the early support for small-scale hydropower was slightly different. The legislation of 1981 explicitly stated that tariff payment should allow for the profitable operation of hydropower plants, indicating a cost-based approach (BOE 1981a; 1981b).

The Spanish tariff calculation methodology for renewable electricity was rather complicated in 1994. In fact, the tariff calculation for renewable electricity producers resembled the calculation of remuneration for conventional power plants in monopolistic power markets. The remuneration level was determined by a complex formula, including a large number of tariff components and coefficients. This included the assured provision of power based on the installed capacity, a premium payment for electricity production in line with general demand patterns and a premium payment for the provision of reactive power. Besides, the tariff payment was reduced if power generation forecasts were not fulfilled.

In the electricity law of 1997, tariff payment for renewable energy sources was justified by reference to the costs of diversifying the electricity portfolio and security of energy supply (BOE 1997b). Besides, further factors were cited for tariff calculation, including the tension level at which electricity is fed into the grid, the positive environmental impact, energy saving and energy efficiency. Most interestingly, one article explicitly refers to a cost-based calculation approach, stating that generation costs shall be taken into account when fixing tariffs, thus guaranteeing reasonable profitability depending on the cost of capital in the financial markets. This approach also explains why the law allows for setting much higher tariffs for solar photovoltaics than for all other renewable energy technologies. Thus, the law of 1997 can be interpreted as a first step away from the value-based approach towards the cost-based approach. For the first time, the law made reference to the generation costs of different renewable energy technologies.

The cost-based tariff calculation methodology was fully developed in 2004. In 2003, the National Energy Commission (CNE) tabled a proposal for an objective methodology for tariff calculation, which was sent to the Ministry of the Economy (CNE 2004).[45] The methodology was applied for the first time for the feed-in tariff scheme of 2004 and also for the amendment of 2007.[46]

In line with the cost-based tariff calculation methodology, the Spanish feed-in tariff guarantees reasonable profitability for renewable energy projects, with the generation costs and specific requirements of each technology taken into account. The income for power producers must provide an internal rate of return similar to that for regulated financial activity. On average, the Spanish feed-in tariff levels are based on an internal rate of return of seven percent. In the case of the premium feed-in tariff option, the profitability varies between five percent and nine percent, depending on the market price (L. Temblenque, personal communication, 13 February 2008). The following parameters are incorporated in the calculation:

---

45   The 2003 proposal did not make any suggestions regarding the specific tariff level of renewable energy technologies for the 2004 amendment. At a later stage, however, the concrete tariff levels resulting from the CNE methodology were evaluated at frequent meetings of representatives of the ministry and the CNE (CNE 2004).

46   The RD 661/2007 explicitly states that all tariffs are based on the calculations of 'reasonable rates of return' (BOE 2007c).

- operation hours per year
- performance of reference plants
- economic lifetime of projects and amortisation period for investments
- investment costs (with potential subsidies taken into account)
- tax burdens and benefits
- income from regional support programs
- operation and maintenance costs (fuels, O & M costs, costs for insurance, rentals for land use, and others)
- other income revenues. (CNE 2004)

To conclude, feed-in tariffs in Spain are fixed by Royal Decrees which are of a lower legal rank than laws and are therefore prepared only by the responsible ministry. Normally, the Spanish ministry tables a proposal. Officially, only the National Energy Council (CNE) has to give a viewpoint on the proposal. The response of the CNE is not binding; that is, the responsible ministry may, but need not, take the suggestions for modifications into account. However, experience shows that in most cases the disapproval of the CNE moved the ministry to modify the proposed Royal Decree (M. Bechberger, personal communication, 14 August 2009). Non-governmental actors, such as industry associations and environmental groups, are sometimes invited to give their viewpoint. In the recent past, however, the Spanish legislator has frequently issued new regulations in the field of renewable energies without consulting these non-governmental actors (APPA 2009).

4.3.3. France

The French feed-in tariff scheme was established by the law on the modernisation and evolution of public services in the electricity sector in 2000 (J.O. 2000a). However, the actual regulations for single technologies (including tariff levels) are published in terms of legally lower ranking arêtes. These by-laws are drafted by the responsible ministry alone. After the Ministry has autonomously decided upon the tariff level, the decree is passed on to two consultative bodies, the *Conseil supérieur de l'énergie*[47] and the CRE. The views expressed by each of the bodies are not legally binding and the Ministry can freely decide whether the recommendation will be taken into account or not. In contrast to the viewpoint of the *Conseil supérieur de l'énergie* the opinion of the CRE has to be officially published.

France first determined a pre-defined tariff calculation methodology by the electricity law of 2000 (J.O. 2000a), the legal foundation of the French feed-in tariff mechanism. Accordingly, feed-in tariff levels are determined by two

---

47   The Conseil supérieur de l'énergie – replacing the Superior Council on Electricity and Gas of 1946 in 2005 – consists of Members of the Parliament, representatives of the Ministry in charge, Unions, representatives of regional entities, representatives of electricity consumers and environmental NGOs and representatives of the gas and electricity industry.

combined factors. First, the long-term avoided costs for electricity generation from conventional energy sources. Second, an extra payment for achieving additional, objectives in the field of public services in the electricity sector which have been defined by law (J.O. 2001c). According to the French electricity law from 2000 public services should contribute to 'energy independence and security of supply, air quality and green house gas emissions, the optimal management of national resources, the management of future technologies and rational use of energy' (J.O. 2000a).[48] According to a decree already issued in 2001, feed-in tariffs should reflect the avoided cost of conventional electricity generation plus the above mentioned additional payment (J.O. 2001c). In short, feed-in tariff shall be calculated based on the avoided costs for conventional power generation plus an increase for additional social and environmental benefits.[49] Due to the rather vague definition of the additional objectives the tariff calculation methodology left a lot of space for interpretation. Therefore, different actors in France have used different tariff calculation methodologies.

The approach for tariff calculation was slightly changed by the energy law of 2005 (J.O. 2005b). Feed-in tariffs should now reflect the avoided private costs for electricity generation (e.g. avoided investment costs plus avoided exploitation costs) plus the above mentioned extra payment for achieving the additional environmental or social objectives. This new tariff calculation approach entered into force in April 2006 (CRE 2006, Lesieur 2005).

Despite this new approach for tariff calculation, the government organisation CRE remained very critical of the ministry's proposals. In 2008, the CRE pronounced itself opposed to the tariff for proposed wind energy tariffs (J.O. 2008a). The criticism was founded in the high profitability of projects and the extra cost for final consumers. Interestingly, the CRE explicitly contrasted the situation in France with the framework for electricity generation in other European countries, namely Germany, Spain and Denmark. The CRE viewpoint states that in the case of other countries, the large-scale promotion of wind energy is 'rational' because it replaces fossil fuel based electricity generation. According to the CRE, this is not the case in France, where '78 percent of electricity is produced by nuclear power plants and 10 percent by hydro power plants' (J.O. 2008a). In other words, if applying the avoided cost approach the reference power source for conventional power generation has to be taken into account. The CRE's viewpoint was not taken into account.

---

48   Author's translation of '... l'indépendance et la sécurité d'approvisionnement, la qualité de l'air et la lutte contre l'effet de serre, la gestion optimale et le développement des ressources nationales, la maîtrise de la demande d'énergie, la compétitivité de l'activité économique, la maîtrise des choix technologiques d'avenir, et l'utilisation rationnelle de l'énergie' (J.O. 2000a).

49   The responsible ministry calculates the reference costs of power generation every 3 to 5 years. The last study was issued by MEEDDAT in 2008 (MEEDDAT 2008b).

Interestingly, however, the last viewpoint of the CRE on the new proposal for solar photovoltaic promotion no longer refers to the notion of 'avoided costs'. On the contrary, the CRE is criticising the tariff payment because of its high internal rates of return for solar photovoltaic producers. According to their analysis, internal rates of return of more than 15 percent might be realised in the case of some solar photovoltaic systems, namely building integrated photovoltaic modules for residential houses (CRE 2009a).

Table 4.1   CRE's calculation of internal rates of return for solar photovoltaic systems

	Average internal rate of return of project (IRR) (in brackets: IRR for entirely equity based projects)		
Building integrated	Residential (3 kWp)	Community (250 kWp)	Other usage of buildings
Complex integration (Tariff: 60 €cent/kWh)	15.2 % (91%)	12.2 % (46%)	12.2 % (46%)
Simple integration (Tariff: 45 €cent/kWh)	15.2 % (90%)	9.7 % (35%)	9.7 % (35%)
Free-standing (Tariff: 32 €cent/kWh)	10.5% (40%)		

Source: Author based on CRE 2009a

In fact, the initially value-based tariff calculation approach has been bit by bit replaced by a tariff calculation methodology based on the actual generation costs. This change can first and foremost be observed with respect to solar photovoltaics. Until 2005, the Ministry of the Economy has set a tariff of 15 cent/kWh, arguing that this would equal the avoided external costs plus the additional benefits (R. Claustre, personal communication, 22 October 2008). Only with the new solar photovoltaic tariffs in 2006, which came as a surprise even for experts in the field of solar photovoltaics in France, the tariff level was increased considerably, now reflecting the actual generation costs.

This shift towards a cost based tariff calculation methodology can partially be explained by the fact that the governmental actor ADEME has established an alternative, cost based tariff calculation methodology, the so called 'Profitability Index Method' which is applied for all technologies (J.-L. Bal, personal communication, 24 October 2008).[50] This approach grounds on the notion of 'fair and sufficient' tariffs and was developed by one of the former employees of ADEME, Mr Chabot (Chabot 2008, Chabot et al. 2002). Generally, an internal rate of return of 8 percent is used by ADEME for the tariff calculation.[51]

Interestingly, the responsible Ministry never disclosed its (internal) methodology for tariff calculation. The Ministry's 'methodology' for setting feed-in tariff has never been disclosed. It consists of a consultation process with a large number of stakeholders and the Ministry's proper calculations. For the different technologies, different internal rates of return have been targeted. The responsible Ministry is not bound to take the suggesting of any specific actor into account. Even the proposals of other governmental organisations such as ADEME are taken into account just like any other proposal from non-governmental actors, such as industry associations or NGOs.

In sum, France does not (always) make use of one transparent tariff calculation methodology for all technologies. Even though some actors are using openly disclosed methodologies, the responsible Ministry's approach is not fully clear. This is certainly one of the reasons why an anonymous expert of the French renewable energy market has referred to the tariff setting procedure as 'political tariffs', meaning that tariffs reflect political preferences inasmuch as economic necessities. However, interviews have revealed that the Profitability Index Method of ADEME was used for biogas, solar photovoltaics and wind energy tariffs in 2000 (J.-L. Bal, personal communication, 24 October 2008). Later, in 2002, the responsible Ministry still referred to the avoided cost approach when justifying the low tariff payment for solar photovoltaics. In the recent years, almost all tariffs were in the range of technology specific generation costs. Consequently, empiric evidence suggest that the initial, value based approach for tariff calculation was abandoned or it has at least been gradually replaced by a cost based approach.[52] Today, even the CRE, one of the strongest opponents to feed-in tariffs in France, refers to the internal rates of return of renewable energy projects when criticising tariff payment.

---

50   For a general evaluation of the 'Profitability Index' in economic literature refer to Mian (2002), pp. 280–86.

51   However, a representative of the governmental organisation ADEME stated that currently a higher profitability margin would have to used for tariff calculation, since capital costs have increased due to the global financial crisis (J.-L. Bal, personal communication, 24 October 2008).

52   This general change of the tariff calculation methodology was confirmed by several interviewees (R. Clauste, personal communication, 22 October 2008; M. del Épine, personal communication, 7 August 2007).

4.3.4 Preliminary Conclusion

As shown in the three case studies, initially all three countries started with a value based tariff calculation methodology, directly or indirectly referring to the avoided costs of conventional power generation (France and Spain) or the avoided external costs of these technologies (Germany). In all three countries, this approach was step by step replaced by a cost based tariff calculation methodology, taking the actual generation costs of renewable energy technologies into account (see Figure 4.2).

Despite using the same tariff calculation methodology, the terminology in use differs in all three countries. In Germany, the notion of 'cost-covering remuneration' guides tariff calculation, in Spain the policy maker speaks of 'reasonable rates of return' and in France the so called 'Profitability Index Method' of ADEME guarantees 'fair and sufficient' tariffs (Mendonça et al. 2009b: 19). All three countries have in common that they guarantee a certain internal rate of return.

Tariff calculation methodologies in Germany, Spain and France reveal a number of similarities. These are especially striking when comparing the published documents in Germany and Spain.[53] First and foremost, it is obvious that both countries offer similar internal rates of return for renewable energy producers, ranging from four to nine percent.[54] Besides, the relevant documents reveal that similar cost factors have been factored in when calculating tariffs. Despite these obvious similarities, there is no evidence of the influence of transnational communication. There is no proof of direct exchange between governmental or non-governmental actors in Germany, Spain and France regarding tariff calculation methodologies. Equally, regulatory competition had no influence on cross-national similarities.

However, policy learning among these European member states has most likely taken place indirectly. As a cost-based tariff calculation methodology is one of the most important 'ingredients' of a successful feed-in tariff scheme, the effectiveness and efficiency of national support mechanisms can be directly linked the legislator's choice regarding this design feature. In this respect, the European framework – more precisely, the national targets set by the European framework – have certainly influenced legislators' policy choices. Moving away from value-based tariff calculation towards a cost-based approach allows for more targeted support. On the one hand, windfall profits for cost-effective technologies such as hydro power, wind power and certain types of biomass can be avoided. On the other hand, less mature technologies will also profit from the support mechanism. Especially in the case of France, the move from a value-based approach to a cost-based tariff calculation methodology was necessary in order to be able to reach the European targets.

---

53   A detailed comparison with the French tariff calculation methodology is not possible as the relevant documents have never been published.

54   The French Ministry preferred not to disclose the targeted internal rates of return for their own calculations.

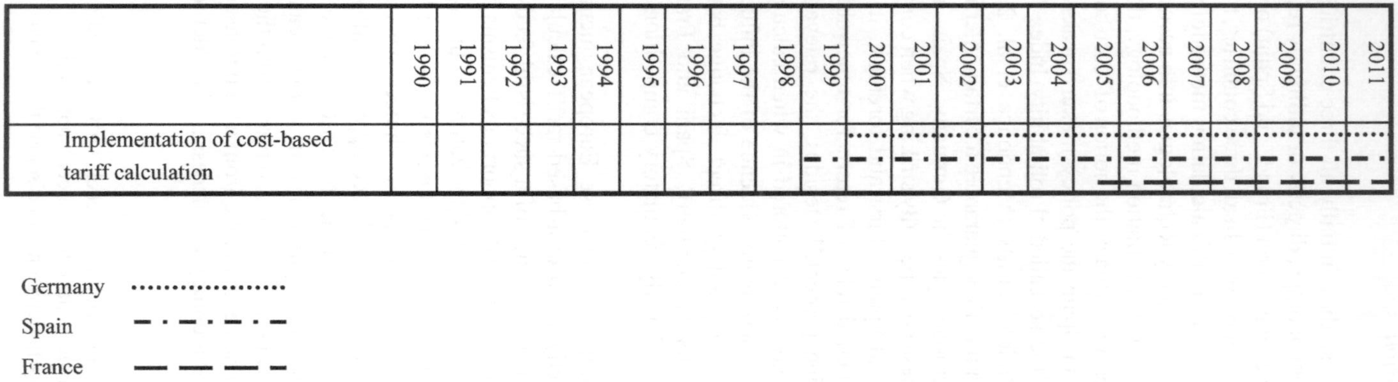

Implementation of cost-based tariff calculation	Germany	Spain	France
2011	⋯	·–·–	–––
2010	⋯	·–·–	–––
2009	⋯	·–·–	–––
2008	⋯	·–·–	–––
2007	⋯	·–·–	–––
2006	⋯	·–·–	–––
2005	⋯	·–·–	–––
2004	⋯	·–·–	
2003	⋯	·–·–	
2002	⋯	·–·–	
2001	⋯	·–·–	
2000	⋯	·–·–	
1999		·–·–	
1998			
1997			
1996			
1995			
1994			
1993			
1992			
1991			
1990			

Germany ⋯⋯⋯
Spain ·–·–
France –––

Figure 4.2   Implementation of cost based tariff calculation methodologies in Germany, Spain and France (1990–2011)

Source: Author

4.4. Duration of Support

Feed-in tariffs are usually guaranteed for a long period of time.[55] This way, the usually high initial investment costs can be recovered over time. Besides, a guaranteed fixed payment over a long period of time significantly increases investment security and therefore facilitates good financing conditions with low interest rates for renewable energy project developers. On average, tariffs are paid for about 20 years because this reflects the general life cycle of many installations for renewable electricity generation. If a legislator opts for a rather short period of guaranteed tariff payment, the tariff level has to be higher in order to assure the amortisation of costs. If tariff payment is guaranteed for a longer period, the remuneration level can be decreased. Some feed-in tariff schemes do not set any time limit for tariff payment; that is, tariffs will be paid during the full lifetime of renewable energy power plants. The main reason for opting for a certain limit is to reduce and control of the overall system cost and to provide an incentive to invest in new technologies after tariff payment has ended.

After the end of the predefined support period, renewable energy producers generally have the right to sell electricity under standard market conditions. Some feed-in tariff schemes also give renewable electricity producers the chance to leave the feed-in contract before the end of the support period. This is especially interesting for producers who achieve higher incomes by selling their electricity on the national spot market at a given moment in time. On the assumption of rising prices for conventionally produced electricity, the market price might become more interesting than the fixed feed-in tariff level in the middle term. This will be the case for the most cost-effective renewable energy technologies, such as wind onshore, in future years. The premium feed-in tariff option (see section 6.1) is often seen as a preliminary stage for renewable electricity producers in which they can learn how to trade electricity on the market.

4.4.1. Germany

The early German feed-in tariff mechanism of 1990 did not explicitly guarantee tariff payment for a longer time period. As described in section 4.3.1, the StrEG followed a value-based tariff calculation methodology, fixing tariffs as a certain percentage of the retail electricity price. These percentage-based payments were related to the retail electricity price of the penultimate legal year (BGBl 1990). Therefore, in every year the tariff payment level was adjusted to the retail electricity price, which, however, was regulated and therefore subject only to marginal annual changes. Nonetheless, the law guaranteed tariff payment for only one year. Guaranteed longer payment duration could be negotiated between power producers and utilities based on longer-term power purchase agreements. This, however, did not effect annual changes in the tariff level.

---

55   In the EU, tariff payment duration ranges from eight to 30 years (Klein et al. 2008).

One of the major innovations of the 2000 amendment was the establishment of a long predefined remuneration period.[56] All renewable electricity producers were now guaranteed a fixed tariff payment for a period of 20 years.[57] The only exceptions were hydro power plants, which, according to the legislator, require longer amortisation periods (BGBl 2000). The amendment of 2000 guaranteed tariff payments for small hydro power plants for the full lifetime of each plant. With the 2004 amendment, the tariff for small-scale hydro power plants (up to 500kW) was increased but at the same time the duration of support was limited to 30 years. Larger-scale hydro power plants (up to 5 MW) received tariff payment for only 15 years (BGBl 2004). With the 2009 amendment, the duration of support for hydro power was changed to 20 years in order to create a unified payment duration for all renewable electricity technologies (BMU 2007b).[58]

4.4.2. Spain

When the feed-in tariff was first established in Spain in 1994, tariff levels were changed every year and no guarantee was given of payment over a long period of time. Similarly, under the 1998 legislation tariff levels were reviewed and changed annually by the responsible ministry. Feed-in rates were adjusted to retail electricity prices, which were fixed by a Royal Decree every year. However, the 1998 legislation for the first time mandated minimum purchase durations of five years. Grid operators were legally obliged to accept these contractual conditions. This provision was the first step towards investment security (BOE 1998a, del Río and Gual 2007). More far-reaching provisions have not been implemented, also because the 1998 legislation primarily targeted market sales.

The need for legislative modifications arose when certain technologies were about to reach the technology-specific mid-term targets. In the case of solar photovoltaics, the tariff payment was guaranteed until the total installed capacity in Spain reached 50 MW. The Royal Decree of 1998 assumed that this capacity would be reached in 2010. Even though this seems rather pessimistic today, it

---

56   Interestingly, the renewable energy associations were initially opposed to the establishment of long payment durations, as proposed by the draft legislation in the parliament. They feared that such major changes of the support instrument could hamper investment security.

57   In fact, the period is usually slightly longer than 20 years, as only the number of complete years that a plant is connected will be counted. If, for instance, a power producer is connected to the grid in February, he will receive tariff payment for 20 years and 11 months.

58   Even though the tariff payment duration is no longer a primary focus of debate in Germany, there is some disagreement regarding the optimal tariff payment duration. In a survey undertaken by the research project Realise Forum, several respondents considered a support period of 20 years for wind power as excessive. By contrast, hydropower and photovoltaics were considered to need longer support (Di Nucci and Mez 2007, Di Nucci et al. 2007).

would have meant that the tariff payment would be granted for a maximum of 12 years or even much shorter periods of time, depending on the year of installation. For this reason, the industry association called for a minimum payment duration of ten years. Referring to the German support scheme, ASIF declared that in the best case tariff payment should be guaranteed for a period of 20 years (ASIF 2002).

The 2004 amendment established the dual remuneration system, comprising the fixed tariff payment option and the market sales option. Renewable power producers opting for the market option were also exposed to the price risk and therefore could not rely on a longer guaranteed payment duration. However, the minimum contractual relations as established in 1998 were still in place (BOE 2004a). The 2004 amendment significantly improved conditions for power producers who opted for the fixed tariff payment, now fixing long payment durations. Tariff payment was guaranteed for a duration of 25 years for solar photovoltaics, solar thermal and hydro power. Wind power, biomass and geothermal power producers[59] even received the tariff payment for the full lifetime of operations, with a slightly higher tariff during the first years of operation. In fact, the lower remuneration, which was granted after 15 or 20 years of operation, was intended to reflect expected future market prices. Therefore, it can be argued that, even in the case of those technologies, above-market tariffs are only guaranteed for the first 15–20 years of operation (M.J. Samaniego, personal communication, 27 February 2008). The 2007 and 2010 amendments did not include any significant changes to the duration of tariff payment.

4.4.3. France

In France, the duration of tariff payment was not defined by the electricity law of 2000 which laid the legal foundations for the French feed-in tariff mechanism. Instead, payment durations were defined within the lower-ranking decrees, which also determine the tariff level for each individual technology. When the first pieces of legislation were issued in 2001 it became apparent that the French feed-in tariff scheme included long tariff payment durations for all technologies. The duration of tariff payment in France varies between 15 and 20 years. Biogas, geothermal, and wind power onshore receive tariff payment for 15 years while solar photovoltaics and wind energy offshore receive a fixed remuneration for a 20-year period. In the case of certain biomass products, the duration of tariff payment was increased from 15 years (2002 legislation) to 2000 years (2009 legislation), thus increasing total revenues for French biomass-based electricity producers.

---

59   This group also includes other less mature technologies, such as tidal and wave power.

4.4.4. Preliminary Conclusion

The analysis of the three case studies shows that feed-in tariff schemes in Germany and Spain have moved from tariff payment based on annual payment variations (Germany until 2000, Spain until 2004) towards longer-term payment durations (see Figure 4.3). It should be noted, however, that, even with annual changes of tariff levels, the early feed-in mechanisms in Germany and Spain offered a certain degree of investment security. This is because the tariff level was linked to the regulated and therefore rather stable retail electricity price. Spain further increased investment security in 1998 by mandating minimum purchase durations of five years. However, these early support mechanisms for renewable electricity in Germany and Spain would not qualify for a feed-in tariff today as they did not guarantee tariff payment over a long period of time. The fact that long tariff payment durations have become an essential part of every feed-in tariff mechanism underlines the importance of payment duration for successful support of green power.

The German feed-in tariff of 2000 was the first legislation to fix the tariff payment for a long period of time. This design feature significantly increased investment security because power producers could now easily forecast their total guaranteed revenues over several decades. This enabled even small actors easily to finance their operations with debt (as opposed to equity).

The implementation of longer payment durations in Spain and France can be attributed to transnational communication based on the positive experience in Germany. Spain followed the same approach in 2004, partly because of pressure from the renewable energy industry, which stressed the success and the higher degree of investment security under the German feed-in tariff scheme (ASIF 2002). When feed-in tariffs were first implemented in France in 2000, the French legislator was already able to profit from experience in other countries, especially Germany. It should be noted that governmental actors in France analyzed the German support instrument as a whole. As a result a number of design features have been adopted in France based on this policy transfer process. Among others, these features include the extended duration of tariff payment.[60] Today, feed-in tariff payment durations in all three countries range from 15 years to the full lifetime of power plants, with 20 years the most widespread payment duration. As Germany was the first country to establish long tariff payment guarantees, the general duration of 20 years became a point of reference for all following countries. Besides, the 20-year period reflected the expected lifetime of a number of technologies. The differences amongst the tariff payment durations in all three countries – ranging from 15 years to the full lifetime of renewable electricity plants – can be explained by the interaction of duration of tariff payment and tariff level: with longer (shorter) tariff payment durations tariff levels can be slightly lower (higher).

---

60 For a more detailed analysis of design options, which can be related to transnational communication between Germany and France, see section 8.1.1.2.

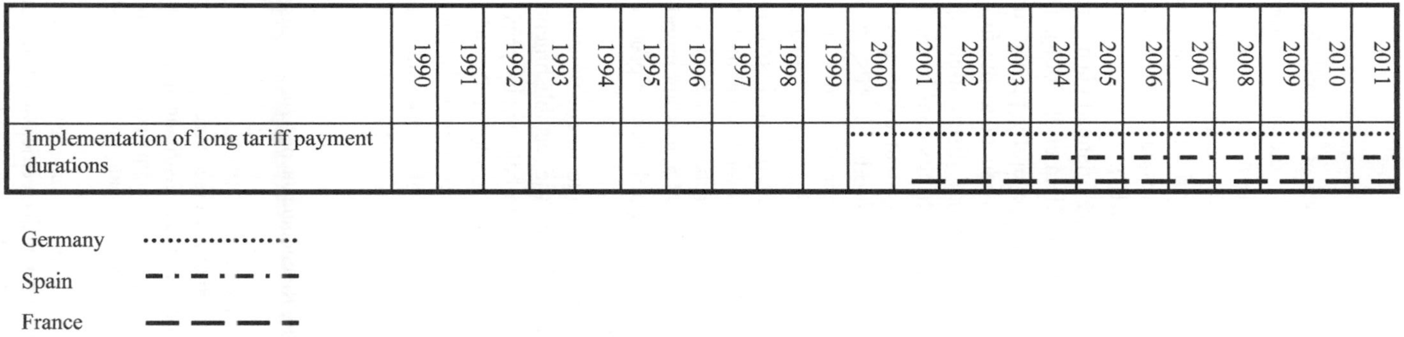

	Implementation of long tariff payment durations
2011	
2010	
2009	
2008	
2007	
2006	
2005	
2004	
2003	
2002	
2001	
2000	
1999	
1998	
1997	
1996	
1995	
1994	
1993	
1992	
1991	
1990	

Germany ·············

Spain — · — ·

France — — —

Figure 4.3   Implementation of long-term tariff payment durations in Germany, Spain and France (1990–2011)

Source: Author

4.5. Financing Mechanism

Supporting renewable energy sources requires finance. In order to cover the additional costs associated with renewable electricity support, the legislator has to set up a financing mechanism for the national feed-in tariff mechanism. The legislator can choose either to distribute the additional costs equally among all final electricity consumers or to finance the additional costs via the general national budget (taxation).[61] These approaches have also been termed 'ratepayer' versus 'taxpayer' financing (Fulton et al. 2010).

In most cases, national feed-in tariffs are financed via a surcharge on the retail electricity price (Mendonça et al. 2009b, Klein et al. 2008). The additional costs for renewable electricity generation are generally passed along the electricity supply chain from the renewable power producer to the final consumer.[62] This financing mechanism allows for the equal distribution of the aggregate of renewable electricity support among all electricity consumers. This burden-sharing mechanism increases the stability of the system as no governmental money is needed to promote renewable electricity. In other words, governmental changes after general elections or a general economic downturn are less likely to have an influence on renewable energy support. Instead of being actively involved in financial activities, in the electricity sector the national government acts only as a regulator of private actors: it sets the feed-in tariff level and the duration of tariff payment, and obliges the grid operator to purchase all electricity produced. Actual revenue streams do not involve governmental actors but merely consist of money transferred between private actors – renewable power producers, grid operators and final consumers (Fell 2009a).

Several feed-in tariff mechanisms include financing exemptions for certain consumer groups. As mentioned above, the establishment of feed-in tariffs normally leads to a small increase in the retail electricity price. Therefore, certain consumer groups that are very responsive to electricity price increases can be exempted from paying the additional amount of money. This is sometimes the case for energy-intensive industries, large shares of whose total 'production costs' are energy related (Mendonça et al. 2009b, Klein et al. 2008).[63] In order to guarantee

---

61   It is also possible to finance the additional costs via a feed-in tariff fund (Mendoca et al. 2009b, Jacobs and Kiene 2009, Reiche and Jacobs 2009). This option has also been discussed as a financing mechanism for renewable energy projects within the international climate regime (Jacobs et al. 2009, Fulton et al. 2010). As this design option is not in use in any European country, it will not be further discussed here.

62   Alternatively, the grid operator can also distribute all costs by a small increase in the pass-through cost of the electricity grid. This approach has been chosen by a number of developing countries in order to reduce additional transaction costs (Jacobs and Kiene 2009).

63   Electricity-intensive industries and companies can be identified by measuring the total amount of electricity consumed, the annual electricity costs in relation to revenues and the ratio of total costs to gross added value (Klein et al. 2008).

the international competitiveness of these companies, some legislators have opted to (partially) exempt them from financing renewable electricity support.[64] A similar exemption could theoretically be implemented for low-income households.[65]

4.5.1. Germany

In Germany, the additional costs of renewable electricity promotion are distributed among all electricity consumers. However, in the 1990s those costs were not distributed evenly. Under the 1990 legislation, only a regional burden-sharing mechanism was implemented. The producer of renewable electricity received the tariff payment from the regional utilities that had a regional monopoly of power supply. The absence of a national burden-sharing mechanism meant that the additional costs of renewable electricity promotion were shared only among the customers of such regional supply monopolies. This increased the costs especially for electricity consumers in the northern part of Germany, were most wind power plants were installed in the 1990s.

In order to eliminate the disadvantage suffered by network operators and final consumers in regions with large shares of renewable electricity, the amendment of 2000 implemented a nationwide equalisation scheme. The distribution system operators were still obliged to pay the tariffs as set under the feed-in tariff mechanism, but they could now transfer them to the level immediately above them – the transmission system operators in their region.[66] At this level, the four German transmission system operators calculate the total costs of renewable electricity promotion and then distribute the costs of the electricity equally among all supply companies. Thus, the additional costs are equally distributed among all German electricity consumers. The fundamental logic of this financing mechanism, as depicted in Figure 4.4, remained unchanged under the 2004 amendment.

---

64 Exemptions for energy-intensive industries have been implemented in a number of European countries, including Austria, Denmark, Germany and the Netherlands (Klein et al. 2008). It should be noted that these practices have often been criticised, as the financing exemption reduces the incentive for the electricity-intensive industry to adopt energy-efficiency measures. Besides, these consumer groups are in any case often able to profit from preferential electricity tariffs.

65 The exemption of low-income households is planned by the Malaysian government for the national feed-in tariff mechanism that is expected to start in 2011. According to the press release from the responsible ministry, only citizens with a monthly electricity consumption exceeding 200 kWh will have to finance the feed-in tariff mechanism (KeTTHA 2010).

66 In exceptional cases, wind power producers are directly connected to the distribution network. In this case, the financial flows are of course slightly different, as shown in Figure 4.4.

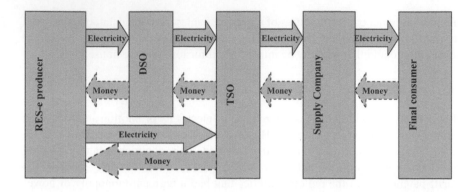

Figure 4.4   Financing mechanism for the German feed-in tariff scheme

Source: Jacobs and Kiene (2009)

In the summer of 2009 the German legislator changed the financing mechanisms for the national feed-in tariff mechanism by detaching the flow of renewable electricity from the financial flow (BGBl 2009a). Previously, supply companies were obliged to purchase fixed amounts of renewable electricity, so-called EEG quotas, from the transmission system operators. The size of these quotas was calculated on the basis of historic data and estimates of the growth of the renewable electricity market. However, because of the fluctuating nature of several renewable electricity technologies, the actual EEG quota often diverged widely from the forecasts, thus imposing additional costs especially on small suppliers, who had to balance these variations though short-term electricity trading. From 2010 onwards, transmission system operators will no longer deliver green electricity to the supply companies. Instead, they will be obliged to sell green power on the on day-ahead market or the spot market (BDEW 2009a).[67]

The exemption for energy-intensive industry, the so-called hardship clause, was implemented for the first time in the summer of 2003 (BBGl 2003). The explanatory memorandum of the 2004 amendment stated that, with regard to the increasing share of renewable electricity, energy-intensive industry had to be partly exempt from the extra payment for renewable electricity support since such companies competed on international markets. As energy costs make up a large share of the total cost of production, these companies were very sensitive to price increases and were consequently exempted (BBGl 2004). In the first year, only 40 companies were eligible for exemption because the criteria of eligibility were rather restrictive. Power-intensive manufacturing companies were eligible for the exemption under the hardship clause only if their electricity consumption was

---

67   For a more detailed analysis of these aspects see Rostankowski and Oschmann (2009) and Ragwitz and Sensfuss (2008).

greater than 100 GWh and the ratio of their electricity costs to gross value added exceeded 20 percent (Deutscher Bundestag 2004b).

With the amendment of 2004, the criteria of eligibility for the exemption became less restrictive so that the exemption could be extended to more companies in the energy-intensive manufacturing industry. Power consumption had to exceed only 10 GWh and the share of the gross value added had to be only 15 percent. In 2007 almost 400 companies were eligible under the hardship clause (BMU 2007a); by 2010 the number had increased to 560 (BMU 2010c). Power-intensive manufacturing companies pay an additional cost of a maximum only 0.05 €cent/kWh to support renewable electricity. However, the EEG amendment of 2004 stated that the additional costs for all other consumer groups should not be increased by more than ten percent, the so-called ten percent cap. Therefore, the electricity-intensive industry had to pay more than the envisaged 0.05 €cent/kWh. With the minor EEG amendment of 2006, this ten percent cap was removed, thus limiting the contribution of energy-intensive industry to 0.05 €cent/kWh (BBGl 2006).[68] After the 2009 amendment, producers who wanted to become eligible for the financing exemption had in addition to prove and certify that energy-saving and energy-efficiency measures had been adopted (BBGl 2008).

4.5.2. Spain

In 1994 the Spanish electricity market had still not been liberalised. As a result, the additional costs for renewable electricity generation simply became part of the overall cost structure in Spain. In 1998 – one year after market liberalisation was introduced by the new electricity law – a national financing scheme for feed-in tariff payments was established. The producer of renewable electricity received the tariff payment from the distribution or transmission system operator, depending on the voltage level of the connection line. Later, the grid operators could pass all the costs to the National Energy Commission, CNE. At this stage, all costs are accumulated and equally distributed among all electricity consumers via the electricity supply companies (del Río and Gual 2007). In Figure 4.5, the arrows indicate the money flow under the feed-in tariff scheme.

However, the feed-in tariff mechanism in Spain is not fully financed via a top-up on the electricity price for final consumers, because retail electricity prices for certain consumer groups are still regulated. In other words, electricity prices for final consumers are fixed by the legislator and therefore do not fully mirror generation costs. Since 2000, the legislator has not increased the regulated tariff sufficiently to cover all costs related to the Spanish electricity system. The additional costs of the national electricity systems that cannot be financed by the

---

68   Overall, the implicit benefits for these companies in 2008 were worth around 650 million €. In 2010 it was already almost 1bn €. Due to the exemption of these industries, the additional cost for all other electricity consumers increased by 17 percent in 2007 (Oschmann and Thorbecke 2006, Wenzel and Nitsch 2008, BMU 2010c).

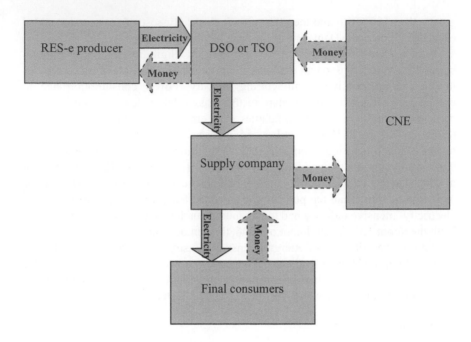

Figure 4.5   Financing mechanism for the Spanish feed-in tariff mechanism

Source: Author based on del Río and Gual 2007

payments of the final consumers have been accumulated in recent years in the so-called 'tariff deficit', that is, the financial debts of the Spanish electricity system.[69]

The refinancing of this deficit was first regulated in 1997. This piece of legislation guarantees full cost compensation for power producers, grid operators and supply companies (BOE 1997c). In 2006, after the debt of the Spanish electricity system had reached a substantial level,[70] the legislator implemented a refinancing mechanism. The extra debt that had been accumulated over a three-month period was now auctioned among actors in the Spanish electricity market, for instance electricity traders or large-scale electricity consumers. In this way the Spanish legislator sought to refinance the debt of the Spanish electricity system via international capital markets. However, with the financial crisis that broke in 2008, fewer and fewer actors were willing to participate in this auctioning mechanism.

---

69   It has to be clear that the tariff deficit does not refer only to the financial deficit related to the Spanish feed-in tariff mechanism. The tariff deficit also originates in other activities of the Spanish electricity system.

70   The electricity deficit has increased rapidly since 2008, reaching more than €10 billion in 2009 (Libertad Digital 2009).

Consequently a new mechanism had to be developed to finance the deficit of the Spanish electricity system.

In 2009, a securitisation fund was established for this purpose (*Fondo de Titulización del Déficit del Sistema Eléctrico*). This fund is intended to enable power companies to recuperate the money they cannot charge to final consumers – the tariff deficit. The money for the fund will be raised via the costs of the Spanish electricity system that eventually have to be paid by the final electricity consumer. For 2009–12, the electricity system deficit was expected to increase further as retail electricity prices are still regulated. However, the Royal Decree 6/2009 had already established annual upper limits for the new deficit. From 2013, electricity prices will have to reflect power generation costs and will no longer be regulated (Morris 2009b). The tariff deficit will be refinanced via an increase in the grid access charges that each electricity consumer has to pay on a monthly basis (BOE 2009a, Bechberger 2009). Refinancing the existing tariff deficit has to be finalised by 2028 (BOE 2010a). In other words, refinancing today's electricity deficit will result in higher future prices for the final consumer, when retail electricity prices will no longer be regulated. It should be noted, however, that the feed-in tariff mechanism in Spain is not financed via the general national budget.

4.5.3. France

The French feed-in tariff scheme is financed by a tax levied on each unit of electricity produced in France. While the tax is raised from electricity producers, the costs are passed on to the final consumers, as they are factored in when electricity tariffs are set. Therefore, the evolving costs for the promotion of renewable electricity are distributed equally among all French electricity consumers. Each consumer pays a certain amount per unit of electricity consumption. The tax in question is the *contribution au service public de l'électricité* (CSPE). This mechanism was established in 2003. For 2009, the tax amounted to 4.5 € for each megawatt-hour of electricity produced (CRE 2009b). However, is should be noted that the promotion of renewable electricity accounts for only a small fraction of this amount. In 2007 only three percent of the revenue from the CSPE was used for renewable electricity promotion under the feed-in tariff mechanism (CLER 2008). Besides, the revenue flows from the CSPE are used for the promotion of co-generation power plants, for financing the additional costs for electricity generation in the French overseas departments, for financing reduced electricity tariffs for low-income households, and for other purposes.

Interestingly, renewable electricity producers receive the tariff payment from the power generation branch of the former monopolist, EDF. The EDF generation branch is legally obliged to guarantee tariff payment to other renewable electricity producers – its natural competitors. The EDF generation branch also purchases all renewable electricity and therefore has the exclusive right to sell this electricity on the market. This procedure has been criticised by competitors of EDF, since, as the former monopolist already owns most of the power production capacity, by buying and selling all renewable electricity EDF strengthens its dominant role in France's power market. EDF itself is not fully satisfied with the current situation, but for a different reason. Although EDF is reimbursed for tariff payment via the CSPE tax, the company has to cover all costs related to the administration of feed-in tariff contracts. According to the utility, these costs amounted to € 10 million in 2008 (P. Balaguier, personal communication, 8 October 2008).[71] This partly explains the company's hostility to the national feed-in mechanism.

In France, energy-intensive industry is partly exempt from the CSPE tax. In 2003 the legislator limited the additional burden for consumers with very high electricity consumption by capping the annual CSPE tax payment at 500,000 € for each consumption site. The exemption of energy-intensive industry was further extended by modifying the aforementioned cap: the electricity law of 2005stipulates that all actors with an annual electricity consumption of more than 7 GWh are partially exempt. The CSPE payment is limited to 0.5 percent of the gross value added of each company (J.O. 2004b, 2005b).

4.5.4. Preliminary Conclusion

Germany, Spain and France all make use of a financing mechanism that distributes the additional costs among all national electricity consumers. Over the years, equal burden sharing mechanisms have been implemented in all three countries analysed (see Figure 4.6). This ratepayer based financing mechanism allows for equal burden sharing and is therefore widely spread in all European countries.[72] The national financing mechanisms have been slightly modified over the years, thus responding to the changing actor constellations due to market liberalisation. In Germany tariff payment is organised through a complex burden sharing mechanism between non-governmental actors; in Spain, the National Energy Commission, a governmental actor, is responsible for organising money allocation; and in France, the necessary financing is generated by a tax and the tariff payment that is curiously guaranteed by the quasi-monopolist EDF.

To some extent, the spread of ratepayer based financing mechanism and equal burden sharing has been steered from the European level through the enforcement of EU state-aid rules. Support mechanisms (more precisely, their respective financing mechanisms) have to be approved by the European Commission, which determines whether national actors are indirectly privileged within the internal market. The European treaty of Lisbon defines state aid as 'any aid granted by a Member State or through State resources in any form whatsoever which distorts or threatens to distort competition by favouring certain undertakings or the production of certain goods, in so far as affecting trade between member

---

71   EDF has to deal with about 2,000 feed-in tariff contracts, excluding solar photovoltaics (P. Balagier, personal communication, 8 October 2008).

72   Eighteen out of 21 EU feed-in tariff countries had implemented an equal burden sharing mechanism by 2008 (Klein et al. 2008).

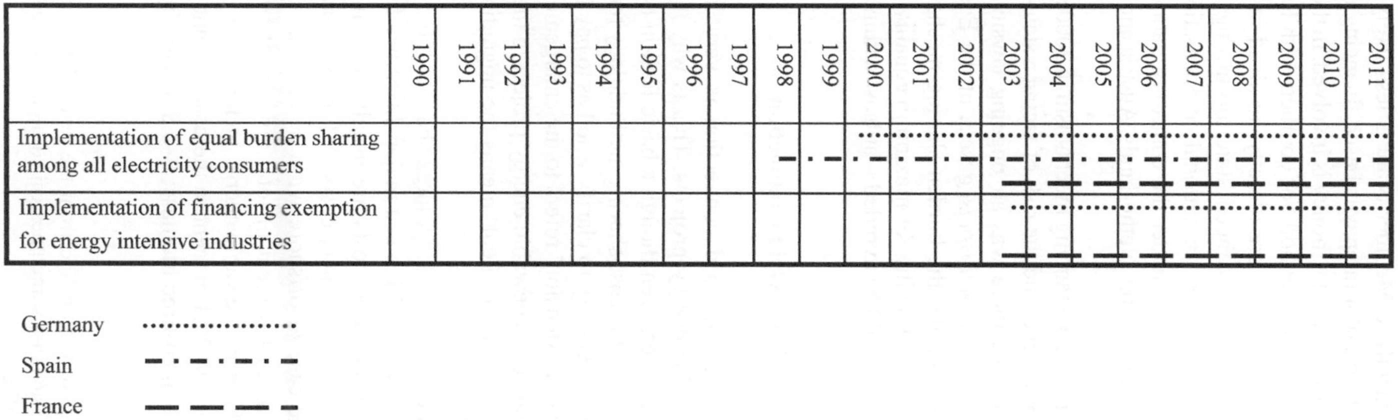

	Implementation of equal burden sharing among all electricity consumers	Implementation of financing exemption for energy intensive industries
2011		
2010		
2009		
2008		
2007		
2006		
2005		
2004		
2003		
2002		
2001		
2000		
1999		
1998		
1997		
1996		
1995		
1994		
1993		
1992		
1991		
1990		

Germany ⋯⋯⋯

Spain ▪ ▪ ▪

France ▬ ▬ ▬

Figure 4.6   Implementation of equal burden sharing in Germany, Spain and France (1990–2011)

Source: Author

states' (EU 2007). Consequently, a Member State is less likely to face state aid infringement procedures in case of a ratepayer based financing mechanisms with equal burden sharing as no governmental money is involved. In the case of taxpayer based financing mechanisms, they likelihood of conflicts with European state-aid rules is higher as governmental actors are directly involved in the allocation of financing. Several European countries faced state aid-infringement procedures as a result of financing feed-in tariffs via national compensation funds which – according to DG Competition – were under the control of national governments. They include Luxemburg, Slovenia, the Netherlands, Austria and Ireland (Fouquet 2006, Maurer 2006).

Germany modified the national financing mechanism in 2000 partly in response to pressure from the European level and the risk of facing a state-aid infringement procedure. Because at that time there was an ongoing lawsuit against the first German feed-in tariff law, the *Stromeinspeisegesetz*, at the European level, the German legislator wanted to design the burden sharing mechanism in the least market-distorting way possible, allowing for maximum equality among all actors. This way, market distortions should be avoided – at least within German borders.

4.6. Cost Sharing Methodology for Grid Connection

The cost-sharing methodology for grid connection is very important for the financial viability of renewable energy projects. This is why this design feature is included in the present chapter, even though a basic feed-in tariff can certainly work without explicitly defining the cost-sharing methodology for grid connection. With this design feature it is necessary to clarify who has to pay for grid extension and grid reinforcement. 'Grid extension' refers to the enlargement of the existing grid in order to connect the new renewable energy project to the already existing grid connection point. 'Grid reinforcement' means the upgrading of the existing grid by increasing overall capacity.[73]

Grid connection costs can account for a large share of the overall investment costs of renewable energy projects. This is a major difference from conventional power plants, whose grid connection costs are usually marginal because of the larger scale and very high overall project costs. A European study on this topic has calculated that, in the case of offshore wind power plants, grid connection costs can account for up to 26.4 percent of total investment costs. Even though the share is smaller for all other renewable energy technologies (3.6–13.6 percent in the case of wind onshore; 9–26.4 percent in the case of wind energy offshore; 2.9–3.5 percent in the case of co-fired biomass; 0–4.8 percent in the case of solar

---

73   Costs related to the connection line from the renewable energy power plant to the extended transmission system are sometimes called immediate 'connection assets' or 'connection works' (Scott 2007).

photovoltaics),[74] the cost sharing methodology for grid connection is often crucial for the profitability of a renewable energy project.

With respect to cost-sharing methodologies for grid connections, that is, cost sharing between renewable electricity producers and the grid operator, one can differentiate between 'deep' and 'shallow' connection charging approaches. The so-called deep connection charging approach assigns all costs to the renewable electricity producer, both for grid connection and for grid reinforcement. This includes the costs of the connection line to the proximate connection point as well as the costs of reinforcing the already established grid infrastructure. In the event of a lack of transmission capacity, the producer is obliged to pay for the necessary upgrading. Under the shallow connection charging approach, the renewable energy producer has to pay only for the new electricity line to the proximate grid connection point, while the grid operator has to cover all costs of potential reinforcement of existing grid infrastructure.[75] In the recent past, some legislators have implemented a so-called super shallow connection charging approach. In this case, even the costs of the new power line to the proximate grid connection point will be covered by the grid operator. This approach has so far been used only for offshore wind power projects, since the proximate grid connection is usually very far away (onshore), which would make grid connection even more expensive (Mendonça et al. 2009b).[76]

4.6.1. Germany

The first feed-in tariff scheme of 1990 included an obligation to connect renewable electricity producers to the grid and to purchase all electricity. The law did not include any further specifications regarding the sharing of costs. Renewable electricity producers had to pay for the grid connection to the proximate grid connection point which was assigned by the grid operator. This provision sometimes meant that – for technical reasons – grid operators chose grid connection points that were far away from the actual production site. In some cases, this made investment in wind power plants unprofitable (R. Bischof, personal communication, 15 September 2008).[77]

The feed-in tariff amendment of the year 2000 was the first law explicitly including provisions concerning grid connection, grid reinforcement and grid extension. Under Article 10, the producer had the right to connect the plant to

---

74   See http://greennet.i-generation.at/.

75   As with all grid-related costs, the grid operator can pass them on to the final consumer in the form of system charges.

76   As well as in Germany, this approach has been adopted in Denmark. Specific planning zones have been established and wind power producers have to pay the connection line only up to the boundary of the zone (Klein et al. 2008).

77   However, wind power plants at that time were so small that they could sometimes even connect to the lowest voltage level.

the 'technically and economically most appropriate grid connection point'. In this way the legislator hoped to prevent grid operators, that is, utilities, from using their dominant market position to exclude potential competitors from power generation. As regards cost sharing, a shallow connection charging approach was adopted. The renewable power producers had to cover the immediate grid extension costs up to the proximate grid connection point. Grid reinforcement, however, had to be covered by the grid operator, who could transfer the costs to the final consumer via the network usage fee.[78] This approach was maintained or even enhanced in the two following amendments.

In 2004 the rights of the renewable shallow grid-connection approach was strengthened by a provision stating that, if the plant was not connected to the 'technically and economically most appropriate grid connection point', the grid operator had to cover the resulting incremental costs (BBGl 2004).In order to further incentivise offshore wind power projects, the German legislator decided to improve grid connection conditions by establishing a super-shallow connection charging approach in 2007. In contrast to the shallow connection charging approach, where the producer of renewable electricity is still financially responsible for grid connection to the nearest connection point, under the super-shallow approach states even this cost has to be covered by the grid operator.[79]

4.6.2. Spain

In Spain, the so-called deep connection charging approach was applied until 2008. This approach was first established under the RD 2366/1994 and confirmed by succeeding legislation in 1998, 2004 and 2007. Accordingly, the renewable electricity producer had to pay for grid extension and grid reinforcement if this is necessary (Knight et al. 2005, A. Ceña, personal communication, 12 February 2008). An effect of the deep connection charging approach was that grid connection makes up a relatively large portion of the overall installation costs in the case of wind energy. A research project commissioned by the wind energy association, AEE, concluded that about 11 percent of total investment costs could be attributed to grid connection (InterMoney Energía 2006). In order to improve investment conditions in Spain, the industry association APPA and Greenpeace called for a shallow grid connection charging approach in 2009 (APPA and Greenpeace 2009).

---

78   The EEG 2000 also guarantees renewable energy producers priority connection to the grid; that is to say, renewable energy projects have to be connected before conventional power plants.

79   The so-called *Energieleitungsausbaugesetz* of 2009 further improved grid connection procedures for renewable energy producers. In particular, the increased use of underground cables was intended to reduce public opposition to wind power plants in touristy, coastal areas. Besides, the new law incentivised the use of pump storage systems by partly excluding them from grid-use charges (BMU 2009c).

In 2008, the Spanish legislator modified the national legislation for grid connection charging, moving a small way towards a shallow connection charging approach. The royal decree 325/2008 differentiated between the 'natural' extension of the transmission grid, meaning grid extension which is necessary to cope with the increasing electricity demand, and grid extension arising from the inclusion of new productive capacity. For the latter, the regulation stated that power plants with a capacity up to 100 kW (in the case of the low tension grid) or up to 250 kW (in the case of the high tension grid) are exempt from paying for grid connection and grid reinforcement (BOE 2008b). This way, small-scale renewable electricity producers can avoid paying the costs of grid reinforcement and extension.[80] Even though the Spanish government claims to have established a shallow grid connection approach (MITYC and IDAE 2010), according to our definition from section 4.6 the new Spanish approach should rather be termed 'shallowish', as larger-scale renewable electricity producers still have to pay for grid extension or reinforcement.

4.6.3. France

France followed a 'deep' connection charging approach until 2001. In 2000, when the feed-in tariff scheme was established for the first time, the law stated that any grid reinforcement that could be directly related to the installation of a new power plant had to be paid for by the plant operator (J.O. 2000a). Conditions for renewable power producers improved in 2001, when a 'shallowish' grid connection charging approach was implemented for both 'green' and 'grey ' electricity. The renewable electricity producer still has to pay for the connection to the proximate grid connection point. With respect to grid reinforcement, the renewable electricity producer has to pay for the reinforcement only of the voltage level he was connected to. The reinforcement of the next-highest voltage level is generally paid for by the grid operator (Knight et al. 2005). As most renewable energy power plants are smaller than 12 MW, they are usually connected to the lower voltage level of 20 kV. Under the aforementioned regulation, these renewable power producers had to pay for grid reinforcement only at the 20kV level. Therefore, the costs of grid reinforcement are shared by the grid operator (high voltage level) and the renewable power producer (low voltage level). If renewable energy producers were obliged to reinforce the existing electricity grid, they could ask other power

---

80   However, grid extensions have to be realised only after a three-year period. In the event of insufficient grid capacity, a renewable power producer is able to decide to feed in a limited amount of electricity so that grid reinforcement is not necessary in the short term. After three years, and in the event that the power plant has been an efficient generation unit, the grid operator is obliged to reinforce the grid based on the Spanish regulation for Binding Transport Planning (L. Temblenque, personal communication, 13 February 2008; BOE 2008b).

producers who later wanted to connect to the same connection point to pay their respective shares (J.O. 2001a).

This regulation was only slightly modified in 2005 (J.O. 2005a), but the cost-sharing methodology was adopted in a major fashion in France in 2007. The grid operator pays a fixed share of all costs related to grid connection and reinforcement. The grid operator pays 40 percent of all costs while the power producer has to pays the remaining 60 percent. This rule applies to all power plants no larger than 12 MW that are connected to the distribution system. This regulation does not, however, apply to all power plants connected to the transmission system. Instead, the share of grid connection costs that power producers have to cover depends on a rather complex calculation, including parameters such as voltage level, the installed capacity of each plant and even the length of the connection line (J.O. 2007b, 2008e). Nonetheless, aspects of the shallow grid connection charging approach are included. Therefore, the French approach is usually referred to as a 'shallowish' approach.

A new regulation is currently being discussed and voted on as part of the second *Grenelle de l'Environnement*. Special cost-sharing methodologies for grid connection will be established for renewable electricity producers. Renewable power producers will still pay a share of grid connection and reinforcement costs, but this share will then be based on a fixed payment for each MW that is connected to the grid. The exact amount of money per megawatt installed has yet to be determined. As well, grid capacity will be reserved for future renewable power generation projects.

4.6.4. Preliminary Conclusion

As shown above, Germany, Spain and France have all moved from a deep grid connection-charging approach towards a shallow or 'shallowish' one (see Figure 4.7). Germany even shifted to a super-shallow approach in the case of offshore wind power. Until 2008, the shallow grid connection-charging approach was not common in the EU member states. An analysis of the same year revealed that only four countries of the EU-15 applied a shallow connection-charging approach, three countries had an intermediate approach and all other countries made use of a deep connection-charging approach.[81] Interestingly, the countries making use of the shallow connection-charging approach are generally those with a high share of decentralised electricity generation units (Knight et al. 2005).

The different design options – shallow and deep connection charging – also reflect the conflict between decentralised and centralised energy systems. The deep connection charging approach is historically grounded and designed around the characteristics of an energy system based on centralised, large-scale conventional power plants. In the light of the high investment costs of a large-

---

81   In 2008 only Denmark, Germany, the Netherlands and Belgium applied a shallow connection-charging approach (Klein et al. 2008).

scale conventional power plant, the additional expenditures on grid connection under the deep approach are negligible. Besides, the location of a conventional power plant does not affect its power output, as in the case of a wind power plant. The deep connection charging approach implies an incentive to optimally use the already existing grid infrastructure, which makes sense in the case of a coal- or gas-fired power plant but not in the case of a wind power plant. The infrastructure of the European electricity grid was mostly established before the liberalisation process was initiated and designed for large scale, centralised power production (EU Commission 2005a). However, wind power plants should be built in the windiest locations and not just in regions with available grid capacity. Therefore, a shallow connection charging approach can be considered more appropriate for a decentralised energy system based on renewable energy sources.

Generally speaking, government are now moving towards a shallow grid connection charging approach because this takes the needs of decentralised power producers better into account. First of all, this approach lowers the overall costs of grid connection. Besides, and even more importantly, renewable energy projects will likely be developed at the locations with the best resources (for example, wind speed) rather than with the best grid availability. The shallow and super-shallow connection charging approaches allow for a strict separation between infrastructure investment and investment in new generation capacity (Auer et al. 2007, Mendonça et al. 2009b).

Besides these more general considerations, the European level has certainly influenced national policy makers' decision to move away from the deep connection-approach and instead apply a shallow or 'shallowish' approach. This approach has been identified and promoted as international 'good' or 'best' practice in official assessments and recommendations of the European Commission (EU Commission 2005a) as well as large-scale European research projects (e.g. Knight et al. 2005, Auer et al. 2007). In order to push national legislators to review their national grid-connection procedures, the new renewable energy directive of 2009 required legislators to review and eventually improve the national framework every two years. The new directive for renewable energy sources of 2009 does not explicitly call for a shallow connection-charging approach. In a more 'diplomatic' way, the directive states that 'Where appropriate, Member States may require transmission system operators and distribution system operators to bear, in full or in part, the costs' related to grid connection and grid reinforcement (EU 2009).

Besides these more general considerations, the European level certainly influence national policy makers' decision to move away from the deep connection approach and instead apply a shallow or 'shallowish' approach. This approach has been identified and promoted as international 'good' or 'best' practice in official assessments and recommendations of the European Commission (EU Commission 2005a) as well as large-scale European research projects (e.g. Knight et al. 2005, Auer et al. 2007). In order to push national legislators to review their national grid-connection procedures, the new renewable energy directive of 2009 required legislators to review and eventually improve the national framework every two

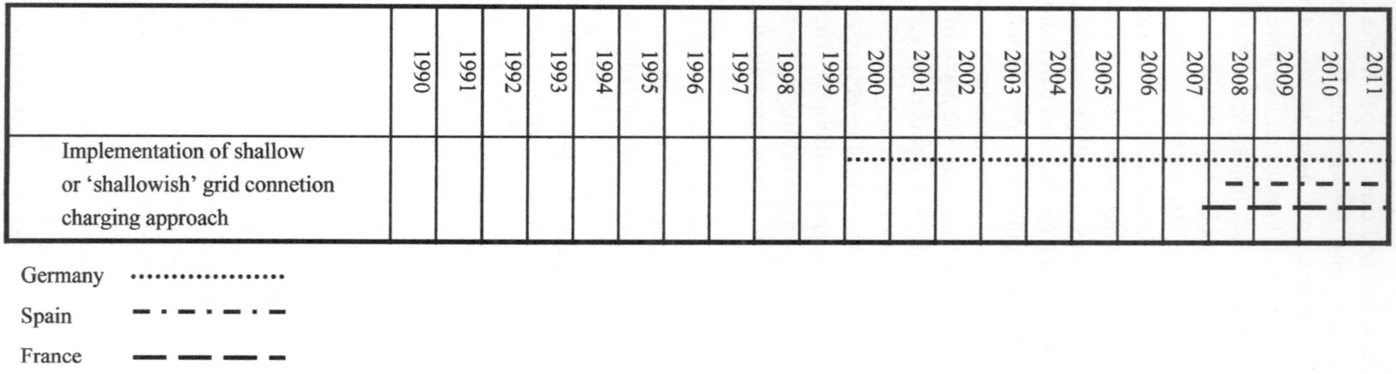

	2011		
	2010		
	2009		
	2008		
	2007		
	2006		
	2005		
	2004		
	2003		
	2002		
	2001		
	2000		
	1999		
	1998		
	1997		
	1996		
	1995		
	1994		
	1993		
	1992		
	1991		
	1990		
Implementation of shallow or 'shallowish' grid connetion charging approach			

Germany ....

Spain — · — · —

France — — — —

Figure 4.7   **Implementation of shallow or 'shallowish' grid connection charging approaches in Germany, Spain, and France (1990–2011)**

Source: Author

years. The new directive for renewable energy sources of 2009 does not explicitly call for a shallow connection-charging approach. In a more 'diplomatic' way, the directive states that 'Where appropriate, Member States may require transmission system operators and distribution system operators to bear, in full or in part, the costs' related to grid connection and grid reinforcement (EU 2009).

4.7. Conclusion

This chapter has shown that there is a high degree of policy convergence on basic feed-in tariff design features in Germany, Spain and France. In all three countries, the number of eligible technologies has increased over time (see section 4.1), capacity caps have either been extended or removed (see section 4.2), legislators have moved from a value-based tariff calculation methodology to a cost-based approach (see section 4.3), all countries now make use of long tariff payment durations (see section 4.4), the countries in question have all established an equal burden-sharing mechanism for the financing of the national feed-in tariff scheme (see section 4.5) and now make use of a shallow or 'shallowish' grid connection charging approach (see section 4.6).

The historical analysis of basic feed-in tariff design features sheds new light on the evolution of this support instrument. In the 1990s there were not many differences between general pricing policies in monopolistic power markets and feed-in tariffs for renewable energy sources. In line with general pricing policies for small producers of conventionally produced electricity (and small hydro power producers in the case of Spain), tariff payment was guaranteed for only a short period of time (generally one year) and tariffs were calculated on the basis of the avoided cost of existing, conventional power plants. Besides, the 'deep' connection charging approach was the predominant cost-sharing methodology for grid connection – an approach that has been developed on the basis of experience with large-scale, conventional electricity power plants.

National policymakers adjusted general pricing policies only very slowly to the specific needs of renewable electricity investment, thus establishing feed-in tariff mechanisms as we know them today. Tariff payment durations have been extended, tariff calculation is now based on the average generation costs of each technology, and a shallow or 'shallowish' grid connection charging approach has been implemented in order to separate investment in grid infrastructure from investment in renewable power generation capacity. The implementation of all three design features can be considered as truly innovative, if not revolutionary, as all three features clash with historically rooted traditions of power market regulation. In all three cases, national legislators have changed the established rules of the power market in order to take better account of the needs of renewable power producers. Never before have national legislators fixed payment duration over a long period of time in order to increase investment security. Never before has the remuneration for independent power producers been detached from the avoided costs of conventional power generation. Never before have legislators established specific grid connection-charging methodologies for renewable electricity producers. These fundamental changes also explain the fundamental opposition of certain actors to feed-in tariff mechanisms.

The difficulty of changing historically rooted power market rules is most obvious in the case of tariff calculation methodologies. The payment of remuneration above the avoided costs of conventional electricity generation clashes with the political objective of providing electricity at the lowest possible cost – one of the three main energy policy objectives of the EU and its member states. Therefore, the process of switching from a value-based to a cost-based tariff calculation methodology was very slow in all three countries. Only when the other two energy policy objectives – energy security and sustainable energy supply – became more important, legislators were willing to grant tariff payment based on technology-specific generation costs. Besides the obvious fact that renewable power sources have a positive effect on energy security and sustainable electricity supply, the argument that renewable energy support today can help to stabilise energy prices tomorrow has become more and more important.

As well, paying renewable electricity producers a tariff which is based on the production costs of each technology is diametrically opposed to the philosophy of competitive, liberalised electricity markets. Theoretically, different power-generation technologies are openly competing against each other in a liberalised market, and the most cost-effective technology will be most competitive and therefore mostly deployed. In other words, paying renewable electricity producers only the avoided costs of the conventional power system would better conform to the idea of liberalised markets, as this means that the average generation costs of existing technologies set the benchmark. This also explains why certain actors that have promoted the liberalisation of electricity markets, such as the European Commission and the International Energy Agency, have long been opposed to feed-in tariff mechanisms (see section 8.1.1.1). The discussion about the best support instrument at the European level can be fully understood only by recognising that market liberalisation and the intensive promotion of renewable electricity started almost simultaneously at the end of the 1990s. While market liberalisation was primarily triggered in order to create a more cost-efficient electricity system (economic objective), renewable electricity support was primarily strengthened because of increasing environmental concerns (ecological objective).

The implementation of all three design features required a change in the mindset of national legislators, which also explains the slow diffusion of precisely those design features. Spain implemented long tariff payment duration only in 2004, and France switched to a cost-based tariff calculation approach only in 2005. Both countries have (still) not (fully) implemented a shallow grid connection-charging methodology. Nonetheless, all three design features are considered international best practice today, since they significantly improve the investment conditions for renewable electricity producers.

According to our hypotheses formulated at the start of this research project (section 2.3), the convergence of basic feed-in tariff design features in these EU member states can be attributed to a number of mechanisms identified by the theoretical framework of policy convergence. Even though regulatory competition seems to have played no role in fostering policy convergence, independent problem solving and transnational communication have been important drivers of the diffusion of basic feed-in tariff design features in Germany, Spain and France. Besides, a certain influence has been exercised from the European level.

When one discusses the effects of transnational communication, it is important to distinguish between the concrete information exchange regarding specific feed-in tariff design features and more indirect communication via the diffusion of information regarding feed-in tariff mechanisms in general. A concrete exchange of policy design information can be observed only in the case of feed-in tariff payment durations, when the Spanish industry association, ASIF, explicitly referred to German experience in order to win over the national legislator (section 4.4.2). However, certain actors have communicated more general information regarding feed-in tariff design. When one compares the German feed-in tariff mechanism of 2000 with the French feed-in tariff legislation of 2001 and 2002, the similarities are striking. They are largely due to intensive communication processes between German and French governmental actors during this time period.

In those years environmentalist parties were members of the governments of both countries. The leading politicians with respect to renewable energy policies of the two governments – Hans-Josef Fell from Germany (spokesman for energy) and Cochet (minister for the environment) – frequently met in order to discuss the design of national feed-in tariffs and to learn about best practice solutions. During those meetings, Hans-Josef Fell explained the design of the German feed-in tariff scheme, which was crucial at a time when France was still discussing the detailed design of its national legislation (H.-J. Fell, personal communication, 25 February 2010). Besides, the French governmental organisation ADEME prepared a report making recommendations for the French feed-in tariff mechanisms. The recommendations were largely based on an analysis of the German feed-in tariff.[82] This also underlines the importance of the German feed-in tariff – especially the amendment of 2000 – for the diffusion of basic feed-in tariff design features. The German support scheme of 2000 was the first to fully apply a cost-based tariff calculation methodology and to establish long payment durations. Both design features were later taken into account by Spanish and French legislators. As well, the German feed-in tariff of 2000 established further policy innovations, namely, a shallow cost-sharing methodology for grid connection and the abolition of program-size caps. Even though the latter design features have not (yet) been completely implemented in Spain and France, they have considerably improved the

---

82 Both the tariff level for certain technologies and the implementation of location-specific tariffs for wind energy were inspired by the German legislation of 2000 (J.-L. Bal, personal communication, 24 October 2008).

investment framework for renewable electricity producers and are now considered to be international best practice. In sum, the analysis shows that communication between governmental actors was crucial for the diffusion of basic feed-in tariff design features.

The development of renewable electricity technologies has also influenced similar policy decisions in Germany, Spain and France. Based on these technological advances, national legislators have independently adjusted national feed-in tariff legislation. In line with the theory of policy convergence, this type of independent problem solving took place in Germany, Spain and France place with respect to the eligibility of technologies and the extension or abolition of program-size and plant-size caps. In the first case, it was shown that renewable electricity technologies were generally included under national feed-in tariff legislation once they had achieved a certain degree of maturity. While research and development spending is applied by legislation at a very early stage of technological development to foster technological learning, feed-in tariffs help to create larger markets, thus considerably lowering the production cost per unit. Similarly, legislators independently extended or removed capacity caps for renewable electricity producers. The technological advances allowed legislators to set higher national targets (at lower cost), thus extending or removing program-size caps, and the increasing size of renewable energy power plants also forced national policymakers to extend plant-size caps.

Besides these voluntary mechanisms for policy convergence, the European framework for renewable energy also had an impact on national policy decisions in the case of some design features. Most importantly, the national targets set by EU directives – indicative targets for 2010 and mandatory targets for 2020 – influenced national policy decisions. With respect to the eligibility of certain technologies, these directives obliged national legislators to include a larger number of technologies or to promote less mature technologies in order to reach the ambitious mid-term targets. Similarly, legislators had to remove or extend program-size caps to comply with European obligations. To a certain extent, modifications of the tariff calculation methodology can also be explained by the need to reach EU targets, because cost-based tariff calculation helps to avoid windfall profits for cost-efficient renewable electricity technologies and at the same time guarantees sufficient rates of return for less mature technologies. Therefore, cost-based tariff calculation methodologies enable the national policymaker to reach the EU targets in the most cost-effective way. Moreover, EU legislation influenced the design of national financing mechanisms for feed-in tariffs. In order to avoid state-aid infringement procedures, Germany, France and Spain have established equal burden-sharing mechanisms that avoid the use of money from the general national budget.

Chapter 5

The Evolution of Design Options for Tariff Differentiation and Adjustment

As in the previous chapter, the detailed analysis of feed-in tariff design options for tariff differentiation and adjustment is intended to answer our research questions as formulated in section 1.5. We analyse the evolution of feed-in tariffs in Germany, Spain and France in order to understand the reasons for increasing policy similarities in the EU. Why are national support instruments becoming more similar over time, even though two attempts at EU harmonisation of these support frameworks have failed?

An analysis of the development of these design options in three countries over a period of more than two decades will show whether feed-in tariffs have become more similar over time and, if so, which driving factors have been responsible for this. Preliminary results are presented after the analysis of each design option. A summary of these conclusions is provided in Chapter 8.

When designing feed-in tariffs, the legislator generally tries to strike a balance between investment security for producers on the one hand and the elimination of windfall profits on the other hand. Windfall profits, or overcompensation for renewable electricity generators, have to be avoided to keep the additional costs for the final consumer as low as possible. Investment security is primarily guaranteed via a set of basic feed-in tariff design options as described in the previous Chapter 4, namely, the purchase obligation and a fixed tariff payment over a long period of time. In order to reduce windfall profits, policymakers have implemented a number of feed-in tariff design options for tariff differentiation. Tariffs can be differentiated according to each technology (see section 5.1) and the size (see section 5.2) and location of a given power plant (see section 5.3). Besides, feed-in tariff schemes can include design options for automatic tariff adjustment. To achieve decreasing tariffs over time, legislators have invented so-called tariff degression (see section 5.4), which consists of annual reductions of a certain percentage in order to anticipate technological improvements. Tariff degression effects only newly installed capacity. All these design options and their application in Germany, Spain and France are analysed in detail in the following sections.

5.1. Technology Specific Tariffs

The most basic way of assuring tariff differentiation is technology-specific support. Technology-specific tariffs are highly interconnected with the tariff calculation methodology (see section 4.3). If the legislator applies a methodology that is based on the generation cost of renewable electricity technologies – contrary to the avoided costs of conventional electricity generation as a benchmark – different tariffs for different technologies are the logical consequence. Depending on the particular generation cost of electricity from each technology, tariffs can vary significantly. Technology-specific support should not be confounded with the general eligibility of technologies. Even though a technology may be eligible for tariff payment, this does not necessarily mean that the tariff payment will be adapted to the technology in question. Especially in the case of a value-based tariff calculation methodology, when the legislator bases the tariff payment on the avoided costs of conventional power production, even two totally distinct renewable electricity technologies might receive the same tariff level. Therefore, a technology might be eligible for tariff payment under the national feed-in tariff scheme without being granted a technology-specific tariff payment.

5.1.1. Germany

Tariff differentiation according to technologies was not very pronounced in the 1990s, basically because the underlying tariff calculation methodology was largely based on the avoided costs of conventionally produced electricity plus a top-up remunerating the environmental benefits of renewable power sources. The tariff level for the different renewable electricity technologies was one of the most controversial issues in the 1990s. An early version of the first German feed-in tariff included a non-technology specific approach. It suggested a tariff level of 90 percent of the retail electricity price for all renewable energy technologies. However, the working group for Economics and Finances of the CDU/CSU coalition and the Free Democratic Party (FDP) rejected this plan because of the expected high costs for electricity consumers. Thereafter a political agreement was reached to the effect that the remuneration for renewable power producers should be in the range of 75–90 percent of the electricity sales price (Kords 1993).

On the basis of this compromise, the German Ministry for Economy elaborated the details of the first German feed-in tariff scheme, including the tariff levels. Even though a relatively large number of technologies was eligible for tariff payment (see section 5.1.1), several technologies with fundamentally different cost structures received the same tariff payment. Thus, hydro power, landfill gas, sewage treatment gas and biomass all received 80 percent of the retail electricity price. The lack of technology-specific support was even more obvious in the case of equal tariffs for wind energy and solar energy, both of which were granted 90 percent of the retail electricity price. This shows that tariff differentiation according to technologies was not intended to reflect technology-specific costs

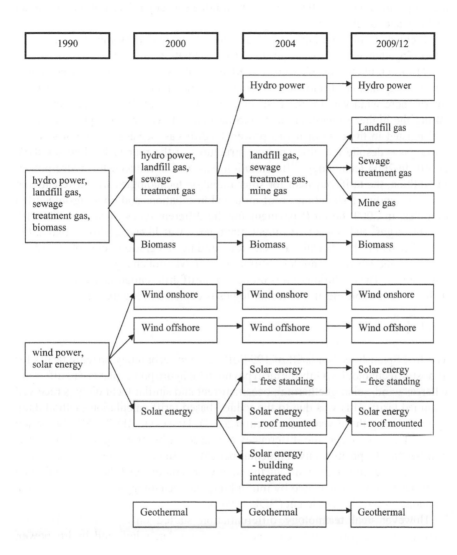

Figure 5.1   Evolution of technology specific tariff differentiation in Germany, 1990–2009/12

Source: Author

but was, rather, based on a political compromise in order to reduce the costs of renewable electricity promotion. Eventually, only two technologies received the higher payment level of 90 percent of the retail electricity price, as envisaged at an earlier stage for all technologies.

Technology differentiation was considerably increased by the amendment of 2000. Together with a change in the tariff calculation methodology (see section 5.3.1), differentiated tariffs were issued for biomass, geothermal power plants, wind energy and electricity derived from solar radiation. As well, the tariff level for offshore wind was for the first time higher than that for onshore wind, thus taking the higher generation costs into account. However, a unique tariff level still existed in the case of hydro power, landfill gas, sewage treatment gas and mine gas – despite the different cost structures of the four technologies (BGBl 2000). In 2004 technology differentiation was also introduced for hydro power. In addition, the tariff payment for solar radiation technologies was differentiated for free-standing installations, roof-mounted installations and building-integrated modules. In 2009 the tariff payment for the different types of biogas – landfill gas, mine gas, and sewage treatment plant gas – was likewise differentiated. On the other hand, the special payment for building integration solar photovoltaic (PV) was abolished.[1] Technology-differentiation was not changed within the 2012 amendment. The evolution of this process of tariff differentiation according to the different renewable electricity technologies is depicted in Figure 5.1.

5.1.2. Spain

The first Spanish feed-in tariff of 1994 offered non-technology-specific tariffs for almost all technologies. Only the tariff payment for hydro power plants was different, whereas wind, solar, geothermal, ocean current and similar technologies received a unified tariff. This was due to the valued-based tariff calculation methodology applied during those years (see section 5.3.2). However, RD 2366/1994 started to take technology-specific differences into account by granting additional tariff bonuses for the positive environmental benefits of renewable energy sources and the societal benefits of diversifying the energy-generation portfolio. This additional tariff payment was calculated by multiplying the technology-neutral tariff with a technology-specific coefficient (BOE 1994a).

However, real technology differentiation started only in 1998. The RD 2818/1998 established differentiated tariffs for solar, wind, and hydro power.

---

1   The bonus of 5 €cent/kWh for building integration in Germany was cancelled by the 2009 amendment. The tariff level for photovoltaics was one of the most hotly debated issues during the political negotiations in 2008. In order to avoid drastic tariff cuts for free-standing and roof-mounted solar photovoltaic plants, the Social Democrats agreed to abandon the special tariff payment for building integration. Consequently, abandoning this type of tariff differentiation has to be understood as a political compromise designed to win the support of photovoltaic critical parliamentarians from the conservative CDU/CSU.

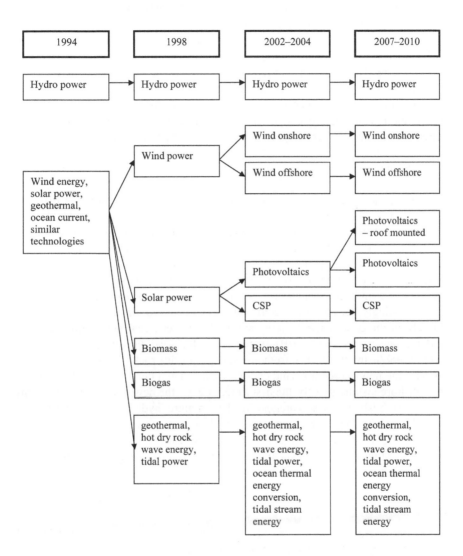

Figure 5.2   Evolution of technology specific tariff differentiation in Spain, 1994–2010

Source: Author

Besides, a differentiated tariff was guaranteed for a fourth category of rather immature renewable energy technologies, including geothermal power, wave energy, tidal power and hot dry rock.[2] In 2002 the RD 841/2002 differentiated between solar photovoltaic plants and concentrated solar power installations, since Spain had the resource potential for both technology types (BOE 2002a). In 2004 onshore and offshore wind power plants received differentiated tariffs for the first time. The group of relatively immature technologies receiving a unique tariff was extended, now comprising geothermal, wave power, ocean current, hot-dry rock, ocean thermal energy conversion and tidal stream energy (BOE 2004a). The evolution of this process of tariff differentiation according to the different renewable electricity technologies in Spain is depicted in Figure 5.2.[3]

5.1.3. France

In France, the legal structure of the national feed-in tariff scheme is different from those in other European countries, such as Germany and Spain. The French legislator established the legal foundation of the feed-in tariff mechanisms via national laws in 2000 and 2005 (J.O. 2000a, 2005b). Based on these more general provisions, the legislator individually issues specific decrees for the single renewable electricity technologies. Therefore, a technology-specific approach is already inherent in the structure of the French feed-in tariff legislation.

Accordingly, technology-specific tariff payment had already started in 2001, namely, in the case of wind energy and hydro power (from lakes, running water and the ocean). In 2002 tariffs were issued for solar power (solar radiation), geothermal (using the energy from aquifers or hot-dry-rock), biomass and biogas. In 2006 the new decree for wind energy differentiated tariff payment for onshore and offshore installations for the first time. In 2007 the new tariff decree for hydro power established a new category of less mature hydro power technologies, including wave power, tidal power and hydrokinetic power plants (J.O. 2007a). Finally, in 2010 the decree for solar energy established new categories depending on the location of solar power plants, namely, free-standing installations, fully building integrated, and systems and power plant where building integration is less

---

2   Besides, tariffs were differentiated for primary biomass product, including all vegetal plants younger than one year which can be used directly or after a transformation process for energy generation (natural resources and energy crops), and secondary biomass products, including biogas.

3   In the penultimate column the time period 2002–2004 is indicated, since the differentiation of tariffs for solar photovoltaics and concentrated solar power was already implemented in 2002 (BOE 2002a). Equally, tariff differentiation according to the location of solar photovoltaic systems, that is, roof-mounted versus free-standing, was implemented only via the extraordinary amendment of 2008 (BOE 2008a). Therefore, the last column refers to the period 2007–2008.

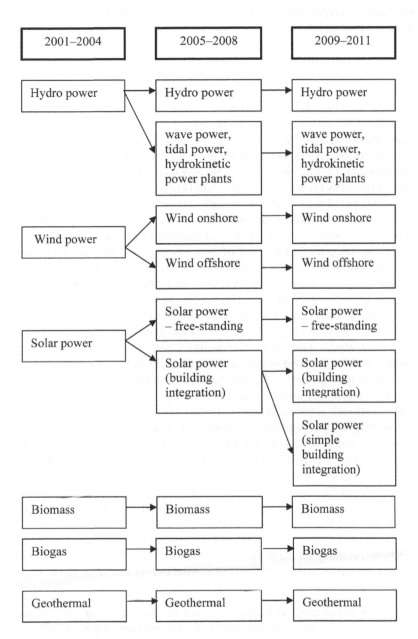

Figure 5.3   Evolution of technology specific tariff differentiation in France, 2001–2011

Source: Author

complicated ('simple integration'). The evolution of technology differentiation is France is visualised in Figure 5.3.[4]

5.1.4. Preliminary conclusion

The three countries in question make use of technology-differentiated tariff payments – one of the main advantages of feed-in tariffs over other support instruments. Technology-specific tariffs enable the legislator to also support less mature technologies while at the same time avoiding overcompensation for more mature technologies. As shown in Figure 5.4 , Germany had already started to differentiate tariffs in 1990, followed by Spain in 1998 and France in 2001.[5] As clearly revealed by Figure 5.1, Figure 5.2 and Figure 5.3, technology differentiation in all three countries has increased significantly over the years.

Even though technology differentiation increased in all three countries, the comparison indicates that this process of differentiation has still not come to an end. In Germany, for instance, all of hydro power technologies, including less mature technologies such as wave power, tidal power, salt gradient and flow energy, all receive a unified tariff level whereas France already pays differentiated tariffs for different hydro power technologies. In Spain the same remuneration is still paid for a rather heterogeneous group of technologies, including geothermal, hot dry rock wave energy, tidal power, ocean thermal energy conversion and tidal stream energy. Once these technologies have reached a certain degree of technological development or the legislator needs them in order to reach more and more ambitious targets, it can be expected that tariff payment will be differentiated according to technology-specific generation costs.

Transnational communication and regulatory competition did not play a decisive role in this convergence process. Instead, the need to reach national renewable energy targets as set out by the European directives led to the increase of technology-differentiated tariffs, as renewable electricity support had to be designed in a more efficient way, thus requiring a cost-based tariff calculation methodology (see section 4.3). The first wave of increased technology differentiation can be related to a change in tariff calculation methodologies in Germany and Spain at the end of the 1990s. The early Spanish feed-in tariff scheme of 1994 is a good example of a non-technology-differentiated support mechanism, granting the same tariff level for almost all technologies. Similarly, in Germany the value-based approach led to wind energy and solar power receiving

---

4   Since feed-in tariffs in France are issued for each technology individually, and given that these tariffs are not issued periodically, the time period between 2000 and 2010 was subdivided into three phases.

5   In fact, all European feed-in tariffs are operating with technology-differentiated tariffs. The only exception until recently was Hungary, offering a unified tariff of 10.5 €cent/kWh for all technologies (Klein et al. 2008:). However, technology-specific tariffs were adopted in 2007.

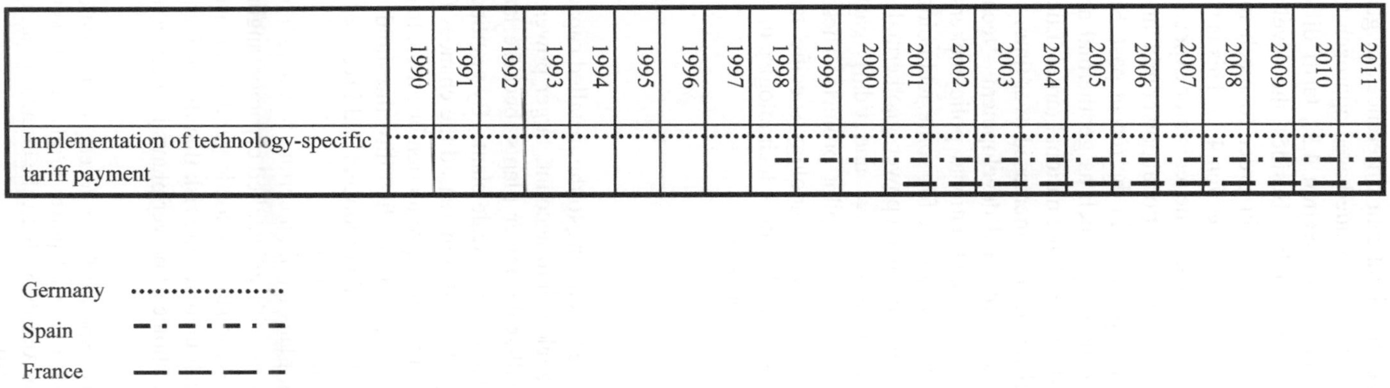

2011		
2010		
2009		
2008		
2007		
2006		
2005		
2004		
2003		
2002		
2001		
2000		
1999		
1998		
1997		
1996		
1995		
1994		
1993		
1992		
1991		
1990		
Implementation of technology-specific tariff payment		

Germany ··········
Spain ▪ ▪ ▪ ▪
France — — —

Figure 5.4 The implementation of technology-specific tariff payment in Germany, Spain and France, 1990–2011

Source: Author

the same tariff payment despite the significant differences in generation costs. When Germany and Spain moved from a value-based approach towards a cost-based tariff calculation methodology (see section 4.3), tariff differentiation for different technologies increased considerably. Besides, the development of less mature technologies has become necessary in order to reach more and more ambitious national and European targets (see section 4.1). This is most obvious in the case of offshore wind power in Germany. The legislator expects offshore wind power to play an important role in the future national power mix[6] and has therefore differentiated (and increased) tariff payment in the recent past. Similarly, Spain wants to develop less mature technologies such as geothermal and less mature hydro power technologies in order to reach the national target (Lucas 2009). This also explains why technology differentiation increased over time.

To a certain extent, general technological development – technology-driven independent problem solving – influenced national policymakers' decisions to implement technology-specific tariff payment for some technologies. Formerly less mature technologies such as offshore wind power, geothermal and alternative hydro power technologies (for instance, wave and tidal) reached a certain technological maturity beyond the stage of pilot projects. These technologies became eligible for tariff payment, and technology-specific tariffs for were implemented to increase the market size for those technologies in order to further reduce generation costs (see also section 4.1).

5.2 Size Specific Tariffs

Feed-in tariffs can be differentiated according to the installed capacity of a given plant. If economics of scale are taken into account, larger power plants usually receive a lower tariff than small-scale power plants, because larger units can produce electricity at lower cost than small-scale plants. By granting higher tariffs for smaller installations, the legislator also increased the chances for small-scale power producers to participate in the power production business. In this case, even smaller-scale projects can reach a certain profitability threshold. Without size-specific tariffs, it is likely that larger-scale producers and investors will profit the most from the national feed-in tariff scheme.

Typically, tariffs are paid for plans within a certain bandwidth of capacity, for instance, different tariffs for plants with an installed capacity ranging from 0–100 kW, 100 kW to 2 MW, 2 to 10 MW, and installation larger than an installed capacity of 2 MW. However, this approach involves the risk that the policymaker's choice of size categories influences the choice of power plants in a given territory. If, for

---

6   The latest lead study for the expected future development of different renewable electricity technologies expects offshore wind power to play an important part in the German power mix in the mid term. Already by 2020, the installed capacity is expected to reach 9 GW (Wenzel and Nitsch 2010).

instance, the tariff for installations up to 100 kW is significantly higher than in the case of the next-highest plant size category, this might encourage the installation predominantly of such small-scale power plants. In order to avoid these effects on the national market, legislators can also opt for establishing formulas which correlate tariff payment with the exact size of a given plant. This way, the size of the power plants installed in a given country is not influenced by the policymaker as tariff level 'jumps' between certain capacity bandwidths are avoided.

5.2.1. Germany

In Germany tariff differentiation in accordance with plant size was first introduced for hydro power, land fill gas and sewage treatment gas under the 1990 feed-in tariff legislation. Instead of explicitly prescribing different tariffs depending on the installed capacity, the higher tariff of 75 percent of the retail electricity price[7] was granted only for the equivalent amount of electricity produced with an installation capacity of 500 kW. The electricity exceeding this limit was remunerated with a lower tariff of 65 percent of the average electricity price (BGBl 1990).

In 2000 plant-size-specific tariffs were implemented for a large number of technologies, including hydro power (two categories) biomass (three categories), geothermal (two categories), and solar energy (three categories). In 2004 larger hydro power plants (exceeding 5 MW capacity) became eligible for tariff payment. Therefore, the number of hydro power categories increased from two to seven (two categories for small hydro power plants and five categories for large hydro power plants). Besides, two categories were implemented for landfill gas, sewage treatment plant gas and mine gas. At the same time, the number of categories for biomass increased from three to four and the number of categories of geothermal from two to four. In 2009 additional categories were included for hydro power plants (now eight categories), mine gas (now three categories) and solar photovoltaics (now four categories). Interestingly, the size classes of geothermal tariffs were reduced from four to two in 2009. The legislator realised that, because of the small market share of geothermal power plants and the limited number of projects, two size categories were sufficient to spur investment in this industry branch (BMU 2007b). In 2012, it was further reduced to just one plant-size category (BMU 2011). Today, plant-size-specific tariff levels exist for all renewable energy technologies except wind energy. Here, the legislator favoured the use of location-specific tariffs (see section 5.3).

The general complexity of plant-size differentiation was further increased by the amendment of 2000. In order to further reduce windfall profits, the tariff payment now depend on the exact size of each plant. The size categories as set out for the different technologies serve only as points of reference for the exact tariff payment. According the EEG 2004, which incorporated this provision for

---

7   The amendment of 1994 increased the tariff level to 80 percent of the retail electricity price (BGBl 1994).

all technologies except wind energy, the tariff level is determined 'according to the share of the plant's capacity in relation to the threshold value to be applied' (BGBl 2004). This means that the tariff level is in fact a linear function among the different remuneration categories. The remuneration depends on the ratio of the total of the kilowatt-hours to be fed into the grid during the calendar year in question to the total number of full hours for the same year (BMU 2008a). This 'gliding remuneration' was implemented in order to avoid jumps in the remuneration level once the producer falls into a new remuneration category. Besides, the legislator intended to avoid insufficient or unnecessary high tariffs (BMU 2004c). This approach was unchanged by the new law of 2009.

5.2.2. Spain

In the early years of the Spanish feed-in tariff legislation the use of plant-size-specific tariff differentiation was limited.[8] The 1998 legislation included size-specific tariffs for hydro power (two categories), biomass (two categories) and solar photovoltaics (two categories).[9] The tariffs for geothermal, ocean and tidal power and wind energy were not differentiated according to the size of the power plant. In 2004 the number of plant-size categories for hydro power was increased from two to three. Besides, tariff differentiation was implemented for wind energy. Wind parks with less than 5 MW installed capacity received a higher tariff than parks between 5 and 50 MW. This regulation also referred to offshore wind power projects. This approach was abandoned again in 2007 as it did not lead to the installation of smaller-scale wind parks. In 2007 a new category was implemented for solar photovoltaic plants. Now three categories existed. Besides, another category for biogas was included. The 2008 amendment for photovoltaic support did lead to the elimination of plant-size categories. However, the maximum size of power plants was defined as 2 MW in the case of roof-mounted systems and 10 MW in the case of free-standing installations.

Even though the Spanish legislator implemented plant-size-specific tariffs, the Spanish feed-in tariff scheme was not sufficiently concise when defining the limits of one power plant. Therefore, power producers were able to divide a large power plant into many small sections, thus becoming eligible for the higher tariff for

---

8   According to the decree of 1994, renewable power plants with an installed capacity exceeding 25 MW received the equivalent of the variable avoided costs of conventional power generation. The tariff for power plants smaller than 25 MW – the primary focus of this piece of legislation – was calculated based on the long-term avoided costs, with the retail electricity price also taken in account (BOE 1994a). However, this approach was not further specified by the RD 2366/1994, and therefore this regulation did not have any effect on the tariff payment level.

9   The RD 2818/1998 offered a significantly higher tariff for small-scale units (39.6 €cent/kWh for installations up to 5 kW and only 21.6 €cent/kWh for larger scale units) (BOE 1998a). Therefore, only small-scale units were installed in Spain.

small-scale installations. In the case of solar photovoltaics, for instance, the RD 661/2007 defined three categories, the first from zero to less than 100 kW, the second from 100 kW to less than 10 MW, and the third from 10 MW to 50 MW. However, primarily plants with a maximum capacity of 100 kW were connected to the grid. In the case of large solar parks, these were often divided into small fractions of 100kW, each running under the names of different owners. These small, 'individual' plants were then connected by a series of transformers.[10]

5.2.3. France

Plant-size-specific tariffs were first implemented in the case of hydro power in 2001, differentiating power plants below or above 500 kW installed capacity. This approach was extended in 2007 to differentiate three plant-size categories of hydro power plants. Interestingly, no plant-size differentiation exists in the case of the other renewable energy technologies, namely, biomass, geothermal and wind power. With respect to the latter, however, the French legislator differentiates tariff payment according to the resource availability at different locations. If a wind power plant is located at a very windy location the tariff payment is substantially lower than that at a location with less favourable resource conditions (see section 5.3.3). In order to differentiate tariffs for wind energy, the concept of linear interpolation is used, that is, a simple linear formula which defines the exact tariff payment between two defined tariff levels. This concept also applies to size-specific tariffs for biogas. The 2006 decree for biogas remuneration sets one maximum tariff of 9 €cent/kWh for small installations up to 150 kW and one minimum tariff of 7.5 €cent/kWh for all installation larger than 2 MW. For all biogas plants with an installed capacity between 150 kW and 2 MW, the tariff payment is calculated based on a linear function between the two remuneration limits. The same approach is used for biogas tariffs in the overseas departments (J.O. 2006c). In the case of solar photovoltaics, sophisticated plant-size differentiation was implemented with the 2011 amendment (J.O. 2011). For both building integrated and roof-mounted systems three different tariff categories were implemented, separating small-scale, medium-size and large-scale systems.

5.2.4. Preliminary Conclusion

Plant-size-specific tariffs are applied in all three countries. Germany started in 1990, Spain in 1998 and France in 2001. However, the level of plant-size differentiation varies significantly. As indicated by Figure 5.5, in Germany tariffs are differentiated according to the size of the power plants for all renewable electricity technologies. In 2000 the complexity of plant-size differentiation was

---

10 The installed capacity of installation below 100 kW grew from 9 MW in 2004 to 140 MW in 2006. In the same period, the installed capacity of larger scale installation only increased from 3 to 5 MW (Calzada et al. 2009).

further increased and tariff payment now depended on the exact size of each plant. In Spain, plant-size differentiation was implemented for the majority of technologies. In France, plant-size-specific tariff payment exists only in the case of hydro power, biogas and solar photovoltaics.

Even though the choice of plant size categories is to a certain degree random, typical plant-size types are taken into consideration by national legislators when fixing size categories. Legislators generally analyse the national market of renewable energy technologies, thus identifying standard sizes of products. Plant-size-dependent tariff differentiation is often implemented accordingly. In the case of solar photovoltaics, Germany has currently implemented four plant-size categories, including plants with an installed capacity no larger than 30 kW, no larger than 100 kW, no larger than 1 MW and all installations that are even bigger. The third category – roof-top installations between 100 kW and 1 MW – was implemented in 2004 when it became clear that the size of roof-top installations had considerably increased over the years (G. Stryi-Hipp, personal communication, 15 September 2008). When plant-size categories were defined for solar photovoltaics in 2008, a discussion was held on whether to implement a new category for small-scale, roof-top installations on private houses, where the maximum capacity normally does not exceed 3, 4 or 5 kW. However, the idea was finally rejected because it would have further increased the complexity of the support instrument (Oschmann, personal communication, 1 September 2008).[11]

In the case of plant-size-differentiated tariffs, there is no evidence to show that transnational communication or regulatory competition was the causal mechanism for policy convergence. However, the implementation of plant-size-specific tariff payment is to a certain extent related to the technological development of renewable energy generation facilities. Over time, renewable power plants grew in size, moving from small-scale plants developed via pilot project towards larger-scale power plants and mass markets. Consequently, the legislator had to reflect this wide range of plant sizes in the national legislation. In order to support both small-scale and large-scale power plants of the same technology, plant-size-differentiated tariff payment was introduced. This sort of policy adjustment was taking place independently in the three countries under analysis.

Besides, similarities in Germany, Spain and France can be explained by the aim of the national legislators to design the national support mechanism in the most efficient way. One of the primary reasons for the implementation of plant-size-differentiated tariff is to avoid windfall profits – as in the case of most design options discussed in this chapter. As a matter of fact, lower tariffs for larger power plants can help to avoid this type of overcompensation. Therefore, at first

---

11   A category for small-scale photovoltaic installation up to 10 kW had already been suggested by the industry association BSW in 2004. However, the German legislator finally implemented a first category of 0–30kW in order to also incentivise the installations of slightly larger photovoltaic plants on rooftops (G. Stryi-Hipp, personal communication, 15 September 2008).

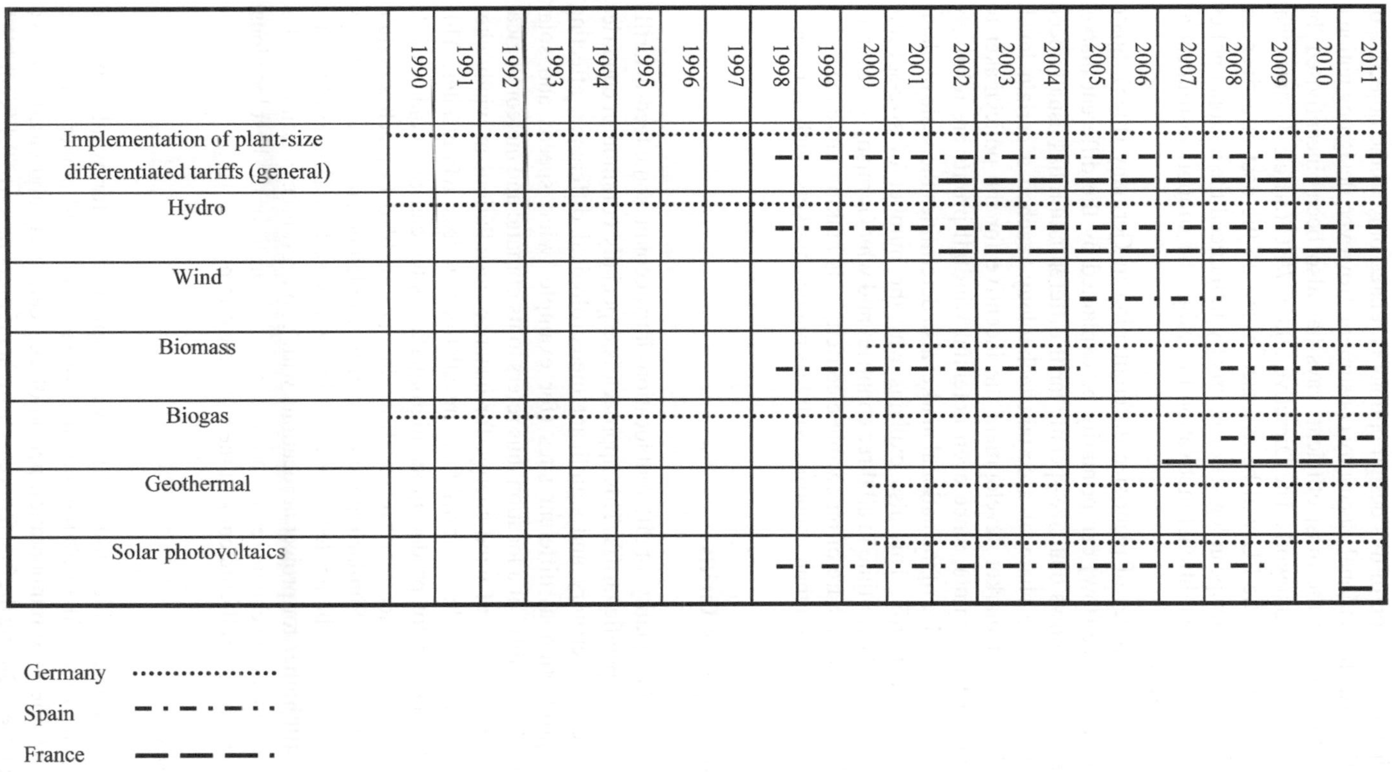

| 2011 |
| 2010 |
| 2009 |
| 2008 |
| 2007 |
| 2006 |
| 2005 |
| 2004 |
| 2003 |
| 2002 |
| 2001 |
| 2000 |
| 1999 |
| 1998 |
| 1997 |
| 1996 |
| 1995 |
| 1994 |
| 1993 |
| 1992 |
| 1991 |
| 1990 |

Implementation of plant-size differentiated tariffs (general)

Hydro

Wind

Biomass

Biogas

Geothermal

Solar photovoltaics

Germany ·············
Spain · — · — · —
France ▬▬ ▬▬ ▬▬

Figure 5.5   Implementation of plant-size differentiated tariff payment in Germany, Spain and France, 1990–2011

Source: Author

sight the implementation of this design options is related only to the objective of reducing the costs for national electricity consumers. However, designing national support instruments in the most efficient ways has also been 'incentivised' by the European policy framework. The Directive 2001/77/EC (article 4) states that support mechanism in the EU will be assessed based on the cost effectiveness. By those means, the European legislator indirectly influenced the national policy framework, giving an additional motivation to design the support instrument as efficiently as possible.

The different degrees of plant-size differentiation in Germany (high), Spain (medium) and France (low) can primarily be explained by the different market sizes. The analysis shows that size-specific tariff differentiation has usually been implemented when the share of a certain technology reached a certain level. At an early stage of market development the incentive for the policymaker to differentiate tariffs is minor, since even potential windfall profits are relatively small. Therefore, the size of the national market also determines the scale of plant-size-differentiated tariffs. This also explains why the number of size-specific tariffs has increased over time in all three countries and why Germany – a country with a relatively large share of renewable electricity – is making more extensive use of this design option than countries with less mature markets, such as France.

5.3. Location Specific Tariffs

In order to take the 'quality' of different locations into account, some feed-in tariffs include location-specific tariffs. This approach is especially common in countries with an extended territory and significant meteorological difference, affecting electricity production at different sites (for example, wind speeds and solar radiation). The 'quality' of a location thus refers to its climatic and meteorological conditions. In the case of wind energy, for instance, installations at windy sites accordingly receive a lower tariff than installations at less windy sites. This way, windfall profits for producers at 'high-quality' sites can be avoided while at the same time producers at 'low-quality' sites are enabled to reach a certain profitability threshold. Similar regulations can be implemented for solar energy technologies. Despite the implementation of location-specific tariffs, profitability is usually still higher for project at locations with good resource conditions. This way, the feed-in tariff scheme retains an incentive to search for the best locations, thus assuring the overall efficiency of the support system.

5.3.1. Germany

The first location-specific remuneration was implemented for wind power in 2000. In order to take the different qualities of sites for wind power generation into account, the total remuneration period of 20 years was subdivided into two payment phases. During the first five years, a higher remuneration of 9.1 €cent/kWh was granted for all wind power plants. Under the so-called reference yield model (*Referenzertragsmodell*), for the following 15 years the remuneration depended on the location-specific conditions in comparison with a pre-established reference yield. If in the first five years of operation the plant achieved 150 percent of the reference yield, the installation received only the lower tariff payment of 6.19 €cent/kWh for the following 15 years.[12] For all installations with a lower yield, the period of the higher initial remuneration was extended, thus increasing the remuneration level over the full lifetime.[13] In the case of offshore wind power plants, the higher initial tariff payment was guaranteed for nine years (BGBl 2000). The purpose of this provision was to prevent windfall profits for wind power plants at very good sites. At the same time, the legislator intended to counteract the concentration of wind power plants along the German coastline in order to avoid resistance from the local population. The new remuneration scheme also intended to make investments in wind power plants more attractive at inland sites (Bechberger and Reiche 2004). In addition, the German legislator wanted to avoid the overcompensation of wind power producers at the very windy coastal areas, thus counteracting the fundamental criticism of the German feed-in tariff scheme by the European Commission (Deutscher Bundestag 1999).

The 2004 amendment only slightly modified this ruling.[14] As politicians from the opposition parties (CDU and FDP) feared that this regulation might lead to inefficiencies in the system, the new law stated that wind power plants at locations with very little wind should no longer be eligible for tariff payment. This effects all installations that do not reach at least 60 percent of the pre-established reference yield. The implementation of this regulation was based on a political compromise between the upper house of the German Parliament (the *Bundesrat*), with a political majority of the SPD and Green Party, and the lower house (the Bundestag), with a political opposition of the CDU and FDP (Lauber and Mez 2007). Even though the German renewable energy law does not require a consensus between the two houses, the governing coalition nonetheless wanted to reach an accord with the opposition parties for better long-term stability. However, this regulation was deleted with the 2012 amendment (BMU 2011). It had become clear that installing wind power plants at locations with less than 60 percent of the pre-established reference yield does not make economic sense anyway.

The 2004 amendment also changed the regulation for wind energy offshore. Similar to the approach for onshore wind projects, the remuneration period is subdivided into a higher, fixed tariff level for the first 12 year of operation and

---

12   The reference yield was calculated according to the reference installation at a reference site, that is, wind speed of 5.5 m per second at a height of 30 meters.

13   The period of higher remuneration was prolonged by two months for every 0.75 percent the actual yield remained below the 150 percent reference yield.

14   The higher initial remuneration was reduced to 8.7 €cent/kWh and the lower follow-up remuneration was reduced to 5.5 €cent/kWh (BGBl 2004).

a lower tariff for the succeeding years, depending on the quality of the actual site. In contrast to the regulation for onshore wind power projects, the higher initial tariff payment is not extended based on the wind speed (reference yield model) but instead based on the actual geographic positioning. The period can be extended only for offshore wind power plants which are located at least 12 nautical miles out to sea and at a water depth of at least 20 meters. It is extended by 0.5 months for each additional full nautical mile beyond 12 and by 1.7 months for each additional full metre of water depth (BGBl 2004). These regulations were not touched by the 2009 amendment (BGBl 2008). Only the tariff level was modified once again.

Based on the national experience of wind energy, the German solar industry association, BSW, proposed a similar approach for photovoltaics in 2003. Depending on the solar radiation intensity, the lobby group wanted to persuade the legislator to grant a slightly higher tariff in northern Germany, in view of the fact that most solar photovoltaic plants were installed in the southern state of Bavaria. The BSW suggested differentiating three zones in Germany according to the degree of solar radiation intensity. However, the German legislator rejected this proposal as it was considered to significantly increase the complexity of the support instrument (G. Stryi-Hipp, personal communication, 15 September 2008).[15] Besides, the primary political objective of the support for solar photovoltaics is not to generate electricity but rather to stimulate market growth. Therefore, it does not matter whether solar photovoltaic installations will be uniquely installed in one region in Germany or whether the installation will be evenly spread throughout the country. The issue was again debated in 2008, though less intensely, since the increasing costs related to the promotion of solar photovoltaics dominated the political debate and left little space for demanding higher remuneration for less sunny locations.

5.3.2. Spain

The Spanish feed-in tariff scheme does not include provisions for location-specific tariff payment, neither in the case of solar photovoltaic systems nor in the case of wind energy. In 2004, Iberdrola – the largest player in the Spanish wind business – proposed the implementation of a location-specific tariff for wind energy, that is, tariff differentiation according to the full load hours at a given location. However, the legislator was reluctant to implement this design feature. The national energy council, CNE, refused to implement location-specific tariffs for wind energy, as it feared that this would undermine the efficiency of the overall system. Wind power producers would no longer have an incentive to look for the sites with the best resources (L. Tembleque, personal communication, 13 February 2008).

---

15 It should be noted that in 2003 the complexity of the tariff regime for solar photovoltaics was already increased by the implementation of plant-size-specific tariffs (see section 5.2).

As well, the ownership structure of the Spanish wind energy market influenced the legislator's decision. In Spain, wind power parks are mostly owned and operated by utilities. Therefore, one utility normally operates a large number of wind power plants, some at very windy locations and others at less windy regions. However, by owning a large number of wind parks the utilities can compensate for the minor revenues for wind parks at less windy locations via the increased income of wind parks at high-quality locations. In other words, by owning a large number of wind parks utilities can balance the revenues from wind parks at good locations and those at bad locations within their wind-park portfolios (A. Ceña, personal communication, 12 February 2008).

5.3.3. France

France first implemented location-specific tariff differentiation for wind energy in 2001 (J.O. 2001b). This approach was only marginally modified in 2006 (J.O. 2006a).[16] Tariffs are differentiated according to the full load hours of a given wind power plant at a given location. In France, tariff payment is guaranteed for a period of 15 years. During the first ten years of operation, all wind power producers receive a fixed tariff payment. During this period, the output and accordingly the full-load hours of the wind power plants are measured. Eventually, these measurements serve as a basis to calculate the tariff payment for the following five year of operation. In order to calculate the average value over the period of ten years, the best and the worst years are excluded from the calculation. According to the legislation from 2006, the tariff payment can vary between 2.8 and 8.2 €cent/kWh, depending on the numbers of annually full-load hours. The decree fixed certain reference values (2400, 2800 and 3600 full-load hours). In between those reference values the tariff payment is calculated via linear interpolation (see table 5.1). Similar regulations exist for offshore wind power plants and wind parks in Corsica, the only difference being that tariff payment is slightly higher. In 2006 a unique tariff level of 11 €cent/kWh was implemented for all overseas departments, independent of the location of each wind power park (J.O. 2006a).

In the case of solar photovoltaics, location-specific tariffs were first implemented in 2010. France was the first country worldwide to implement this type of tariff differentiation for solar photovoltaic systems. Location-specific tariffs for this technology were first been suggested in a report from the French parliament in 2009. The report argued that, contrary to other European countries like Germany with a rather even distribution of solar radiation intensity across the national territory, sunshine intensity in France varies greatly. According to M. Poignant, the rapporteur of the French parliament, the country is divided into two parts, one

---

16   The 2006 legislation was slightly more generous in terms of tariff payment. The reduced tariff payments started only when the wind park reached more than 2400 full-load hours, not 2000 full-load hours. Besides, the higher tariff payment was initially guaranteed for only five years.

similar to the area in Germany that experiences intense solar radiation and the other experiencing sunshine conditions similar to those in the north of Spain. In order to limit the complexity of tariff differentiation, the study suggested dividing the French territory into two, with solar photovoltaic installations in the northern part receiving a tariff payment 20 percent higher than those in the southern part (Assemblée Nationale 2009).

Table 5.1   Location-specific remuneration for onshore wind power in France (mainland)

Annual duration of reference operating time	Tariff for the first ten years of operation (€cent/kWh)	Tariff for the following five years of operation (€cent/kWh)
2400 hours and less	8.2	8.2
Between 2400 and 2800 hours	8.2	Linear interpolation
2800 hours	8.2	6.8
Between 2800 and 3600 hours	8.2	Linear interpolation
3600 hours and more	8.2	2.8

Source: Author, adopted and modified from J.O. 2006a

The recommendations of the French parliament were taken into account in the new legislation for solar photovoltaic promotion, issued in early 2010 (J.O. 2010a). The new decree stated that free-standing solar photovoltaic systems with an installed capacity of more than 250 kW were subject to location-specific tariff differentiation. Installations in the sunniest regions in the southern part of the country received the tariff levels as indicated by the legislation. Installations in less sunny regions received the same tariff multiplied by a coefficient. The coefficient was calculated based on the average annual sunshine intensity in a given region (kWh/m^2). Accordingly, the country was subdivided into 20 different zones. As a result of this regulation, tariffs in the northern part of the country are up to 20 percent higher than those in the very southern part. This way the legislator intended to incentivise a more even distribution of free-standing solar photovoltaic installations throughout the country. This had become necessary in order to avoid competition within the areas used for other purposes in southern France. In several cases, the installation of solar photovoltaic plants had caused trouble with the well-established winery industry, which feared the loss of land available for winegrowing (Hirshman et al. 2010, MEEDDAT 2010a).

5.3.4. Preliminary Conclusion

The convergence of feed-in tariffs in Germany, Spain and France is limited with respect to location-specific tariffs. Of the three countries under investigation, only two – Germany and France – have implemented this design option. Both countries implemented location-specific tariffs for wind energy in 2000 and 2001, and France even applied this design feature in support of solar photovoltaics in 2010 (see Figure 5.6). In Spain, the national legislator considered this design option in 2004 but opted against it.

Despite the limited diffusion, transnational communication played a crucial role in all three countries. The implementation of location-differentiated tariffs for wind energy in France can be related to direct policy transfer from Germany to France. When preparing the first feed-in tariff for wind in France, a study of the governmental organisation ADEME was of special importance. ADEME had analysed the German support framework for wind energy, including the regulation for location-specific tariffs. Therefore, the location-specific remuneration for wind in France was strongly inspired by the German system (J.-L. Bal, personal communication, 24 October 2008). This direct policy transfer also explains some striking similarities in the detailed designs in the two countries. In both countries, the quality of the site for wind power production is measured during the first years of operation. In this time, the legislator in both countries guarantees a fixed and higher tariff payment. The potentially lower tariff payment in the following years is fixed according to measurements made during the first years of operation. In fact, the origin of the implementation of this specific design can be traced to direct policy transfer from German to France (Lesieur 2005). Besides, Iberdrola's proposal to implement location-specific tariffs in Spain was clearly inspired by the design of feed-in tariffs in Germany and France, the only countries at the time that had already implemented this design feature. The fact that Spain refused to implement this design option is related to a number of factors.

Primarily, the Spanish legislator declined to implement location-specific tariffs mainly in order to maintain a strong incentive for searching for the best possible locations. It was feared that the implementation of location-specific tariffs would undermine the efficiency of the whole support system (M.J. Samaniego, personal communication, 27 February 2008, L. Tembleque, personal communication, 13 February 2008). The high importance of efficiency is also reflected in the tariff-setting approach in the case of wind energy. While Germany and France are reducing tariffs over time based on the concept of tariff degression (see section 5.4), Spain is increasing tariffs for wind energy from time to time. This way, the Spanish legislator is trying to establish an incentive to use the best locations first. Once all the locations where a certain number of full-load hours can be achieved are all taken, the legislator will increase the tariff and the wind power producers will start to develop the locations with fewer wind resources (M.J. Samaniego, personal communication, 27 February 2008).

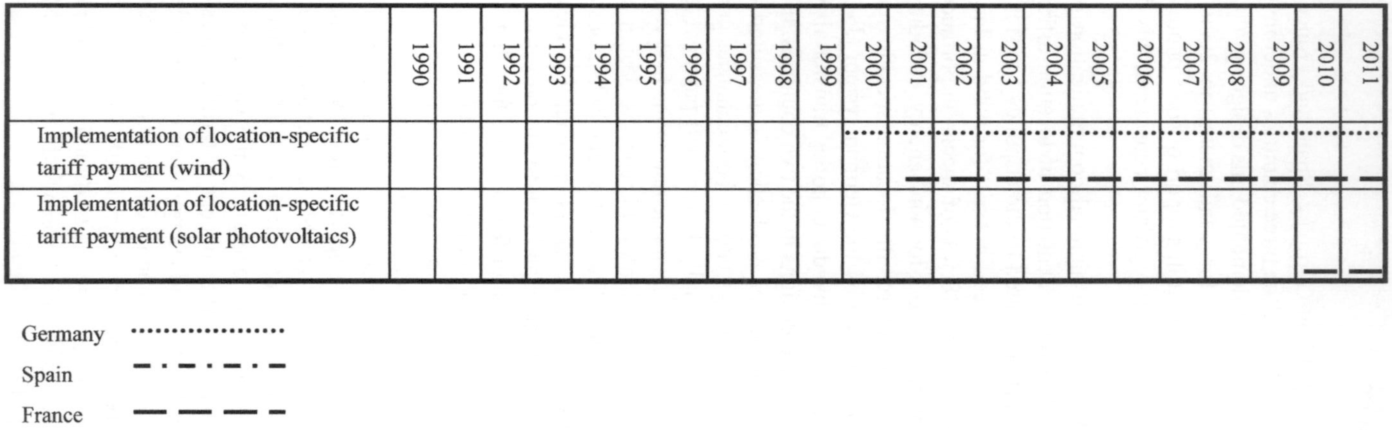

	Implementation of location-specific tariff payment (wind)	Implementation of location-specific tariff payment (solar photovoltaics)
2011		
2010		
2009		
2008		
2007		
2006		
2005		
2004		
2003		
2002		
2001		
2000		
1999		
1998		
1997		
1996		
1995		
1994		
1993		
1992		
1991		
1990		

Germany ⋯⋯⋯⋯

Spain ▬ · ▬ · ▬

France ▬ ▬ ▬

Figure 5.6 Convergence of location-specific tariffs in Germany, Spain and France, 1990–2011

Source: Author

5.4. Tariff Degression

Several countries have implemented so-called tariff degression, an automatic, annual reduction of feed-in tariff payment. Through annual tariff degression the legislator aims to anticipate technical progress, economies of scales, rationalisation and the overall learning potential of a given technology. Economic analyses of various industry branches have found that with every doubling of cumulative output of an industry branch the cost per unit decreases by a certain percentage (Klein et al. 2008). In the case of solar photovoltaics, for instance, a doubling of the globally installed capacity has decreased photovoltaic module prices by 22 percent annually in recent decades (EPIA 2009c).

At the same time, tariff degression serves as an incentive for technology learning and cost reduction in the renewable energy industry. If the legislator reduces the tariffs annually by a certain percentage, the manufacturing industry is 'forced' to constantly improve its product in order to pass cost reductions on the installer and the purchaser of renewable energy technologies. Finally, tariff degression also has stimulated investors to speed up the planning process, because the sooner the power plant is connected to the grid, the higher will be the tariff payment.

Since feed-in tariffs are revised not yearly but only periodically after a given number of years, the implementation of tariff degression is important for automatic adjustments in the meantime. This way, constant changes to the legislative text because of technological improvement can be avoided. The degression rate varies according to the maturity of each technology. Technologies with a relatively steep learning curve, that is, a high potential for technological improvement, usually have a high rate of tariff degression (for example, photovoltaics). More mature technologies, like wind energy, usually have a much lower or no degression rate. The tariff degression affects only new installation, since technological improvements have no effect on installations that have already started operating.

As an alternative to the aforementioned fixed tariff degression, several countries have opted to link the degression rate to the market growth of a certain technology. This so called 'flexible tariff degression'[17] is a relatively new design option (Jacobs and Pfeiffer 2009). As described above, the cost reduction potential of a certain technology partly arises from economies of scale. If the overall installed capacity increases significantly, one can expect production costs to decrease significantly. Therefore, costs can be expected to decrease faster and the tariff can be lowered when market output augments.

The implementation of flexible degression requires the legislator to define a standard degression rate as in the case of normal tariff degression. Alterations from this standard degression rate can be linked either to a predefined pathway of market growth (for example 20 percent growth per year based on the installed capacity)

---

17   Couture et al. (2010: 41) use the term 'responsive degression' to reflect the self-adjusting nature of the mechanism.

or to a given capacity target (for example, 1,500 MW newly installed capacity per year). If those targets are exceeded the degression will be automatically increased; if they cannot be achieved the degression rate will be automatically reduced. The evaluation of the actual market development requires the establishment of a national registry that all new projects must sign. The statistics gathered on market growth will provide the basis for setting the annual degression rate (Jacobs and Pfeiffer 2009).[18]

5.4.1. Germany

In 2000 Germany became the first country worldwide to implement tariff degression. From 2002 onwards, the tariffs were to decrease according to the expected learning curve of each technology. Accordingly, the law of 2000 fixed degression rates for a number of technologies, namely, biomass (1 percent), wind (1.5 percent) and solar photovoltaics (5 percent). Other technologies, including small hydro power, gas from landfills, mines and sewage treatment plants, and geothermal power were not subject to tariff degression.

This new design option was primarily implemented to confront the general criticism of a price-based support instrument. Opponents of feed-in tariffs frequently argued that by fixing a 'price' technology learning is not incentivised and that the dynamics of market development are not taken into account. However, reducing the tariff on an annual basis in line with the expected cost-reduction potential for each technology can encourage technological learning. To a certain extent, this new design option was also implemented to better comply with the European state-aid guidelines and to be prepared for a potential legal confrontation with the European Commission (Bechberger and Reiche 2004). As described in Chapter 3, the European Commission had criticised the German tariff scheme for offering excessively high remuneration for wind energy producers.

In 2004 tariff degression was extended to all renewable energy technologies eligible under the feed-in tariff mechanism except small hydro power (less than 5 MW). The Germany legislator perceived tariff degression as being successful in tapping the learning potential of renewable energy technologies and making renewable energy sources competitive (BMU 2004c). Geothermal power producers, now receiving a considerably higher tariff, were confronted with a degression rate of 1 percent from 2010 onwards. Landfill gas, sewage treatment gas and mine gas tariffs were now reduced by 1.5 percent yearly. The degression for wind energy onshore was increased from 1.5 to 2 percent while offshore wind energy plants were exempted from facing reductions until 2008. As from 2006, the

---

18   The volatility of the degression rate can be limited by setting maximum deviations from the standard degression at both the top and the bottom end. For instance, if the standard degression rate is 5 percent, the deviation at the top or bottom is limited to +/– 3 percent, creating an overall 'corridor' of 2 to 8 percent.

degression rate for solar photovoltaics was increased to 6.5 percent in the case for free-standing photovoltaic plants.

The 2009 law brought further changes to degression rates, while the degression rate remained stable for biogas, biomass, geothermal and small hydro power, the rate for wind energy was reduced from 2 to 1 percent annually. The reduction of the wind energy degression had become necessary because of the increasing costs for steel and other raw materials that are needed for turbine production (Deutscher Bundestag 2008b). The German wind energy industry association BWE even called for interrupting tariff degression for a number of years (BWE 2008). In 2012, the tariff degression for wind energy will again be increased to 1.5 percent (BMU 2011). All other modifications are depicted in Table 5.2.

Table 5.2   Degression rates under the German feed-in tariff law of 2012

Renewable energy technology	Annual degression rate
Hydropower (more than 5 MW)	1 percent
Landfill gas	1.5 percent
Sewage treatment gas	1.5 percent
Mine gas	1.5 percent
Biomass	2 percent
Geothermal	5 percent
Wind power offshore	7 percent (from 2018 onwards)
Wind power onshore	1.5 percent
Solar photovoltaics	9–24 percent (depending on the market growth)

Source: Author based on BMU (2011)

In the case of solar photovoltaics, the 2009 amendment brought about further changes. As described above, legislators have to option to link the annual degression rate to the market growth in a given year. This so-called flexible degression was implemented in 2009 for solar photovoltaics. During the political decision-making process in 2008, the German Green Party suggested linking the degression rate to a 15 percent standard market growth rate as there was no register for renewable energy plants and thus only limited knowledge about the actually installed capacity. If the real growth was to be below or above this expected rate, each percentage point would be translated into an increase or decrease of the degression of one thousandth.[19] Under this first proposal of the Green party, the standard degression

---

19   For instance, if the national market grew by 20 and not 15 percent, that is, 5 percent more than expected, the degression rate would increase by 0.5 points (Jacobs and Pfeiffer 2009).

rate was set at 6.5 percent and the maximum deviation was plus or minus 2.5 percent. This first proposal of the German Green party was well perceived by large parts of the government and members of the parliament. However, the debate on feed-in tariff remuneration was highly politicised and controversial and therefore the proposal did not enter into the political debate at an early stage. Only after an agreement between the two governing parties was questioned by some members of the Conservative Party was flexible tariff degression hastily implemented (Krägenow and Kade 2008).

Eventually, the legislator decided to adjust the degression rate in relation to the installed capacity, rather than using the market growth rate expressed in percentages as the benchmark. In order to facilitate measurement of the installed capacity, the law established a register for all renewable energy producers who wanted to be eligible under the feed-in tariff scheme. The maximum deviation from the normal degression rate is plus or minus 1 percent per year. The legislator set a predefined pathway of market growth until 2011, expecting the installed capacity to reach 1,500 MW in 2009, 1,700 MW in 2010 and 1,900 MW in 2011. Depending on the actual capacity installed in those years, the degression rate can be either reduced or increased by 1 percent.

The mechanism for flexible degression was further extended and modified by the regulatory amendment to solar photovoltaic promotion in 2010. Under the new regulation, the government targets an annually installed capacity of 3.5 GW. If the market overshoots this target, the tariff will be lower by one additional percentage point for each GW. In other words, up to an installed capacity of 4.5 GW the degression rate will increase by one percentage point, up to 5.5 GW the degression rate will increase by two percentage points, and so on (BGBl 2010). The system was further modified with the 2012 amendment. The degression rate in January 2012 can range from 9 to 24 percent, depending on the actual market growth.

Table 5.3   Flexible degression for solar photovoltaics in Germany (2012)

Scenarios	MW to be installed in 2011 (projected)	Resulting degression in January 2012
Base case	3500	9%
+1 GW	4500	12%
+2 GW	5500	15%
+3 GW	6500	18%
+4 GW	7500	21%
> +4 GW	>7500	24%

Source: Author based on BMU 2011

5.4.2. Spain

By mid-2011, tariff degression as being applied only to solar photovoltaic installations. The implementation of tariff degression was debated in 2007. However, it was believed that the cost reduction potential for all other technologies was difficult to estimate, and consequently tariff degression was not implemented (A. Ceña, personal communication, 12 February 2008). However, the legislator decided to index the tariff only partly to the national inflation rate in order to achieve a similar effect.

Even though the solar industry association ASIF had already promoted the idea of tariff degression in 2002 (ASIF 2002), tariff degression for solar photovoltaics started as late as 2008. The reason for the relatively late implementation is partly the relatively small market. Even in 2006, the accumulated installed solar photovoltaic capacity was of 146 MW (see section 7.1.2). However, when it became apparent that a new piece of legislation was needed, since the 2010 target had already been reached in 2007, the renewable electricity associations ASIF and APPA actively started to lobby for a support framework including tariff degression. The industry associations proposed a degression rate of 5 percent starting from 2010 (ASIF and APPA 2007). In December 2007 the Spanish minister for industry, Juan Clos, announced his intention to include tariff degression in the remuneration scheme for the Spanish photovoltaic sector. In order to better reflect actual generation costs in tariff payment, the minister wanted to adopt 'the German system' of tariff degression (Energelia 2007). Accordingly, the intention to implement tariff degression for solar photovoltaics in Spain was clearly based on an analysis of the German feed-in tariff scheme (T. Diaz, personal communication, 25 February 2008).

Several months later, in March 2008, ASIF tabled another, more detailed proposal that aimed at reducing overall system costs related to solar photovoltaic promotion while at the same time avoiding the implementation of a technology-specific capacity cap. This two-sided objective was to be achieved via the implementation of flexible tariff degression. ASIF wanted to reach sustainable growth in the photovoltaic sector by coupling the level of tariff payment to annual market growth. Therefore, the rate of tariff degression was dependent on annual market growth. The targeted annual market growth was to be of 20 percent. In this case, the annual tariff degression would be of 5 percent. If market growth deviated from this expected rate, the degression rate would automatically be adjusted (see Table 5.4) (ASIF 2008).[20]

---

20   The industry association Asociación Empresarial Fotovoltaica (AEF) came up with a similar but simplified proposal. They called for a general degression rate of seven percent which was to increase to ten percent once the annual installed capacity exceeded 800 MW (Ristau 2008).

Table 5.4   Interdependency of market growth and tariff degression based on the ASIF proposal

Market growth (%)	Degression rate (%)
<= 5	2
10	3
15	4
20	5
25	6
30	7
35	9
>= 45	10

Source: Author based on ASIF (2008)

In the succeeding political decision-making process, the proposal for a flexible degression scheme was considerably modified. Instead of taking annual market growth as a point of reference, the ministry (MITYC) decided to base all variation on the newly installed capacity since this was more in line with the Spanish system of technology-specific targets and caps (see section 4.2.2). As all solar photovoltaic power producers had to sign a national registry, it was easy to keep track of the annually installed capacity. Even though ASIF managed to persuade the ministry to implement flexible tariff degression in order to control market growth, the second objective of the photovoltaic industry – the elimination of the capacity cap – was not achieved. Instead, the responsible ministry decided to combine flexible tariff degression with a capacity cap.

The new legislation, Royal Decree 1578/2008, establishes trimestrial capacity limits for the different types of photovoltaic systems. The total limit of 400 MW for 2009 was further divided according the system size, that is, small-scale roof top installations (< 20 kW), large scale rooftop installations (> 20 kW) and free-standing photovoltaic systems. More precisely, two thirds (267 MW) are allocated to rooftop installation (of which 10 percent are reserved for small-scale installations) and the remaining 133 MW to free-standing photovoltaic systems (BOE 2008a). Tariff degression takes place if the installed capacity in the previous trimester exceeds 75 percent of the target. In this case the tariff is reduced in accordance with a given formula.[21] If the target is fully reached in one trimester the maximum degression rate is of 2.6 percent every four months.[22] Tariffs might

---

21   $((1 - 0.9^{1/4}) * (P0 - P) / (0.25 * P0) + 0.9^{1/4}$, P0 being the capacity target for the previous trimester, and P the pre-registered capacity in the previous trimester.
22   $100 * (1 - 0.9^{1/4}) = 2.6\%$

also increase according to the same formula. This happens if in two consecutive trimesters not even 50 percent of the targeted capacity is installed.[23]

5.4.3. France

France applied tariff degression straightaway when the feed-in tariff scheme was first established in 2000/2001. Tariff degression applied both to rather mature renewable energy technologies (for instance, wind energy) and to less developed technologies (for instance, solar photovoltaics). The first decree for wind energy from 2001 stated that wind power tariffs were to decrease by 3.4 percent annually as from 2003 (J.O. 2001b). In 2006, when the decree for wind energy was amended for the first time, the degression rate was reduced to 2 percent per year (J.O. 2006a), even though the French wind industry association called for the complete cancellation of tariff degression (Rigaud 2006).

In the case of solar photovoltaics, the first decree included an annual tariff degression rate of 5 percent starting in 2003 (J.O. 2002a).[24] Surprisingly, the 2006 tariff scheme for solar photovoltaics – which offered considerably higher tariffs (see section 7.1.3) – no longer included tariff degression. An anonymous expert of the French solar photovoltaic industry commented that when the solar photovoltaic tariff was approved in 2006 it was planned to keep the tariff in place for a maximum of two years. For this reason, the implementation of automatic tariff reduction, that is, tariff degression, was not considered to be of primary importance. Because of replacement of employees and the restructuring of the responsible ministry, no new tariff scheme for solar photovoltaics was implemented until 2010.

A 2009 report from the French parliament suggests re-implementing tariff degression, referring to the negative experience of the Spanish photovoltaic feed-in tariff, where excessively high tariffs finally had disruptive effects on the support framework as a whole. The report explicitly refers to the 'German model' of tariff degression. Based on the experience in Germany and Spain, it recommends a degression rate of 7 to 8 percent. The report even suggests relating the degression rate to market growth (flexible tariff degression), as is done in Germany and Spain

---

23 Flexible tariff degression applied for the first time in February 2009. Mostly due to administrative problems, the applications for roof top installations did not reach 75 percent of the reserved capacity in the first quarter of 2009. Therefore, the tariffs remained at 34 cent/kWh (small scale) and 32 cent/kWh (large scale) for the second quarter of 2009. However, tariffs were reduced for free-standing photovoltaic installations from 32 to 30.72 cent/kWh as more than 75 percent of the allocated capacity were reached. Similar result were observed in the following degression rounds. The tariff level for roof mounted solar photovoltaic systems remained stable due to the fact than less than 75 percent of the targeted capacity were installed while the tariffs for free standing systems were further reduced to 29 €cent/kWh (MITYC 2009b).

24 However, this design feature did not have an effect on the French market. Soon after the tariffs of 2002 have been published it became clear that the tariff level was far too low in order to spur any investment.

(Assemblée Nationale 2009: 75). Accordingly, the new French tariff scheme for solar photovoltaic (2010) includes a 10 percent tariff degression starting from 2012 (J.O. 2010a).

The concept of flexible tariff degression was implemented in March 2011 (J.O. 2011). Based on the policy design in Spain, the French legislator adjusts the degression rate according to the trimestral quota issued for each plant-type category. As mentioned in section 4.2.3, the newly installed annual capacity for small-scale building integrated systems is limited to 100 MW. If in a given trimester the demands for grid connection exceed 65 MW, the tariff will be reduced by 9.5 percent. If the demand for this market segment is lower, the degression rate will also be lower as depicted in table 5.5.

Table 5.5   Flexible degression in France as of July 2011

Demand for grid connection for building integrated systems for individuals in a given trimester	Tariff degression scenarios based on the installed capacity in the previous trimester
More than 65 MW	9.5%
≤ 65 MW	7.5%
≤ 55 MW	6%
≤ 45 MW	4.5%
≤ 35 MW	3.5%
≤ 25 MW	2.6%
≤ 15 MW	1.5%
≤ 5MW	0

Source: Author based on J.O. 2011

5.4.4. Preliminary Conclusion

Germany, Spain and France have converged in implementing tariff degression since 2000 (see Figure 5.7). The German legislator invented this design option in 2000 to stimulate technological learning, to confront the criticism of the European Commission regarding a price-based support mechanism, and to meet the objections of national actors who argued that renewable energy sources would rely on subsidies for ever – similar to the German experience in the coal sector. France implemented this design option for the first time in 2001 and Spain in 2008.

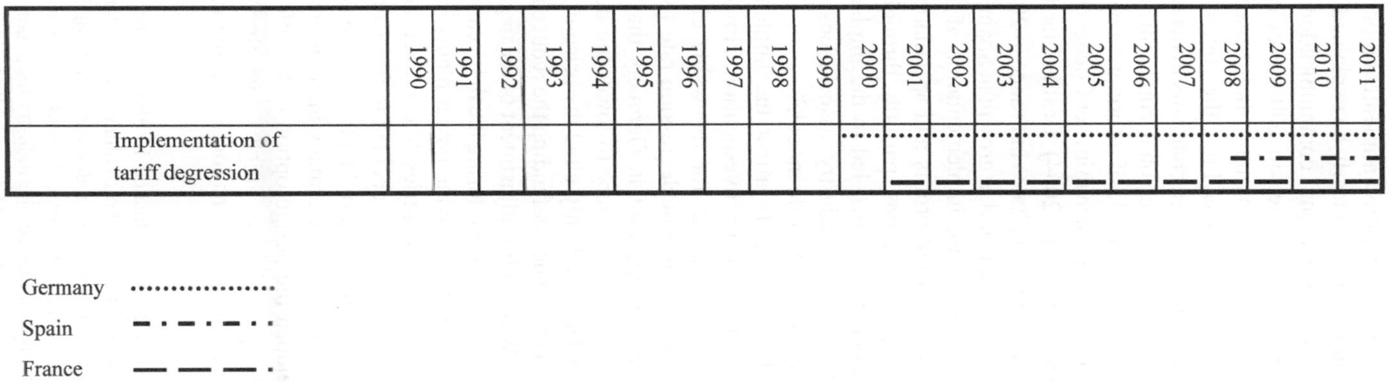

	Implementation of tariff degression
2011	
2010	
2009	
2008	
2007	
2006	
2005	
2004	
2003	
2002	
2001	
2000	
1999	
1998	
1997	
1996	
1995	
1994	
1993	
1992	
1991	
1990	

Germany ·········

Spain ▪ ▪ ▪

France ▬ ▬ ▬

Figure 5.7 Implementation of tariff degression in Germany, Spain and France, 1990–2011

Source: Author

Why did all three countries under investigation implementation tariff degression? Its implementation in France and Spain can clearly be explained in terms of policy transfer from Germany based on transnational communication. In France, tariff degression for wind energy was implemented in 2001. Previously, France had learned from the German experience via communication between governmental actors, who were frequently exchanging ideas. First, the governmental institution ADEME analysed the German feed-in tariff mechanisms in detail when designing the French support instrument (J.-L. Bal, personal communication, 24 October 2008) and second, there was frequent exchange of information between member of the German Green Party and the former French minister Cochet (H.J. Fell, personal communication, 25 September 2008, Riepe 2004). Besides, tariff degression was re-implemented in the case of solar photovoltaics in France in 2010 partly because of the negative experience of exponential market growth in Spain. A report from the French parliament explicitly stated that the implementation of this design option was necessary in order to avoid windfall profits and speculation. Spain learned from the German and the French regulations, partially through communication between national industry associations which led to the implementation of this design options in 2008. The Spanish industry associations promoting tariff degression clearly viewed Germany as the role model.[25]

Regulatory competition did not directly affect the implementation of this design feature in the three countries under investigation. However, the need to be competitive on the international market for renewable electricity products influenced the German legislator's choice of degression rate for wind energy in 2008. In 2007 and 2008 it became apparent in Germany that rising prices for certain raw materials were making it necessary to lower the degression rate for wind energy. The German legislator explicitly cited the price increases of copper and steel on the international market when defending the reduced tariff degression (Deutscher Bundestag 2008b). Without the adjustment of the degression rate – and a simultaneous increase of the tariff payment during the first years of operation (see section 5.4.1) – the incentive for installing wind power plants in Germany would have been insufficient. Besides, other countries like France and Spain offered higher tariffs, and so wind power developers decided to focus on markets outside Germany. Consequently, the German legislator had to increase the tariff level and reduce the degression rate in order to keep Germany internationally competitive.

Germany and Spain developed the design option of flexible tariff degression almost simultaneously. Although this fact might give the impression of a coordinated approach or the result of transnational communication, a detailed

---

25   The direct transfer process is also indicated by the fact that all countries used or proposed similar degression rates for solar photovoltaics. When tariff degression was first implemented for solar photovoltaics in 2000, a degression rate of 5 percent was fixed (BGBl 2000). France implemented exactly the same degression rate in 2002 (J.O. 2002a). Based on the experience in Germany and France, the industry associations ASIF and APPA proposed a degression rate of 5 percent in 2007 (ASIF and APPA 2007).

analysis and interviews with the national actors in the two countries reveals that this was not the case. Both the individuals in German and Spain who 'invented' and promoted this design option within the political decision-making process confirmed that they had not heard of one another's effort. Therefore, the almost simultaneous implementation of flexible tariff degression has to be related to 'independent problem solving' through parallel problem pressure. Both Germany and Spain implemented the flexible tariff degression in order to control market growth (Deutscher Bundestag 2008b, BOE 2008a). Therefore, flexible tariff degression can be considered to be a design option for relatively large solar photovoltaic markets (Mendonça et al. 2009b, Jacobs and Pfeiffer 2009). This also explains why France implemented this design option once the demands for grid connection started to grow exponentially.

5.5. Conclusion

There is a high degree of policy convergence in design options for tariff differentiation and adjustment in Germany, Spain and France. All countries under investigation have implemented tariff differentiation by reference to the technologies and the sizes of power plants (see section 5.1 and 5.2). In the case of location-specific tariffs, only France and Germany opted for this design feature (see section 5.3). All countries have implemented tariff degression for automatic and annual tariff adjustment and even the concept of flexible tariff degression (see section 5.4). Their primary reasons for the implementation of design options for tariff differentiation and tariff adjustment are to avoid windfall profits, to increase investment security and to promote certain specific producer groups.

The implementation of the three design options for tariff differentiation – technology-, size-, and location-specific tariffs – has significantly increased the complexity of national feed-in tariff mechanisms. In Germany, the number of tariffs increased from three in 1990 to 32 in 2009, in Spain from three in 1994 to 72 in 2007 and in France from eight in 2000 to 21 in 2009. The increasing number of tariffs is also reflected in an increasing number of articles and paragraphs in national feed-in tariff legislation. The German feed-in tariff law, for instance, started with only five articles in 1990. In successive amendments the number increased to 13 in 2000, to 21 in 2004, and to an astounding 66 articles in 2009 and 2012.[26] Generally speaking, increasing tariff differentiation reflects a trend away from technology-specific support towards more project-specific support, with the size and the location of different power plants also taken into account.

---

26   The same tendency can be observed in France and Spain. However, since Spanish legislation also regulates combined heat and power generation, and the French feed-in tariff legislation is fragmented into a large number of ministerial orders and decrees, the exact number of articles cannot be estimated.

Testing the hypotheses of this research project, we have established that transnational communication was decisive for the spread of certain feed-in tariff design options. This was the case for the diffusion of location-specific tariffs for wind energy. As shown in section 5.3, the communication between governmental actors in Germany and France led to direct policy transfer from one country to the other, which explains the striking similarities. Even though the implementation of this design option was unsuccessful in Spain, the national debate about the advantages and disadvantages of this design feature was clearly fuelled by the experience in Germany and France, the only countries operating with this design option at the time.

Transnational communication was also important for the spread of tariff degression (see section 5.4). Once again, the direct exchange of information between the German and the French legislator was the basis for the implementation of this design feature in France. Interestingly, the re-implementation of tariff degression in France in 2010 was primarily based on the negative experience with solar photovoltaic promotion in Spain. The French legislator, wanted to avoid the exponential growth of solar photovoltaics such as occurred in Spain during 2007 and 2008, reintroduced tariff degression for solar photovoltaics in 2010 (after it had been excluded in 2006) in order to better control market growth and the overcompensation that might lead to speculation. This case shows that policy learning can also be based on negative experiences in other countries.

Regulatory competition played a minor role in the convergence of design options for tariff differentiation and adjustment. Only in the case of tariff degression in Germany did we detect a competition-related influence. The German legislator's decision to lower the degression rate from 2 percent to 1 percent was made against the background of decreasing installation rates in Germany. Obviously, the internal rates of return in other countries – including France and Spain – were higher at that time. This is why wind power developers decreased their engagement in Germany. Consequently, the German legislator had to increase the tariff level and lower the degression rate in order to regain competitiveness with other European markets.

Besides, independent problem-solving was a crucial driving force for the similar development of national support mechanisms. As shown in section 5.1, increasing technology differentiation reflected general technological development, especially the development of certain sub-technologies such as offshore wind power and integrated solar photovoltaic systems. As technologies developed, national legislators enshrined more specific definitions of the eligible technologies. Besides, the increasing size of renewable energy power plants prompted legislator to take the differences between small-scale producers and large-scale producers into account, thus establishing plant-size-specific tariffs. This occurred without cross-national communication or regulatory competition. National legislators made these decisions independently because of international technological development.

An interesting instance of independent problem-solving was revealed in case of flexible tariff degression in Germany and Spain (see section 5.4). Even though the two countries developed and implemented this design feature in the same year, and even though the details of their designs strongly resemble each other, apparently no information exchange took place between them. The people who initially 'invented' this design option in Germany and Spain were both interviewed, and both confirmed that they did not know of each other's efforts. Therefore, the implementation of this policy innovation in the two countries was a true case of independent problem-solving based on similar problem pressure. In both countries flexible tariff degression was primarily implemented in order to control market growth and the cost of the national solar photovoltaic sector. This is also why France implemented this design option in 2011.

As already identified in section 4.7, the European level also had a certain influence on the development of national feed-in tariff design. In the previous chapter we showed that the need to reach the European indicative targets for 2010 forced national legislators to include a larger number of technologies under national feed-in tariff schemes. This was managed by applying a new tariff calculation methodology which takes technology-specific generation costs into account (see section 4.3). In this chapter we revealed that the application of a cost-based tariff calculation methodology also affected the degree of technology differentiation. In section 5.1 it was shown that the number of technology-specific tariffs increased once the country under analysis adopted a cost-based tariff calculation methodology. In other words, the need to reach the national targets that were set at European level prompted national legislators to opt for a more efficient tariff calculation methodology and consequently to increase technology-differentiated tariff payment. Besides, national governments were eager to design their support instrument as efficiently as possible, because the European Directive 2001/77/EC stated that support instruments would be assessed on the basis of their cost-effectiveness.

Chapter 6

The Evolution of Design Options for System and Market Integration

As in the previous two chapters, the detailed analysis of feed-in tariff design options for system and market integration will serve to disclose national policy similarities. We want to understand the causes of increasing policy similarities of feed-in tariffs – despite the failure of harmonisation. The extended timeframe of the analysis of more than two decades allows us to determine whether feed-in tariffs have become more similar over time and, if so, which factors have been responsible for this. Therefore, we can test our hypotheses as presented in section 2.3. Preliminary results are presented after the analysis of each design options. A summary of these conclusions is provided in chapters 8.

In recent years, legislators have implemented an increasing number of design options that facilitate the integration of renewable electricity into the existing power system and market. 'System integration' refers to all regulations improving the technical integration of renewable power into the national grid. 'Market integration' refers to all measures affecting the framework for renewable electricity sales. These design options usually improve the 'quality' of electricity supply, meaning that electricity supply better matches electricity demand and that grid stability is enhanced. Design options for improved system and market integration include alternative sales options (market sales, premium feed-in tariff and direct-consumption), tariff payments for improved system integration (auxiliary grid services, demand-oriented tariff payments, tariff payments for steady supply of renewable electricity) and regulations for controlling power output (forecast obligations and remote-controlled power output).

6.1. Alternative Sales Options

Over the years, legislators have implemented a number of alternative sales options for the producers of renewable electricity, primarily targeting improved integration into the existing electricity market. Under standard feed-in tariffs, producers receive a fixed remuneration per kilowatt hour independent of electricity demand. Alternative sales options include the possibility of selling renewable electricity on the grey power market (the market sales option) and market sales plus the payment of a reduced tariff (the premium feed-in tariff option). In order to facilitate system integration, a special tariff payment can be guaranteed for electricity that is not fed into the grid but consumed by the producer itself (direct-consumption).

Market sales of electricity are possible in liberalised electricity markets. Generally speaking, power can be traded within different frameworks. Producers can sell power based on long-term bilateral agreements (so called over-the-counter deals), on a power exchange market and balancing markets (for balancing power supply and demand), or on the forward markets (for future power demand). Additionally, electricity can be traded within different time frames (month ahead, week ahead, day ahead, intra-day, etc.). Generally speaking, market sales are expected to improve system integration because power producers will respond to the price signals of the market – the result of the interaction of supply and demand.

Power producers who are able to control the timing of power output (for instance, biomass-based generation units) will try to produce power during periods of high demand and low wind because they will be able to generate higher income during peak-demand periods. Power producers using fluctuating renewable energy technologies that are not able to control the timing of power output (for instance, wind and solar photovoltaic producers) will try to better forecast their generation, as national market design usually penalises short-term deviations, and find partners on the supply and demand side. Because of the price spread between off-peak and peak electricity prices, the development of storage technologies might also be incentivised through market participation. The price spread need to be rather large so that producers will decide to invest in storage technologies in order to shift the provision of electricity from low-demand periods to high-demand periods. The necessary price spread might need to be smaller in the future, once the costs of storage technologies have fallen.

Usually, renewable electricity producers cannot compete with other, conventional power technologies on the market because of their higher costs – which is why a feed-in tariff or any other support instrument becomes necessary in the first place. If the negative external effects of conventional energy sources are not internalised, the generation costs of renewable energy sources are usually higher. However, certain renewable energy technologies have managed to decrease generation costs considerably over the recent decades. Depending on the (hourly) market price, selling electricity on the market is sometimes already financially more rewarding for green power producers than receiving the fixed feed-in tariff. These time periods are expected to occur more frequently in the future because of decreasing costs for renewable energy sources and increasing costs for conventionally produced electricity.

In order to encourage market participation, some feed-in tariffs give renewable electricity producers the opportunity to leave feed-in contracts before the end of the guaranteed tariff payment period. In this case, the legislator will also have to clarify to what extent producers will be allowed to switch between market sales and the fixed feed-in rate. Theoretically, legislators have three options. First, they can mandate that the predefined duration of tariff payment (see section 4.4) has to be 'fulfilled' and that renewable electricity producers do not have the right to sell their power in the market beforehand.[1] In this case, market participation will be postponed to the end of the tariff payment period. Second, renewable electricity producers can be given the right to leave the fixed feed-in tariff prematurely. However, once they have decided to leave the fixed feed-in tariff scheme, they have no right to re-enter it.[2] Finally, the legislator can give producers the opportunity to switch between the fixed-tariff option and the premium feed-in tariff, allowing renewable electricity producers to gain first-hand experience in the power market without being exposed to the full risk related to volatile market prices. In this case, the policymaker has to determine a time period (for instance, one month or one year) in which the producer is obliged to remain with one remuneration option before it is allowed to switch to another option. Generally speaking, very short periods might incentivise cherry-picking as producers will opt for the market option in times of high market prices and fall back on the fixed-tariff option in times of low market prices (Mendonça et al. 2009b).

A more elaborated form of market sale for green power producers is a premium feed-in tariff (premium-FIT).[3] In the case of premium feed-in tariffs, the producer has to sell the renewable electricity on the market. As the market price of electricity alone will often not suffice to reach a certain profitability threshold, a reduced feed-in tariff, that is, a premium payment on top of the market price, is paid to the producer of renewable electricity. The premium payment depends on the technology-specific generation costs. This way, the premium feed-in tariff payment covers only the average cost difference between the market price and the average generation cost for each renewable technology. The combination of these two payment components – the electricity market price and the reduced feed-in tariff –allows the renewable electricity producer to operate its power plant profitably.

In most cases, premium feed-in tariffs are introduced as an optional alternative to the fixed tariff payment. In other words, renewable electricity producers can choose to receive the 'normal' fixed feed-in tariff payment or to sell their electricity on the market and then receive a reduced tariff. As with the general market sales option, the legislator has to define the extent to which producers are allowed to switch between the fixed and the premium feed-in tariff option. A premium feed-in tariff is sometimes considered to be more 'market-based' than a fixed feed-in tariff and less problematic in the light of article 34 of the Treaty on the Functioning of

---

1   The positive effect of this approach is lower electricity costs for final consumers, once the power price for conventional power exceeds the guaranteed tariff level. In this case, the feed-in tariff will stabilise and lower the average electricity price. However, such a policy could delay the integration of green electricity into the conventional power market as developers will be getting less for their renewable electricity (Mendonça et al. 2009b).

2   This would in essence increase the risk of renewable electricity producers participating in the conventional grey market as future prices might be difficult to anticipate.

3   In North America, the term 'market dependent' is sometimes used (see Couture and Gagnon 2009).

the European Union regulating the free movement of goods (Klein et al. 2008, EU Commission 2005a).[4]

However, a premium feed-in tariff is also more complicated to establish. The combination of two remuneration components – the market price and the reduced tariff payment – makes it harder for the legislator to set the tariff levels 'correctly' (that is, in line with the technology-specific generation costs) when market prices are fluctuating. The development of the market price has to be anticipated in order to avoid windfall profits (in the case of high market prices) and to guarantee sufficient returns of investment (in the case of low market prices) (Mendonça et al. 2009b).

In the case of direct-consumption, the producer 'directly consumes' its own decentralised electricity without feeding it into the grid. This option usually applies to small-scale power plants, such as solar photovoltaic plants on private houses, as local power supply should not exceed local power demand. Direct-consumption becomes financially interesting when the retail electricity price exceeds the generation costs of a decentralised power generation unit (grid parity). If this is not (yet) the case, legislators can incentivise direct-consumption by granting additional tariff payment to compensate for the cost difference. Therefore, direct-consumption can be considered an additional 'sales option', even though it does not involve selling (all) electricity to the national grid.

The rationale of additional tariff payments for direct-consumption is that the extra costs of the feed-in tariff scheme can be reduced. Only the cost-differences between the retail electricity price and the actual generation costs have to be covered via the financing mechanism of the feed-in tariff. As explained in more detail in section 4.5, these costs eventually have to be distributed between all final electricity consumers. Moreover, this additional sales option can help producers take better advantage of the decentralised nature of renewable electricity generation, thus facilitating system integration. The national electricity grid has to absorb less (fluctuating) electricity from renewable energy sources and is therefore indirectly stabilised. Theoretically, costs of grid reinforcement can be reduced. On-site generation also reduces line losses, because less power has to be transmitted and distributed. Besides, legislators can already prepare power producers for the post-'grid parity' conditions, that is, the time when direct-consumption will be cheaper than purchasing electricity from the grid.

6.1.1. Germany

In Germany, the implementation of both a market sales option and a premium feed-in tariff has been debated together since 2007. A framework for market sales of green power producers was established within the amendment of the German

---

4   Therefore, the implementation of premium feed-in tariffs has often been considered as a means to overcome resistance to price-based support mechanisms in general (see Mendonça et al. 2009b).

feed-in tariff scheme in 2009, granting electricity producers the right to relinquish the fixed tariff option and sell their electricity on the market (BGBl 2008).[5] Under the article 17 of the same law, renewable electricity producers may also sell only a certain share of their production on the market. This provision was explicitly implemented to take into account the needs of small and medium-sized companies, as this option helps to reduce price risks (BMU 2008b). For instance, small producers could sell 90 percent of their electricity via the fixed tariff scheme and ten percent on the market in order to gain first-hand experience without being exposed to the full price risk.

Unlike with the premium feed-in tariff, there was wide consensus among German actors for implementing the market sales option. But different points of view were expressed regarding the time frame during which producers had to remain with either option. Inspired by the Spanish regulation (see section 6.1.2), the responsible ministry first called for a rather restrictive approach, allowing producers to switch between the two sales options on an annual basis (BMU 2007c).[6] When the first proposal was tabled it was still believed that the premium feed-in tariff would be established as well – one factor explaining this rather restrictive approach. The Ministry for the Economy argued that an annual choice might obstruct market participation, in response to which the Ministry of the Environment agreed to reduce the period to six months (Bröer 2007). Because of further political pressure to facilitate market participation, the minimum period available to a producer to sell its electricity on the market was further reduced to one month. A shorter time period for switching between the fixed tariff and the market sales options was opposed by the Federal Grid Agency. Some actors called for an approach that allowed power producers to switch hourly between the remuneration options, but the Federal Grid Agency argued that this approach would be incompatible with the financing mechanism of the German feed-in tariff scheme (Bundesnetzagentur 2007).

Besides the aforementioned provisions, article 37 of the 2009 feed-in tariff amendment provides another indirect incentive for market sales. However, it has to be noted that only utilities can profit from this regulation, not renewable electricity producers in general. It states that utilities will be exempt from payment under the nationwide equalisation scheme if at least 50 percent of the electricity delivered to the final consumer stems from renewable energy sources that are eligible for tariff payment. Instead of receiving tariff payments, the green electricity – which would otherwise be eligible under the German feed-in tariff – can be directly sold to the final consumer. This sales option will be attractive if the retail electricity

---

5   The legislation of 2000 and 2004 did not regulate market sales of green power. In practice, renewable electricity producers were allowed to switch between the fixed tariff payment option and market sales as often as they wanted.

6   The proposal from 2007 required producers to indicate at the end of September whether they wanted to sell their electricity on the market or profit from the fixed tariff regime for the whole of the following year.

price plus the avoided cost of the national equalisation scheme is higher than the actual tariff payment. This is already interesting for a number of technologies with a relatively low tariff level, including small hydro power plants and wind energy producers, which receive a reduced tariff payment at the end of the remuneration period. This option will become even more attractive if the cost of renewable electricity promotion increases further. While the additional costs per unit of electricity amounted to 1.1 €cent/kWh in 2009 it reached reach 2.0 €cent/kWh in 2010 (BDEW 2009d). In 2011, the additional cost per kilowatt-hour reached 3.5 €cent/kWh. Therefore, the new feed-in tariff law of 2012 included a provision which limits the potential savings to 2 €cent/kWh. Additionally, utilities selling renewable electricity directly to final consumers will have to include 20 percent of electricity from fluctuating sources such as wind power and solar energy (BMU 2011).

During the political decision making process in 2007 and 2008, a premium feed-in tariff was devised to further incentivise market participation. This intention was first expressed at an international conference in Brussels in June 2007. A representative of the BMU stated that Germany would implement a premium feed-in tariff based on the Spanish experience. A later report of the German ministry of the environment on improved integration of renewable energies in the electricity sector explicitly envisaged the possible future implementation of a premium feed-in tariff (BMU 2008b). However, in autumn 2007 it had already become clear that including all necessary provisions in the 2009 amendment would be difficult because the political decision-making process was accelerated (Solarthemen 2007, Bröer 2007). Therefore, the amendment of 2009 included only an authorisation for the government to issue and ordinance for the implementation of premium feed-in tariff at a later stage (BGBl 2008).

Subsequently, the Fraunhofer ISI research institute conducted a study of the specific design of the German premium feed-in tariff scheme on behalf of the German ministry of the environment (Ragwitz and Sensfuss 2008). This document acted as the scientific basis for the subsequent political decision-making process. Under this proposal, the premium feed-in tariff was to be a voluntary alternative to the existing fixed tariff payment scheme. The premium feed-in tariff level would be defined *ex post*, that is, after the market price for electricity of the previous month was calculated in order to avoid windfall profits and to increase investment security. The proposal for a premium feed-in tariff no longer included a purchase obligation for green power production. Therefore, producers of green electricity would be responsible for forecasting future power generation. Possible deviations from this prognosis have to be levelled out via the purchase balancing power on the market. Therefore, the proposal also included an extra premium payment for market participation, in order to reimburse producers for the related additional costs (Sensfuss and Ragwitz 2009, Ragwitz and Sensfuss 2008).

However, in 2008 no political consensus was reached regarding the implementation of a premium feed-in tariff. The Christian Democrats (CDU) preferred premium feed-in tariff while the Social Democrats (SPD) opposed this design option and wanted to improve system integration via an additional tariff payment for a more steady electricity supply (see section 6.2.1). In the absence of a political compromise, ultimately neither design option was implemented (CDU/CSU 2009). Besides the opposition from the SPD, the implementation of a premium feed-in tariff was contested by large sections of the renewable energy industry, including the all major renewable industry associations headed by the BEE. The opponents primarily questioned the effectiveness of this approach in bringing about better market integration and developing storage technologies.[7] However, it should be noted that a part of the renewable energy industry supported this additional incentive for market participation.[8] Besides the Christian Democrats, this approach was supported by the German industry association BDEW (Bröer 2007a, BDEW 2009b) and a number of research institutes (for instance, Sensfuss et al. 2007). Also within the responsible ministry BMU, the implementation of a premium feed-in tariff was a matter of controversy.

Nonetheless, in summer 2011, the premium feed-in tariff was eventually implemented under a new government coalition. System and market integration of renewable energy sources was declared to be a primary objective of the latest amendment. This also explains why the new law consisted of a new section including all provision direct marketing of renewable electricity (§33a to 33i). The design of the premium feed-in tariff is largely based on the aforementioned proposal by the Fraunhofer ISI. In additional, the new law offered a so-called flexibility premium for biogas producers which manage to shift their power production to peak-demand periods via investment in storage capacities (BMU 2011).

Since 2009, the German feed-in tariff mechanism has offered an additional incentive for the direct-consumption of photovoltaic electricity (BGBl 2008). All producers of electricity based on solar photovoltaic power plants up to an installed capacity of 30 kW can directly consume the electricity generated and receive a reduced feed-in tariff. Either the operator of the installation or a third party has to consume the power in the immediate vicinity of the plant. The tariff is reduced from 43.01 €cent/kWh to 25.01 €cent/kWh. The draft of the law for solar photovoltaic promotion of 2010 even proposed increasing the additional incentive for direct consumption of solar electricity from 3.6 €cent/kWh to 8 €cent/kWh for all installations up to 800 kW (BMU 2010a).[9] However, the final

---

7   Opponents of premium feed-in tariffs also point to the German experience with co-generation power plants. In practice the power plants sell their electricity under a premium feed-in tariff (market sales plus payment of an additional remuneration). However, this policy did not lead to the development of storage technologies.

8   Eleven medium-sized companies, mainly from the wind power business, produced an independent position paper in early 2009. In contrast to the industry associations' position, these companies called for the fast implementation of a premium feed-in tariff based on the proposal of Fraunhofer ISI (Natenco et al. 2009).

9   As the effects of this new regulation are hard to predict, this sales option will be in place only until the end of 2011, when the government will re-evaluate its impact (Podewils 2010b).

version of the law offered fewer incentives for direct consumption than the first draft. The legislator feared that the generous framework elaborated in the original proposal would create a loophole in the legislation and undermine the government's efforts to reduce costs for final consumers. This 'sales option' is now limited to installation with a total capacity of no more than 500 kW.[10]As before, the new legislation offers power producers that directly consume their electricity an average additional remuneration of 3.6 €cent/kWh. This incentive is paid in the event that direct consumption is below 30 percent of the total power consumption of a given consumer. Once direct consumption exceeds this 30 percent threshold, the additional tariff payment will rise to 8 €cent/kWh (BGBl 2010). The German legislator hopes in this way to incentivise the use and development of storage technologies. Moreover, the new regulation can be understood as an incentive to install smaller-scale units, since it is easier to reach a share of directly consumed electricity of above 30 percent with these power plants (Wenzel and Nitsch 2010). The law of 2012 limited this sales option to installations with maximum 100 kW. In addition, the higher tariff rate of 8 €cent/kWh was cancelled (BMU 2011).

6.1.2. Spain

Shortly after the liberalisation of the national electricity market in 1997, the Spanish legislator allowed renewable power producers to sell their electricity on the market. As early as 1998 the Spanish legislator passed a new piece of legislation that not only enabled market sales but also provided additional tariff payments in line with the aforementioned premium feed-in tariff mechanism. With this regulation, Spain was the first country to establish a premium feed-in tariff. Under RD 2818/1998 producers were given the opportunity to sell electricity to the distribution companies at the average hourly market price. On top of that, producers would receive a technology-specific premium payment as defined by the Royal Decree (BOE 1998a).[11] The premium payment was updated annually in line with the variations in the average electricity sales price (del Río and Gual 2007). Despite the early implementation of the premium feed-in tariff option, selling renewable electricity on the market was more of a theoretical option and did not have any effect on the choice of actors in the renewable energy industry.

---

10   Interestingly, the Spanish legislator had implemented a very similar definition of direct consumers in 1998. The feed-in tariff legislation of the same year defined direct consumers as persons who primarily produced electricity for their own use and whose installations were no larger than 25 kW. However, this definition of direct consumers applied not to solar photovoltaic installations but to small-scale cogeneration units.

11   Besides, producers under the premium option were eligible for additional payments for the provision of reactive power (see section 6.2.2).

All renewable electricity producers opted for the fixed tariff option because of its greater investment security and higher profitability (AEE 2004).[12]

Royal Decree 436/2004 considerably improved the conditions for participating in the conventional electricity market in terms of the premium feed-in tariff. Tariffs were increased, the forecast obligation for renewable electricity generation was established (see section 6.3.2), and legal uncertainties were removed. Besides the premium feed-in tariff, the legislator offered an additional incentive for the participation of renewable electricity producers in the conventional power market.[13] Shortly after the implementation of the amended decree of 2004 renewable electricity producers shifted from the fixed tariff option to the premium feed-in tariff option. This movement was especially pronounced in the case of wind energy, where already by the end of 2005 more that 90 percent of all producers were remunerated under the premium feed-in tariff (AEE 2006). In recent years, the number of wind power producers opting for the premium feed-in tariff option has decreased marginally. Wind power producers opting for the fixed tariff payment are either small-scale producers or producers on the Spanish islands such as the Canary Islands.[14] The latter are obliged by law to opt for the fixed tariff option (A. Ceña, personal communication, 12 February 2008).

The average cost of electricity in the Spanish electricity market increased considerably after the legislation was implemented in 2004 (see Figure 6.1). Accordingly, the profitability of renewable energy projects selling their electricity under the premium feed-in tariff option increased as well. Internal calculation of the CNE suggested that the internal rate of return for wind power projects rose to 16.3 percent in January and February 2005, in contrast to the targeted five to nine percent (CNE 2007a).

---

12   The Royal Decree of 1998 did not manage to incentivise renewable electricity producers to participate in the wholesale market as some regulations of the valid electricity law of 1994 conflicted with the new Royal Decree (BOE 1994b). It was implemented before the liberalisation of energy markets in Spain and did not accommodate the possibility of third parties selling electricity on the wholesale market (AEE 2004).

13   This additional incentive was offered from 2002 to 2007. Royal Decree 841/2002 already granted this additional remuneration for certain technologies, namely, installations based on fossil fuels (BOE 2002a). Other technologies, such as wind energy, became eligible for this extra payment in only 2004. In the case of wind power, RD 436/2004 guaranteed producers 40 percent of the average electricity price under the premium feed-in tariff option. In addition, producers of wind-powered electricity received an extra ten percent of the average electricity price for market participation (Jacobs 2008). This extra payment amounted to an additional tariff payment of 0.7 €cent/kWh in 2006 (AEE 2008a). This incentive was also intended to compensate for the additional transaction costs related to activities in the market.

14   In 2008, the capacity of wind power installed on the Canary Islands was 140 MW. All these producers were obliged to sell their electricity under the fixed tariff option.

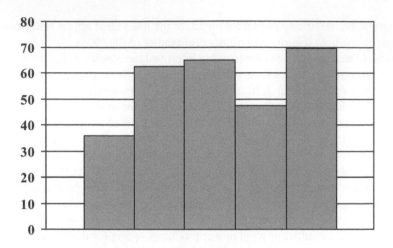

Figure 6.1   Development of average electricity price at the Spanish power market (€/MW – 2004 to 2008)

Source: Author based on REE 2009: 13

Consequently, the 2007 amendment to premium feed-in tariffs was complemented by a 'cap' and a 'floor' to the total remuneration. The cap prevents the combined remuneration (market price and reduced feed-in tariff) exceeding a certain limit in the case of high market prices, while the floor prevents the remuneration falling below a minimum threshold in the case of falling market prices. For onshore wind power, for instance, the premium was fixed at 2.9 €cent/kWh, the combined tariff payment (market price plus premium tariff) cannot fall beneath 7.1 €cent/kWh. If the electricity price reaches or exceeds 8.5 €cent/kWh, the premium feed-in tariff is no longer paid.

A special tariff payment for direct-consumption has not (yet) been implemented in Spain. However, the Spanish industry association for solar photovoltaics, ASIF, has proposed a similar approach. According to a study commissioned by ASIF, grid parity for solar photovoltaic will be reached in 2016, that is to say, electricity from decentralised solar panels will in that year be as expensive as Spanish retail electricity. Therefore, ASIF is pushing for a legislative amendment which would reduce tariffs and establish additional incentives for direct-consumption of solar photovoltaic electricity. This so-called 'feed-in compensation' should compensate for the difference between the retail electricity price and the cost of solar photovoltaic electricity generation until grid parity is reached. According to ASIF, electricity production exceeding direct-consumption could still be fed into the grid and remunerated in line with a so-called 'FIT-Net-Metering Tariff' (ASIF 2009). ASIF claims that the responsible ministry has already expressed an interest in this design option (Saravia et al. 2010). The proposal was based on an analysis of existing regulations for direct-consumption in other countries, including Germany, Italy, Belgium and the United States (T. Diaz, personal communication, 15 March 2010).

6.1.3. France

Market sales of renewable electricity are not a political priority in France. Theoretically, renewable energy producers can leave feed-in tariff contracts which they have signed with EDF before the end of the contract. However, penalties for leaving the contract are so great that no producer has ever left a feed-in tariff contract prematurely (J.-L. Bal, personal communication, 24 October 2008). Therefore, market sales are a merely theoretical option in France. In 2012 the first feed-in contracts signed by hydropower producers will come to an end. It is still not clear whether these power producers will be offered a follow-up contract for a feed-in tariff or whether the will have the opportunity to sell their electricity on the market (J.-L. Bal, personal communication, 24 October 2008).

France has not implemented either a special remuneration scheme for direct-consumption or a premium feed-in tariff. EDF has internally debated the implementation of a premium feed-in tariff, which it would back as it is interested in integrating renewable energy sources into the conventional energy market (P. Thérond, personal communication, 22 October 2008). SER, the industry association, opposes an immediate implementation of premium feed-in tariffs because it considers the fixed tariff option to be sufficient for the moment. However, in the mid-term future this option might have to be reconsidered, given that market electricity prices are expected to rise continuously in the future (M. Lettry, personal communication, 9 October 2008). The implementation of this sales option has never been debated at governmental level (J.-L. Bal, personal communication, 24 October 2008).

6.1.4. Preliminary Conclusion

There is limited convergence on the implementation of alternative sales options. Even though market sales of green electricity are theoretically possible in all three countries, only the Spanish feed-in tariff scheme currently manages to incentivise large-scale market participation, primarily via the premium feed-in tariff option. Of the three countries under study, Spain and recently Germany have established this alternative sales option (see Figure 6.2). Spain was the first country in the world to implement a premium feed-in tariff. In that sense, the legislation from 1998 was truly innovative, even though this remuneration option became attractive for renewable electricity generators only after the 2004 amendment.[15]

---

15   The same approach was followed by a number of other European countries, namely Denmark, Slovenia, the Czech Republic, Italy, Estonia and the Netherlands. Finland also

In Germany, the implementation of a premium feed-in tariff has been debated at governmental level since 2007. After a controversial debate, this design option was eventually implemented in 2011. In France, market sales are practically impossible before the end of a feed-in tariff contract because of the high penalties for leaving the contract prematurely. The implementation of a premium feed-in tariff has so far been debated internally only by EDF, never at the political level. Interviews with French stakeholders revealed that some actors did not even know about this remuneration option and that knowledge of its advantages and disadvantages was very limited.[16]

The implementation of the market sales option in European countries was largely technology-driven. Over the years, power generation costs of renewable electricity technologies decreased considerably and consequently the cost difference vis-à-vis the electricity price on the conventional power market decreased. Today, it is sometimes financially more interesting for certain renewable power producers to sell their electricity on the market rather than receiving the fixed remuneration via feed-in tariff mechanisms. Therefore, policy makers in Germany and Spain (and theoretically as well in France) allowed renewable electricity producers to sell their power on the market. It should be noted that national legislators implemented this design option independently, that is, without transnational communication, regulatory competition or any influence from European level.

The fact that Spain excluded solar photovoltaics from participation under the premium feed-in tariff shows that the actual differences between the technology-specific generation costs and the electricity price on the grey power market matters. However, for solar electricity the cost comparison with the retail electricity price is almost equally important. Because of technological learning, the generation costs of solar photovoltaics are now approaching the retail electricity price in several countries – so-called 'grid parity'. This is why Germany implemented yet another remuneration option: direct sales. Although the German legislation includes some features of existing net metering approaches (see section 6.1.1), the 2009 legislation can be considered as truly innovative. This design option can be seen as an incentive not to feed electricity into the grid but rather to consume it at the actual production site. In other words, this design option is intended to prepare renewable electricity producers for grid parity. ASIF, the Spanish industry association, proposed a similar approach shortly afterwards partly because of transnational communication but also because Germany and Spain are the two biggest markets for solar photovoltaic (see section 7.1). Legislators therefore have a common interest in reducing the costs that are passed on to the final consumer.

Two factors can explain the limited diffusion of market sales and premium feed-in tariffs in Germany, Spain and France, namely, the share of (supported) renewable electricity and the national ownership structure. The size of the national

---

plans to implement this remuneration option for wind energy in 2011 (Mendonça et al. 2009b).

16   It has to be noted that most interviews in France were already done in 2008.

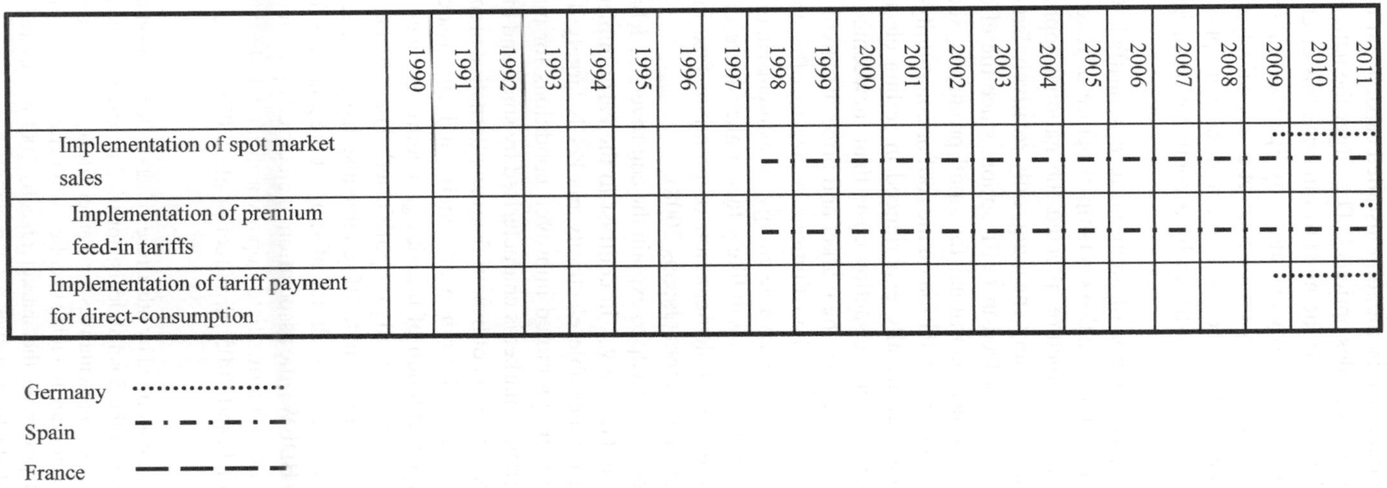

	1990	1991	1992	1993	1994	1995	1996	1997	1998	1999	2000	2001	2002	2003	2004	2005	2006	2007	2008	2009	2010	2011
Implementation of spot market sales									I	I	I	I	I	I	I	I	I	I	I	I	I	I
Implementation of premium feed-in tariffs									I	I	I	I	I	I	I	I	I	I	I	I	I	I
Implementation of tariff payment for direct-consumption																						

Germany ·········
Spain I · I · I
France I — I — I

Figure 6.2   Implementation of additional sales options for renewable electricity in Germany, Spain and France (1990–2011)

Source: Author

renewable electricity market, that is, the total installed capacity, affects the policy makers' approach to alternative sales options.[17] The larger the national market for renewable electricity, the stronger is the need to integrate this electricity into the conventional power market. This partly explains why Spain gave power producers the opportunity to sell their electricity on the market at a relatively early stage. In fact, as shown below, the stronger need to implement design options for system and market integration in Spain was driven by a combination of market growth and grid constraints.

The relatively late implementation of premium feed-in tariffs in Germany can primarily be explained by the national ownership structure of renewable energy power plants. The ownership structure affected the national approach towards incentives for market sales. Generally speaking, only relatively large players opt for market sales or the premium feed-in tariff option, since the effort of selling electricity on the market is disproportional for small producers, such as private households with a photovoltaic installation on the roof or community-owned wind farms. In contrast, utilities are usually experienced in selling electricity on the market and therefore have lower transaction costs than 'newcomers'. Moreover, large players can more easily handle the market and price risk involved in selling electricity on the market. Financial institutions prefer to offer loans to large players who can cross-subsidise renewable energy investment via other activities in the energy business. Small players will likely face higher interest rates for their potential loans. Therefore, the implementation of premium feed-in tariffs suits larger players better than small actors (Jacobs 2008).

These general considerations help to explain the controversial debate regarding premium feed-in tariffs in Germany – in contrast to the early implementation in Spain. While the Spanish renewable electricity market is largely dominated by utilities that have proactively promoted improved conditions for market sales,[18] the German renewable energy market is characterised by small and medium-sized companies. These companies will not benefit from an alternative premium feed-in tariff option. This is why the major renewable energy industry associations have been opposed to the implementation of this design feature. Besides, it should be noted that the debate in Germany is highly politicised. The reluctance of several German actors to implement a feed-in tariff scheme also has to be seen in the light of the general debate concerning the choice of support mechanisms for renewable electricity. VDEW (now BDEW), the association of electricity producers, proposed a premium feed-in tariff as an intermediate step towards a quota-based mechanism in 2005 (Bleuel and Hillebrand 2005, di Nucci et al. 2007). Interestingly, many

---

17   These findings were confirmed by other researchers (Klessmann et al. 2008).

18   It should be noted that the Spanish legislator tried to ease market entry for small producers by establishing legal representatives for market sales – so called 'comercializador de último recursos'. These actors are intended to collect offers from several small producers and to sell the combined power on the market (Jacobs 2008). This reduces the total transaction costs and potentially the penalties for non-compliance.

of the actors that were in favour of quota-based mechanisms are now in favour of a premium feed-in tariff. Similarly, actors that were opposed to a quota-based mechanism are now opposed to a premium feed-in tariff as they fear that it will be a first step towards a support instrument that favours utilities and large players.

The German debate was clearly triggered by policy developments in other European countries, namely Spain (Rostankowski and Oschmann 2009). The exchange of information via the International Feed-in Cooperation was of special importance for this transfer process. The above analysis reveals how far the policy was initially 'copied' and later modified during the (ongoing) political decision-making process. First of all, Germany, based on the analysis of the Spanish support scheme, planned to give producers the opportunity to switch between the fixed tariff and the premium feed-in tariff options on an annual basis. Later, the time frame for selling electricity on the market was reduced to a minimum of one month.

The comparison between the existing Spanish legislation and the new German legislation reveals that the two countries followed different approaches to a common problem: insecurity about the future market price of electricity. While Spain opted to prevent under- and overcompensation via the implementation of a cap and a floor, Germany fixes the premium tariff payment *ex post*, after ascertaining the exact market price for electricity in a given period, in order to reduce as much as possible the price risk for small and medium-sized producers. Here we have another example of how policy design differs in order to take the need of national actors into account. As well, fixing the premium tariff payment *ex post* would simultaneously tackle the problem of reduced revenues caused by the merit-order effect.[19]

6.2. Tariff Payment for Improved System Integration

Besides the establishment of alternative sales options, market integration and grid stability can be improved via special tariff payment for improved system integration. This includes special tariff payment for auxiliary grid services (the provision of reactive power[20] and the capacity to support voltage dips) and

---

19   The merit-order effect occurs when large amounts of renewable electricity are generated at the same time, for instance because of high winds, thus lowering the hourly price for electricity on the market. Even though from a macroeconomic perspective this effect might be desirable, revenues for renewable electricity producers who cannot control the timing of power output might be reduced considerably (Sensfuss et al. 2008, Bode 2006a, de Miera et al. 2008, Pöyry 2010).

20   Reactive power is the power consumed in an alternating current power systems because of the expansion of magnetic and electrostatic fields. The capacity of wind power plants to provide reactive power for stabilising the power grid has been discussed in more detail by Braun (2008).

increased or decreased tariff payments depending on electricity demand or residual load patterns (demand-oriented tariff payment).

Auxiliary grid services can be provided by modern renewable electricity generation plants. State-of-the-art wind power plants, for instance, often have the capacity to support voltage dips[21] and can provide reactive power for grid stability. Policymakers can incentivise the use of these technologies by paying an addition incentive, compensating either for the necessary upgrading of already existing power plants or for the higher costs for the initial purchase.

Demand-oriented tariff payment means that feed-in tariffs vary with the demand for electricity at different moments of time. They can vary with the time of day or with the seasons. In periods of high demand (peak periods), renewable electricity producers receive higher tariff payments. In periods of low demand (off-peak periods), producers receive a lower tariff. This way, producers are given an incentive to align their production to general electricity consumption patterns. This approach is most suitable for renewable electricity technologies that have the capacity to control power output at a given moment in time, for instance biomass and certain types of hydropower. Renewable energy technologies with fluctuating power output, such as wind energy and solar photovoltaics, are generally excluded from these provisions.

Demand-oriented electricity supply can also be provided by combining fluctuating with non-fluctuating renewable energy technologies or by using storage technologies. Basic feed-in tariff schemes usually do not incentivise demand-oriented power production as the purchase obligation guarantees power purchase independently of electricity demand. By paying an additional premium for more steadily power generation in line with the national residual load, legislators can incentivise the combination of several renewable electricity technologies or the use of storage technologies.

6.2.1. Germany

The German feed-in tariff scheme grants additional tariff payment for auxiliary grid services. The regulations were first implemented in 2009. As early as 2006 BWE, the German wind energy association, commissioned a study analysing the additional remuneration for auxiliary grid services of wind power plants in Spain. This study analysed not only the Spanish feed-in tariff scheme but also the detailed regulation for additional tariff payments in the case of reactive power

---

21   Formerly, wind power plants had to be disconnected when so-called voltage dips occurred. Voltage dips are short reductions of the alternative current main voltage, normally lasting no longer than a few seconds. When wind power plants were disconnected it normally took some time to restart and reconnect them. Today, modern technical equipment allows wind power plants to stay connected, thus supporting the stability of the grid (Mendonça et al. 2009b).

and grid ride-through capacity in the case of voltage dips (Jacobs 2008).[22] BWE used this information when lobbying for the regulation in Germany in 2009. The so-called system service bonus (*Systemdienstleistungsbonus – SDL*), issued by an ordinance in July 2009, grants an additional tariff payment of 0.5 €cent/kWh for new wind power plants if they complied with certain technical requirements relating to the capacity to provide reactive energy and grid ride-through in the case of voltage dips.[23] The retrofitting of old wind power plants is voluntary, and attracts a slightly higher remuneration of 0.7 €cent/kWh (Weber 2009). This bonus will be paid only for wind power plants that are installed before 2014 and for a maximum of five years. From July 2010 onwards, new wind power plants without certain technical requirements will no longer be eligible for tariff payments (BGBl 2009b).[24] According to the feed-in tariff law of 2012, the additional incentive for retro-fitting will be extended until the end of 2015. New wind power plants will no longer receive this SDL-Bonus starting in 2012 (BMU 2012).

Extra tariff payment for demand-oriented electricity supply was debated in 2008 and 2009. When the amendment of the EEG was prepared in 2008, certain employees of the Ministry of the Environment, the renewable energy associations and the Social Democrats favoured the implement a so-called *kombi* bonus.[25] This would have guaranteed an additional tariff payment in the event that two power producers with different technologies combined their output or used storage technologies, thus making their supply more steady and predictable. The first proposal for the *kombi*bonus, which was prepared by a number of independent research institutes on behalf of the BMU, the ministry for the environment, foresaw two additional streams of revenue, one for demand-oriented power generation and the other for the technological development of electricity storage systems.

The incentive for demand-oriented power generation was based on time-differentiated tariff payment with the aim of shifting power generation from periods of high renewable electricity penetration to periods of low renewable electricity penetration in line with general demand patterns. Under this first proposal, the day was subdivided into three eight-hour periods, reflecting typical residual load patterns; the additional remuneration for demand-oriented power production would be of 2 €cent/kWh. If power were provided through the use of storage technologies, the additional tariff payment could rise to 3.3 €cent/kWh (Schmid et al. 2009). This additional tariff payment would be granted for electricity provided during the eight hours with the highest electricity demand (peak load) or stored during the eight hours with the lowest electricity demand (off-peak load) (Schmid et al. 2009). In other words, this regulation was intended to incentivise the shift of

---

22   The study was written by the author of this research project.

23   A first draft proposal was tabled by the BMU in March 2009 (BMU 2009a).

24   The deadline for retrofitting was postponed in July 2010. New wind power plants now need to comply with the technical requirements starting from March 2011.

25   The implementation of this design option was debated together with the implementation of the premium feed-in tariff (see section 6.1.1).

renewable electricity to the time of day when conventional power sources are most used (Rostankowski and Oschmann 2009). An additional payment was envisaged to further incentivise the development of certain storage technologies. The level of support depended on the storage unit and the size of the power plant, and ranged from 21 to 198 €/kW annually for the first ten years of operation.

The renewable energy industry associations supported this approach and even tabled similar recommendations. The so-called integration bonus or '4,000 full load hours approach' was intended to provide renewable electricity more steadily. Generally, wind power plants in Germany operate for an average of 2,000 full-load hours annually. Through the combination with other technologies, such as biomass, hydropower or other storage technologies, the total amount of full load can be doubled and, more importantly, electricity output can be planned.[26] All wind power plants which arithmetically reach 4,000 full-load hours should be eligible for tariff payment. According to the BEE's calculations, an additional tariff of 2 €cent/kWh could cover the related costs (BEE 2008b).[27]

6.2.2. Spain

With respect to auxiliary grid services, Spain was the first country worldwide to grant additional tariff payment for capacity to support voltage dips. In 2004, the Spanish feed-in tariff scheme granted an extra tariff payment of five percent of the average electricity price for the first five years of operation (BOE 2004a). The producers had to provide certification to the effect that their wind turbines were capable of supporting voltage dips. In 2007, when the percentage-based payment was abandoned, the additional tariff payment was fixed at 0.38 €cent/kWh for all installations that were connected before January 2008. The necessary changes to wind power plants had to be made before January 2010. This regulation is being amended as of 2010. The deadline for wind energy installations will likely be postponed to 2011. Besides, the new regulation will also oblige large-scale solar photovoltaic producers to install the necessary technological equipment in order to support voltage dips. New installations are expected to provide these services without any extra payment. As for the provision of reactive power, Spain implemented a 'bonus-malus' system as early as 1998, rewarding producers that fulfil the requirements of the grid operator and penalising producers that do not comply (BOE 1998a). This provision was not modified by the 2004 and 2007 amendments (Jacobs 2008).

---

26   The combination of different renewable electricity technologies is already possible today. However, no higher tariff payment is currently granted under the German feed-in tariff legislation.

27   Critics of special tariff payment provisions for the development of storage technologies argue that this might go beyond the requirements of the German electricity system (Rostankowski and Oschmann 2009).

The combination of different renewable energy technologies was first entertained in the amendment of 2007. The new Royal Decree allowed for renewable electricity generation in hybrid installations, explicitly referring to the combination of solar thermal power plants with certain types of biomass for steady electricity supply (BOE 2007c). APPA, the industry association, appreciated this approach and called for it to be extended to other technologies such as solar photovoltaics and small wind energy (APPA 2007).

Interestingly, elements of demand-oriented tariff payment were implemented as early as 1994. The first feed-in tariff scheme of the same year calculated the tariffs based on a number of parameters including the hourly provision of electricity – determining the quality of electricity fed into the grid. If the power was provided during peak demand periods, the tariff payment was multiplied with a certain coefficient (BOE 1994a). From 1998 to 2007, demand-oriented tariff payment was no longer included, probably because it was expected that the premium feed-in tariff and market sales would create an incentive for adjusting electricity production to general demand patterns. Since this is not the case with the fixed-price option, demand-oriented tariff payment was re-implemented in 2007. Under the royal decree 661/2007, renewable electricity producers can opt for time-differentiated tariff payment (*discriminación horaria*). Depending on the time of day and the respective demand for electricity, producers receive (marginally) lower or higher tariffs. In the case of wind energy, for instance, the fixed tariff is multiplied by a factor of 1.0462 during peak-demand periods and by a factor of 0.9670 in times of off-peak demand. Producers choosing this remuneration option have to stay with it for at least one year (BOE 2007c, Jacobs 2008).

6.2.3. France

The French feed-in tariff scheme does not include any provisions concerning additional tariff payment for auxiliary grid services. The French grid operators require wind energy generators to meet certain technical requirements, such as grid codes (J.O. 2008b, 2008c, 2008d), without granting any additional tariff payment. However, the latest regulation relating to the use of public power grids – TURPE 3^{28} (J.O. 2009a) – gives the grid operators RTE and ERDF permission to offer financial incentives in the event that power producers are capable of delivering reactive power. A working group is currently analysing the potential design of these regulations. Besides, certain lobby groups are arguing for additional tariff payments for auxiliary grid services, as occurs in Germany and Spain. The French wind power industry calls for the implementation of an additional tariff payment for the capacity to support voltage dips. Additional tariff payments for the combination of several renewable energy technologies have not been discussed or implemented.

---

28   Troisième Tarif d'Utilisation des Réseaux Publics d'Électricité

Demand-oriented tariffs have been implemented in the case of hydropower and waste energy. For hydropower producers, tariff payment is differentiated according to both the time of year and the time of day since 2001 (J.O. 2001d, 2007a). From November until April, hydropower producers receive an increased tariff. Likewise, the tariff is slightly lower from 10 p.m. to 6 a.m. all year round. From December to March, two peak demand hours in the morning and in the evening are remunerated with a higher tariff payment (J.O. 2007a). Hydropower producers can opt into demand-oriented tariff payment. Accordingly, they can demand one single tariff (no differentiation), remuneration based on two components (summer and winter), four components (summer, winter, off-peak and normal demand) or five components (summer, winter, off-peak, normal demand and peak-demand periods). Time-differentiated tariffs apply to installations on the French mainland, not to those in the overseas territories. They do not apply to installations with more than 3 MW capacity.

Table 6.1   Time-differentiated tariff payment for French hydropower producers

Single-component tariff	No differentiation	6.07 €cent/kWh
Two-component tariff	Summer	8.38 €cent/kWh
	Winter	4.43 €cent/kWh
Four-component tariff	Winter, normal demand	10.19 €cent/kWh
	Winter, off-peak demand	5.95 €cent/kWh
	Summer, normal demand	4.55 €cent/kWh
	Summer, off-peak demand	4.25 €cent/kWh
Five-component tariff	Winter, peak demand	17.72 €cent/kWh
	Winter, normal demand	8.92 €cent/kWh
	Winter, off-peak demand	5.95 €cent/kWh
	Summer, normal demand	4.55 €cent/kWh
	Summer, off-peak demand	4.25 €cent/kWh

Source: Author based on J.O. (2007a)

Time-differentiated tariffs have also been implemented for a number of other renewable electricity technologies. The decree on 'methanisation' allows biogas producers to opt into a special tariff payment scheme for steady electricity provision. Those tariffs are calculated based on the capacity that a producer can guarantee over a certain period of time. This fixed capacity has to be provided either for the full year or only for a six-month period during winter (J.O. 2002b).

Similar regulations exist for biomass resources from animals and plants (J.O. 2002c).

6.2.4. Preliminary Conclusion

The convergence of additional tariff payment for improved system integration is limited. Only two out of three countries have implemented one and the same design feature. Tariff payments for improved system stability have been implemented in Spain and Germany (the provision of reactive power and the capacity to support voltage dips). France and Spain makes use of time-differentiated tariff payment. Spain can be considered a front runner with respect to improved system integration; it issued the first regulatory incentives for reactive power provision as early as 1998. The first elements of demand-oriented tariff payment were already included from 1994 to 1998. Incentives for the capacity to support voltage dips were introduced in 2004 (see Figure 6.3).[29]

The existing similarities with respect to the implementation of tariff payment for the provision of reactive power and the capacity of renewable power plants to support voltage dips in Spain and Germany can be explained by transnational communication. There are clear indications of policy transfer from Spain to Germany with respect to these design options. The German industry association for wind power, BWE, commissioned an analysis of the Spanish regulation for use in the political decision-making process in Germany. The similarities between the regulations in Germany and Spain are striking. However, the fact that both countries have implemented provisions for the capacity to support voltage dips and to provide reactive power also has to be seen in the light of technological advances. Modern wind power plants, as they are produced in both countries, have the technical capacity to provide those services. As described above, the additional financial incentive primarily targets the retrofitting of old wind power turbines.[30]

That fact that Spain implemented all of the discussed design options for improved system integration at a very early stage can primarily be explained by national grid constraints. First, the Spanish electricity grid is weaker than the other national systems on the European continent, and second, Spain's interconnection capacity with neighbouring countries is very limited. Interconnectors facilitate cross-border electricity transmission. As early as 2002 the EU agreed to increase minimum interconnection levels between EU member states to ten percent of

---

29   Demand-oriented tariff payment is also used in a number of other European countries, including Slovenia, Portugal and Hungary (Klein et al. 2008).

30   An anonymous observer has noted that, by incentivising such technological innovations for improved system integration (the capacity of wind power plants to support voltage dips and to provide reactive power), Germany and Spain are implicitly supporting the national wind industry. Both countries are home to leading world-class wind energy companies which dispose of the technological know-how to produce wind turbines that are able to meet these technical requirements.

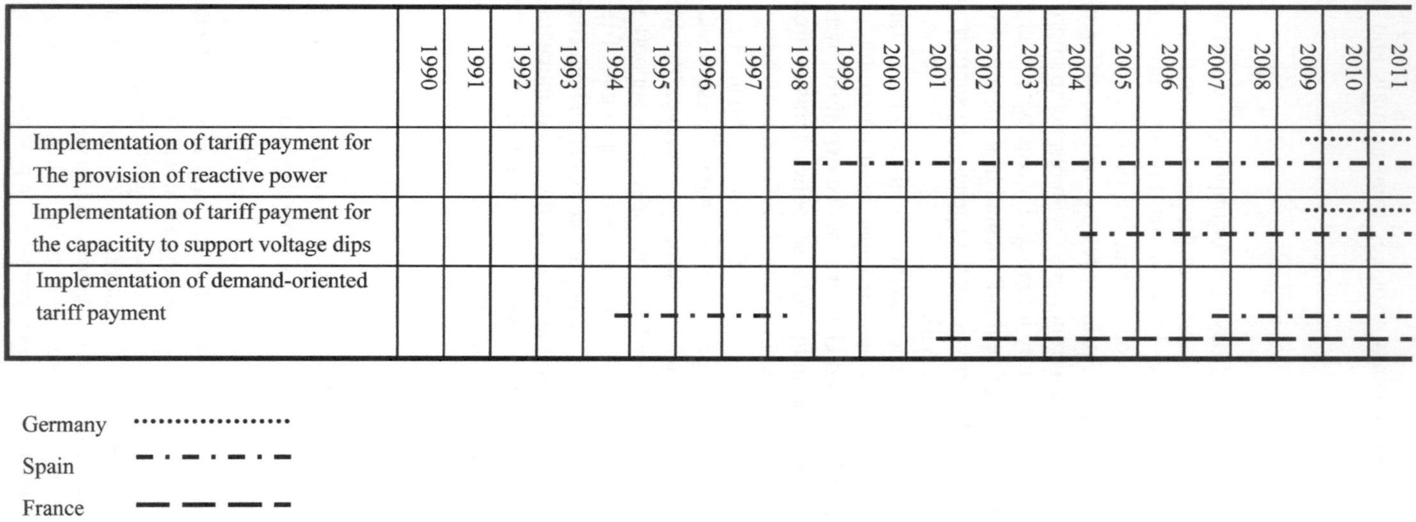

Year	Implementation of tariff payment for The provision of reactive power	Implementation of tariff payment for the capacity to support voltage dips	Implementation of demand-oriented tariff payment
2011			
2010			
2009			
2008			
2007			
2006			
2005			
2004			
2003			
2002			
2001			
2000			
1999			
1998			
1997			
1996			
1995			
1994			
1993			
1992			
1991			
1990			

Germany ············

Spain — · — · —

France —— ——

Figure 6.3 Implementation of design options for improved system integration in Germany, Spain and France (1990–2011)

Source: Author

the total power output. Many member states have still not achieved this target (EU Commission 2007e).[31] The situation is especially precarious in Spain, where interconnection levels have reached little more than four percent (Bechberger 2009, Jacobs 2008). Therefore, the provision of auxiliary grid services from wind power producers is very important for the Spanish electricity system. The growth of the renewable electricity sector is primarily based on the wind sector. Accordingly, wind power producers managed to provide more than 50 percent of the total electricity demand on 8 November 2009 (at least for a couple of hours) (AEE 2009). Thus, it is crucial that wind power producers play a major role in stabilising the Spanish power system.

Besides the national grid infrastructure, differences at national level can also be explained by the share of national renewable energy markets. Obviously, countries with a larger share of renewable electricity have a stronger incentive to implement design options for improved system integration. When the German feed-in tariff was amended in 2009, system integration was one of the top issued on the political agenda because of the steadily increasing market share of renewable electricity – and because of the ambitious national mid- and long-term targets. Therefore, in 2009, the German legislators implemented a number of design options for improved system integration, including additional tariff payment for wind power producers that have the capacity to provide reactive power and to support voltage dips. Other possibilities, such as an extra payment for the combination of technologies and demand-oriented tariff payment, are still being discussed. France has a relatively low share of new renewable energy sources (excluding hydro) and a relatively good grid infrastructure, making the implementation of design options for improved system integration a less urgent issue.

The German legislator's intention to implement the so-called *kombi* bonus can be explained by the specific ownership structure of the German renewable electricity market. The thousands of small-scale power producers need an incentive to combine their power output in order to make it steadier. The German legislator considers paying an additional incentive for producers of different technologies to enter into cooperation. The issue of granting additional tariffs for the combination of certain renewable energy technologies seems to be less important for the Spanish and French legislators, as their renewable energy markets are dominated by large-scale producers. National utilities own a large number of power plants using different technologies that can be combined when electricity is sold on the market.

Interestingly, a comparison of all three case studies reveals that France is the leading country with respect to demand-oriented tariff payment. France makes use of a rather complex system of time-differentiated tariff payment in the case of

---

31   The last assessment of interconnection capacity between EU member states dates back to 2007. The European Commission stated that Poland, the United Kingdom, Spain, Ireland, Italy, France, Portugal, Bulgaria and Romania had not yet achieved the 10 percent target (EU Commission 2007e).

hydropower, even though the contribution of renewable electricity capacity is still rather limited. The need for system integration largely arises from certain factors related to the structure of the national electricity system. First, France's electricity system is largely based on nuclear power. As nuclear power plants have a limited technical ability to follow the electricity demand pattern, the French legislator offers additional incentives for renewable electricity producers to fulfil this task. Second, many French citizens make use of electric heating systems during winter. For this and other reasons, peak electricity demand in winter is more than three times higher than off-peak electricity demand in summer. The capacity to shift hydro-based power generation to peak-demand periods is therefore crucial for the stability of the French electricity system. As shown in Table 6.1, the tariff payment during peak demand in winter is four times higher than during off-peak periods in summer. This reveals in how far the national electricity mix also influenced the national design of renewable energy support.

6.3. Regulations for Controlling Power Generation

In order to integrate renewable electricity into the overall electricity system, legislators can adopt certain measures in order to better forecast (forecast obligation) and control (remote-controlled regulation) green power output. As the market share of renewable electricity in the national electricity portfolio increases, it becomes more and more important for the grid operator to know how much power will be provided at a given moment in time. Just as all conventional power plants are obliged to forecast their power generation, some legislators impose this obligation also on renewable electricity generators. This forecast is especially important for intermittent sources of power generation, such as wind energy and solar photovoltaics.[32] Besides, state-of-the-art renewable energy power plants often include remote-control electronic equipment, which improves the so-called 'dispatch ability'[33] of renewable energy power plants. In this case, the grid operator can decrease or increase the power output of a power plant from a national or regional control centre. This technical provision, also called 'curtailment', can considerably increase grid stability in the event of high renewable electricity penetration (Mendonça et al. 2009b).

---

32   Future power forecasting has not yet become a big issue with solar photovoltaics as the share of electricity from this technology has remained relatively marginal. For a more detailed analysis of state-of-the-art wind power forecasting methods, see Monteiro et al. (2009).

33   Dispatchable power plants can be turned on and off according to power demand.

6.3.1. Germany

In Germany, forecasting wind power generation and providing balancing power are in the hands of the four transmission system operators, each operating in its respective control zone. Generally speaking, *amprion* controls the zone in the western part of Germany, *50Hertz* the eastern part, *EnBW* the south-west and *transpower* the rest of the country. Based on the provisions of the German feed-in tariff scheme, single wind power operators are not obliged to forecast their power output.[34]

All four German transmission operators work with the Wind Power Management System (WPMS), developed by the research institute ISET. This online monitoring system is used to determine wind power generation in each control zone. The respective values are exchanged between the four transmission system operators, thus reducing the need for balancing power in the country as a whole (Lange et al. 2006).[35] To provide balancing power, transmission system operators traded power on the day-ahead market and used reserve power. In recent years, the option of trading electricity in intra-day markets has become increasingly important. In this case, power is traded on the market some hours in advance based on updated, short-term generation forecasts.[36] The forecasting errors for wind power production have diminished considerably in the past (Lange et al. 2006).

As for the remote-controlled power output of renewable energy plants, the German feed-in tariff scheme of 2009 requires all new installation with more than 100 kW to dispose of the necessary technical equipment. For old installations, the deadline for possibly necessary upgrading of power plants is 2011.[37] According to the feed-in tariff law of 2012, smaller scale solar photovoltaic systems with up to 30 kW installed capacity will have to reduce their actual power output to 70 percent of the actual installed capacity (BMU 2011).

---

34   The feed-in tariff of 2009 obliged single power producers to provide information only about their real-term power production at each moment (BGB 2008: Art. 6.1). Although transmission system operators play the primary forecasting role, a number of private actors offer forecasts for wind power producers for optimised operation and maintenance of wind parks (R. Bischof, personal communication, 15 September 2008).

35   Up until 2004, communication between the four transmission system operators was limited, and immediate horizontal wind exchange did not take place (Lange and Focken 2005).

36   Up until 2009, the transmission system operators balanced wind power generation and passed it on as a constant power delivery to the distribution grid operators. Balancing was achieved by selling power on the day-ahead power market and the use of reserve power (Lange et al. 2006).

37   These strict conditions reflect the fact that Germany will produce about 30 to 35 percent of its electricity from renewable energy sources by 2020. In order to ensure grid stability, even small-scale production units need to have the capacity to regulate power output.

In order to protect renewable electricity producers and to limit the power of the grid system operators, the German feed-in tariff legislation clearly defines certain circumstances when remote-controlled power reduction is possible. Art. 11 of EEG 2009 states that this is the case if:

1. The grid capacity in the respective grid system area would otherwise be overloaded on account of that electricity,
2. they [the grid operators] have ensured that the largest possible quantity of electricity from renewable energy sources and from combined heat and power generation is being purchased, and
3. they [the grid operator]) have called up the data on the current feed-in situation in the relevant region of the grid system (BGB 2008, section 11).

If the grid operator reduces the power output of renewable energy power plants, renewable electricity producers can demand financial compensation. Compensation is mandatory and grid operators even have to state the reason why the power output has been reduced (BGBl 2008). As of 2012, dispatched renewable energy producers will only receive a financial compensation equalling 95 percent of usual expected revenues (BMU 2011). This way, the German policy maker was trying to introduce an indirect incentive for taking the potential national grid constraints into account when planning new renewable energy power plants.

6.3.2. Spain

Spain first implemented a forecast obligation for renewable electricity in 2004. With the amendment of the promotion scheme for renewable electricity in 2004, all wind power producers had to make individual forecasts for their power production regardless of the remuneration option. In the case of a fixed tariff payment, the legislation for the first time included penalties for non-compliance. However, the obligation to forecast the power generated applied only to wind farms larger than 10 MW. Wind power plant operators were obliged to communicate their forecasts 30 hours in advance to the grid operator. Short-term adjustments of the actual amount of power produced could be made up to one hour in advance of power generation. Penalties had to be paid if power generation deviated from the hourly forecast amount by more than 20 percent in the case of solar photovoltaics or five percent in the case of all other technologies.[38] In the case of the premium feed-in tariff, forecasting requirements were even stricter: the general rules of the conventional power market applied, leaving almost no room for deviations from

---

38 Because forecasts could be adjusted up to one hour in advance of generation, relatively simple forecasting methodologies enabled power producers to avoid penalty payments (*Windpower Monthly* 2004). Initially, RD 436/2004 obliged renewable energy producers to make forecasts from 2005 onwards. However, the necessary technical adjustments caused the deadline to be postponed to January 2006 (BOE 2004c).

power generation forecasts (BOE 2004a). As of July 2006, penalty payments due to power output deviations were changed to take account of actual power demand at each given hour (BOE 2006c). In the event that the forecast error coincides with the actual trend of power demand – that is, the grid operator's estimates exceeded the actual amount of power demanded and the renewable generator was producing less than forecast – no penalties were imposed. In case of a genuine lack of electricity, the necessary balancing power had to be purchased on the market (Jacobs 2008).

With the amendment of 2007, the regulation of electricity generation forecasting was further extended. From September 2007 onwards, all wind power plants with an installed capacity of more than 5 MW had to make provisions even within the fixed tariff option. This limit was reduced to 1 MW as of 2009 (BOE 2007c). Penalties already have to be paid if the actual power output deviates by more than five percent from the forecast hourly provision (Jacobs 2008). In order to reduce forecasting errors, renewable electricity producers have the possibility of combining their forecasts. In the case of wind power, deviations have fallen significantly in recent years because of the combination of several power plants and improved forecasting technologies.[39] Besides individual forecasts from each renewable electricity producer, the grid operator, REE, produces its own wind power forecasts for the entire Spanish electricity system.[40]

In 2007, the Spanish legislator also implemented provisions for the remote-controlled power output of renewable electricity plants. With decree 661/2007, the Spanish legislator obliged all wind power producers with more than 10 MW capacity to be technically equipped for remote-controlled operation. As of February 2010, the Spanish grid operator can control the power output of all wind power plants through 31 national control centres. The generation control centres send real-time information to the system operator. Investment and maintenance costs of those centres are covered by all plant operators (Klein et al. 2008, Mendonça et al. 2009b). REE is already making use of remote-controlled dispatch of wind power installation.[41]

6.3.3. France

France enacted a piece of legislation concerning power forecasts of renewable electricity producers as early as 2003 (J.O. 2003b, 2003c, 2003d). The regulation

---

39   Forecasting errors were reduced from 30 percent in 2005 to only 12 percent in 2007 (Jacobs 2008, Lucas 2007).

40   The real-time share of wind power generation can be seen at https://demanda.ree.es/eolica.html.

41   On 2 November 2008, for instance, wind power producers were obliged to disconnect for several hours. Only 2,800 MW of wind power capacity remained connected to the grid. This significant important reduction of wind power output was necessary to assure the stability of the national grid (RTE 2009).

was only slightly amended in 2008 (J.0. 2008b, 2008c, 2008d). Under both pieces of legislation, the French grid operator was legally entitled to demand power production forecasts from all producers with an installed capacity of more than 5 MW. However, it should be noted that this legislation did not impose a general obligation on power producers to forecast their production. It only gave distribution and transport system operators the legal right to demand this information if needed. However, the grid operators RTE and ERDF have never made use of this provision. Instead, RTE estimates the future production of wind power independently for each wind park. The IPES projects,[42] established at the end of 2009, allow the grid operator to forecast power production from wind and solar power producers (G. Duclos, personal communication, 2 April 2010). RTE, the transmission system operator, is currently gathering experience with a new forecasting model for wind power output called PREOLE. In forecasts of wind power output one hour ahead of generation, deviations are as small as three percent. Deviations in forecast three days ahead are about seven percent (RTE 2009b).

6.3.4. Preliminary Conclusion

There is almost no convergence with respect to forecasting obligations and remote-controlled power output in the three countries analysed. Forecast obligations of single power plant operators have been implemented only in Spain (see Figure 6.4). In Germany, forecasting is undertaken at regional level by the four German transmission grid operators. Similarly, in France the transmission system operator makes forecasts for the whole country even though national legislation offers the possibility of passing this responsibility to each individual power producer. The only similarities exist with respect to remote-controlled power output in Spain and Germany. Both have issued related regulations in 2007 and 2009.

Generally speaking, forecasting errors for wind power generation can be reduced by combining the power forecasts of several wind power plants within a given area. Germany, France and Spain follow different approaches for achieving this shared objective. In Germany and France, transmission system operators forecast power generation for all wind power plants within their areas of activity. In Spain, each wind power producer is responsible for forecasting future power output. This implicit barrier to accurate forecasting is overcome by setting up private traders that combine the forecasts of several individual plants within a give area and sell the electricity on the market. The function of these electricity traders was explicitly mentioned by the feed-in tariff legislation of 2004 (BOE 2004a). This option of combining power forecasts is primarily used by small-scale producers (Jacobs 2008).[43]

---

42   The French abbreviation means 'Insertion de la Production Eolienne et photovoltaïque sur le Système' (System integration of solar and wind power).

43   Studies have shown that combining power forecasts can reduce deviations by 50 percent (Ceña 2005).

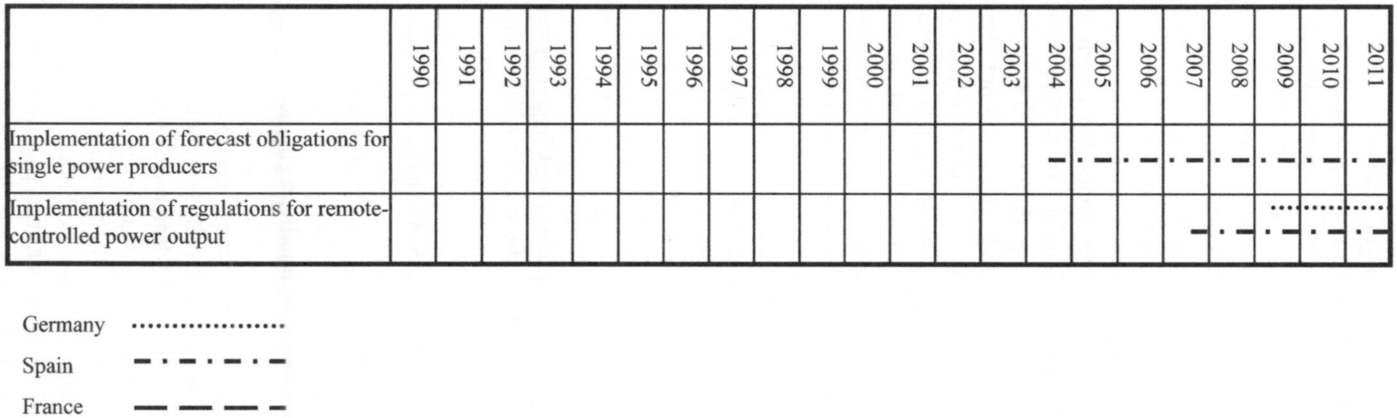

Year	Implementation of forecast obligations for single power producers	Implementation of regulations for remote-controlled power output
2011		
2010		
2009		
2008		
2007		
2006		
2005		
2004		
2003		
2002		
2001		
2000		
1999		
1998		
1997		
1996		
1995		
1994		
1993		
1992		
1991		
1990		

Germany ············
Spain ▪ ▪ ▪ ▪
France ▬ ▬ ▬

Figure 6.4   Implementation of power forecast obligations and remote controlled power output in Germany, Spain and France (1990–2011)

Source: Author

The differences between German, Spanish and French forecasting obligations and provisions for remote-controlled operation of power plants can be explained primarily by the size of the market and the market share of fluctuating renewable electricity. In France, the loose regulations for power forecasting and the absence of legislation for remote-controlled operation primarily reflect the size of the national wind power market – which is why interviewees expect the regulations to be modified in the future. Three test wind parks are already communicating real-time power generation data to the grid operator (G. Duclos, personal communication, 2 April 2010). The increasing share of renewable electricity is the primary reason for the remote-controlled operation of power plants also in Germany and Spain.

Differences between German and Spanish forecasting obligations can be explained by the size of national wind power parks and national actor constellations. In Germany, the obligation to forecast power output has not been imposed upon producers because the German wind power market consists of several thousand rather small wind parks, each of which would find power forecasting difficult (Schlögel 2008). The Spanish market, in contrast, consists mainly of large wind parks which are owned by a limited number of large operators, whose output forecasts could in principle be readily combined. Besides, it should be taken into consideration that without a forecast obligation, the Spanish utilities could artificially influence the national spot market price. Theoretically, they could make wrong power forecast for wind energy and thus artificially increase the price for the conventionally produced electricity.

6.4. Conclusion

In contrast to basic feed-in tariff design options (see Chapter 4) and design options for tariff differentiation (see Chapter 5), the convergence of design options for improved system and market integration has been limited. These design features have been implemented, if at all, in two of the three countries. Of alternative sales options (see section 6.1), Germany and Spain have implemented spot market sales and an optional premium feed-in tariff. As for design options for improved system integration, Germany and Spain have included additional tariff payments for certain auxiliary grid services, whereas France and Spain make use of demand-oriented tariff payment (see section 6.2). Finally, regulations for controlling power generation are mostly used in Spain (see section 6.3).

In fact, the Spanish regulation for renewable electricity includes almost all the design features discussed in this chapter. In this sense, the Spanish legislation can be considered an international model for design options for system and market integration. As shown above, these design features needed to be implemented in Spain largely because of the high share of (fluctuating) renewable electricity, a relatively weak national grid infrastructure and limited interconnection capacities with other European countries. However, the proactive role of the Spanish renewable energy industry was equally important for the implementation of a large number of design options for system and market integration. In 2003 it became clear that the growth of the wind energy sector might be hampered by grid-related barriers. Therefore, the national utilities entered into a fruitful dialogue with the national grid operator in order to overcome these barriers (Stenzel and Frenzel 2008). As a result, certain design options for market integration have been implemented, namely, the tariff payment for auxiliary grid services of wind power plants, regulations for remote-controlled power output and obligatory power production forecasts. In this respect, the royal decree of 2004 can be considered as truly innovative, including a large number of the design options discussed in the above section. Despite of the importance of the Spanish legislation of 2004, it has to be noted that further design options have been developed by legislators in other countries at a later stage.

Design options for improved system and market integration have initially been implemented to better integrate an increasing share of renewable energy sources into the conventional power market. In other words, the design of national electricity markets influenced the design of national support instruments for renewable electricity. However, we have also shown instances where the general market design was changed in order to better integrate renewable sources. For instance, national legislators established a larger number of intra-day markets in order to facilitate market sales of less predictable generation technologies such as wind energy. It remains to be seen in how far the design of national electricity markets needs to be changed in order to incorporate even larger shares of renewable electricity. Some aspects that are currently debated will be presented in section 8.1.2.3.

The few similarities with respect to design features for system and market integration are due to transnational communication. In various occasions we have shown that policy makers in Germany have learned from the experience in Spain. The establishment of the premium feed-in tariffs in 2012 was clearly inspired by the already existing Spanish legislation. In this respect, the frequent exchange of governmental actors in both countries via the International Feed-in Cooperation was crucial (see section 8.1.1.2). Moreover, the implementation of additional support for auxiliary grid services (the provision of reactive power and the capability of wind power plants to support voltage dips) reveals direct policy transfer from Spain to Germany. As depicted in section 6.2.1, the German wind energy association analysed the Spanish legislation in detail and used this information during the political decision making process in 2008. In one occasion, policy learning took place in the opposite direction. In 2009, the Spanish solar photovoltaic industry association promoted the concept of direct-consumption, as it was implemented in Germany in the previous year (see section 6.1.2).

Besides transnational communication, the global technological development of renewable energy equipment caused a certain degree of policy convergence. Independently, national legislators established certain design options to pay tribute to certain technological advances. In the case of the alternative sales option of market sales it was shown that legislators opened this possibility to renewable electricity producers because the cost differences between the generation costs of (some) renewable electricity technologies and the actual market price declined. Therefore, it was sometimes more interesting for renewable electricity generators to sell their power on the market and not in line with the fixed remuneration established by the feed-in tariff mechanism. Consequently, national legislators were urged to establish certain rules regulating market sales of renewable electricity producers. Similarly, the implementation of extra tariff payment for auxiliary grid services of wind power plants (see section 6.2) we depicted that this type of additional remuneration was only possible because suddenly wind turbines were able to provide 'services' such as the provision of reactive power and the capacity to support voltage dips.

However, apparently there were also factors obstructing cross-national policy convergence. First, it should be noted that the diffusion of these design options is probably limited because they have mostly recently been 'invented'. International best practice for improved market integration still needs to be evaluated and assessed; and transnational transmission usually requires a certain number of years to become effective. Nonetheless, the differences in national feed-in tariff design can also be explained by a number of other factors, namely, national differences with respect to renewable electricity market development (the share of renewable electricity), the national grid infrastructure, the national actor constellations and the ownership structure of the national renewable electricity market, and, finally, the overall national regulatory framework of energy markets.

Regarding the share of renewable electricity, it is evident that countries with a larger share of renewable electricity have a stronger incentive to implement design options for system and market integration. This explains why these design options are more frequently used in Germany and Spain than in France. If the substantial amount of hydropower, which is mostly not supported under national feed-in tariff schemes, is excluded, the share of renewable electricity was 13.1 percent in Germany (BMU 2010b),[44] 15.7 percent in Spain (MITYC and IDAE 2010) and only 2.4 percent in France (RTE 2010) as of 2009. Germany and Spain have implemented alternative sales options, additional tariff payment for wind power plants that provide auxiliary grid services and regulations for remote-controlled power output. France has implemented none of these design features.

As well as the overall share of renewable electricity, the share of certain technologies can make a difference. In Spain, a large share of the sector's growth in recent years can be attributed to wind energy. In 2009, wind power provided 12.9 percent of all electricity produced in the country (MITYC and IDAE 2010). Consequently, wind power has to play its role in stabilising the electricity grid via auxiliary grid services. The high share of wind energy in the Spanish system explains why this design option was implemented as early as 2004. Since wind

---

44   In 2009, the share of renewable electricity was 16.1 percent. Seventy-seven percent of all green power was supported under the feed-in tariff scheme. The share of hydropower was of only 19 percent of total renewable electricity production (BMU 2010b).

power is a fluctuating energy source, remote-controlled operation of these power plants has also been of importance at a relatively early stage. The need for improved market integration is even stronger in Spain because of the national grid infrastructure. The Spanish electricity system is less developed than in other European countries and has limited capacity for electricity exchange with neighbouring countries. As a result, design options for better market integration have been especially important in Spain (L. Temblenque, personal communication, 13 February 2008). Although its technical capacity for electricity exchange with other European countries is very limited, Spain managed to integrate large shares of renewable electricity into the national power system by implementing many of the design options for that purpose.

Moreover, the design of the conventional energy market influences the capacity of renewable electricity producers to participate in this market (Mendonça et al. 2009b). In order to facilitate the entrance of renewable electricity producers into the conventional energy market, the Spanish legislator changed the design of the national power market. Today, the spot market includes a number of intra-day markets, offering the option of forecasting wind power on an hourly basis. These intra-day gates a very important for renewable electricity producers in order to make short-term adjustments for power generation and forecasting (Morthorst et al. 2007, Woyte 2007). Besides, special power forecasting rules have been established for renewable electricity producers (see section 6.3.2). Germany only established intra-day markets for power trade on EEX in 2006.[45] France followed in 2007 (TradeWind 2009). This example shows that the political debate on how to design support mechanisms for renewable electricity will have to be extended, also analysing the design of power markets as a whole.

When Germany, Spain and France are compared, significant differences can be observed in national progress towards fully liberalised power markets. Although the liberalisation process of European energy markets was triggered simultaneously at the European level by Directive 96/92/EC, market liberalisation has taken different forms in Germany, Spain and France. The existing national market design also affects the national support mechanism for renewable electricity. Especially with respect to unbundling (the separation of power generation, transmission and distribution), the three countries have adopted different approaches. In Spain REE, the transmission system operator, is no longer owned by the former Spanish utilities. The Spanish state bought the last shares of the Spanish electricity grid in 2005 (Jacobs 2008).[46] The Spanish market is thus characterised by both legal and

---

45   Due to the fact that transmission grid operators are responsible for balancing power, these actors could potentially trade renewable electricity on the day-ahead or intraday market.

46   In addition, the engagement of Spanish utilities in the renewable energy business did not conflict with their asset base in conventional power generation capacity. Contrary to the situation in Germany and France, Spanish utilities no longer operated transmission lines and therefore were not directly confronted with the problems of integrating wind power

ownership unbundling. In France and Germany, the transmission grids are legally separated from power producers but there is (as yet) no ownership unbundling. In Germany three large utilities – RWE, Vattenfall and EnBW – operate large shares of the transmission system. Researchers have pointed to the potential negative effects of insufficient unbundling on the development of renewable energy sources (Ropenus and Jensen 2009). The absence of ownership unbundling can have negative effects on grid connection procedures, as grid operators might wish to exclude potential competitors from the power generation business. The institutional linkage between the German grid operators and the largest utilities is also one reason why the renewable energy industry associations are opposed to the implementation of a premium feed-in tariff. Under a premium feed-in tariff renewable electricity producers can no longer rely on the purchase obligation and are subject to general dispatching rules (Jacobs 2008).

The primary reason for the establishment of the purchase obligation in all three countries has been to protect small-scale, independent power producers operating in monopolistic or oligopolistic markets without unbundling. Otherwise, a grid operator that is operating conventional power plants might be biased when deciding whether renewable electricity can be fed into the grid and, if so, which power plants should be used (Mendonça et al. 2009b). This consideration explains why the German renewable electricity industry frequently emphasises that the purchase obligation is one of the most important components of the German feed-in tariff mechanism. The industry fears that without the purchase obligation the incumbent industry, which for many years was opposed to renewable electricity support via the feed-in tariff mechanism, might use its market power to exclude unwelcome competitors in the power generation business. This also explains why the German industry is opposed to the implementation of alternative sales options, such as a premium feed-in tariff, which would no longer include a purchase obligation (see section 6.1).

The national power mix also influences the design of national feed-in tariffs and general framework conditions for renewable energy sources. In France, almost 80 percent of the electricity is produced by nuclear power plants. As these plants have difficulties regulating power output according to general demand patterns in the short term, this role was traditionally allocated to the national hydropower plants. Accordingly, the French legislator established a complex system of time-differentiated tariffs for smaller-scale hydropower plants in line with general electricity demand patterns. In Spain and Germany, the share of gas-based electricity generation has increased in recent years. Although this is partly due to national climate protection policies – gas-fired power plants emit

---

into the grid. Furthermore, electricity demand in Spain continued to rise, so that investment in new power generation capacity was not going to lead to stranded investments in the conventional power business (Stenzel and Frenzel 2008). In Germany and France, utilities' investment in renewable power generation capacity always creates competition with the existing conventional generation units, which might lead to stranded investments.

less CO_2 than coal-fired power plants – the increasing use of gas also allows for better system integration of renewable energy sources. Gas-fired power plants can provide electricity on a very short-term basis and can therefore level out the supply fluctuations of certain renewable energy technologies, such as wind energy and solar photovoltaics. A recent study analysing the effects of a 47 percent share of renewable electricity on the Germany power market clearly shows that baseload power plants, including coal-fired power plants and nuclear power plants, will be less needed in the future (Saint-Drenan et al. 2009).

Chapter 7

The Evolution of Tariff Levels:
The Case of Solar Photovoltaics

In this chapter we analyse in detail the development of the tariff levels for solar photovoltaics over time. This is the last chapter comparing the policy designs in Germany, Spain and France. In line with the three previous chapters, in this chapter we analyse the causes of policy convergence, focusing on our three hypotheses revolving around transnational communication, regulatory competition and independent problem solving. We want to understand the reasons for increasing policy similarities in the EU.

In view of the complexity of tariff differentiation in Germany, Spain and France (see Chapter 5), the comparison is based only on tariff levels as explicitly prescribed in national legislation. Differences in tariff payment duration and resource intensity are not be taken into account. However, the automatic, annual reduction of tariff levels (tariff degression, see section 5.4) and tariff indexation to inflation and other economic benchmarks are factored into the comparison. It is important to be clear that the comparison of tariff levels in the three countries can by no means serve as a point of reference for investment decision. Equally, policymakers should not be blinded by the comparison of tariff levels, as other factors need to be taken into account to guarantee sufficient rates of return.

In addition, the comparison of tariff levels in Germany, Spain and France is confined to the fixed tariff options. As indicated in section 6.1.1 and 6.1.2, Spain and Germany offer national producers of renewable electricity a choice between a fixed tariff option and a premium feed-in tariff, which consists of the market price of electricity plus a reduced tariff payment. Since this remuneration option includes one fluctuating source of income, namely, market prices, comparisons over a long period of time are difficult. Besides, France has not implemented this alternative sales option. Moreover, the tariff level comparison does not include tariff payments for technological innovations or other bonuses.

The design of national feed-in tariffs, including the tariff level, is sometimes related to the market size in a given country at a given moment in time. Therefore, and in order to give the reader a better understanding of the general market framework, the analysis of tariff level development of each technology starts with an account of the growth of the national market over time. This research project analyses the development of feed-in tariffs over a 21-year period, from mid-1990 to mid-2011. As always, we start with an analysis support mechanisms in Germany, followed by those in Spain and France. Conclusions are presented in section 7.2.

7.1. Solar Photovoltaics

Although solar photovoltaic systems supply less than one percent of total electricity demand on a global scale, the solar market has experienced impressive growth rates since 2000. From 1999 to 2009, the globally installed capacity increased from 1.1 GW to 22.2 GW (EPIA 2009b, 2010). One year later, already more than 35 GW solar photovoltaic capacity were installed world-wide. In the EU, the market grew by 24 percent in the period 1998–2003. In the following five years, the average market growth even reached 39 percent (EPIA 2009c). This highly dynamic global market growth is reflected in the development of the installed capacity in Germany, Spain and France (see Figure 7.1).

In Germany, the installed capacity increased modestly during the 1990s, from about 1 MW in 1991 to 42 MW by the end of 1999. At the end of 2004, the installed capacity had already surpassed the 1 GW milestone; and in the following

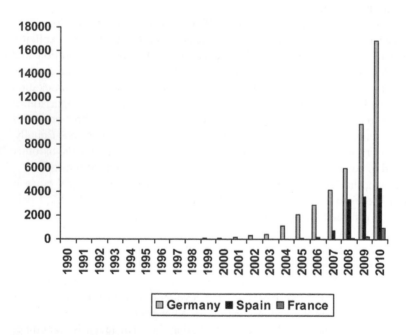

Figure 7.1   Installed solar photovoltaic capacity (MW) in Germany, Spain and France (1990–2010)

Source: Author based on BMU (2010b), BSW (2010), EPIA (2010), CNE[1], ADEME (2006), SER-SOLER (2009, 2010, 2011)

---

1   CNE publishes all data from the so-called special regime on a monthly basis, see http://www.cne.es/

years the capacity almost doubled annually. By the end of 2009 a total of 9.8 GW solar photovoltaic capacity were connected to the German electricity grid. In 2009 alone, the newly installed capacity in German more than doubled, from 1.8 GW to 3.8 GW. The German market represents more than 52 percent of the world market (EPIA 2010). In 2009 solar for the first time experienced the strongest market increase of all renewable electricity technologies. With 3,800 MW, the newly installed solar photovoltaic market capacity was twice as high as the newly installed wind capacity in the same year (1,917 MW). Rapid market growth continued in 2010 with a newly installed capacity of about 7.4 GW. At the end of 2010, with about 17 GW more than half of the world-wide installed solar photovoltaic capacity was located in Germany.

The Spanish solar market barely existed in the 1990s. By the end of the century, little more than 2 MW of solar photovoltaics were installed in the entire country. Even by the end of 2004 only 23 MW were connected to the grid. With the improved support framework under Royal Decree 436/2004, especially in the case of free-standing systems, the market started to grow considerably in the following years. By the end of 2006 the installed capacity had reached 146 MW, followed by 695 MW in 2007 and 3,342 MW by the end of 2008. Consequently, the Spanish market had increased more than twenty-fold in just two years. Because of the comparatively high photovoltaic tariff and very good solar irradiation in Spain, the market grew by around 500 percent in 2006 and 2007. With newly installed capacity of about 2,600MW in 2008, Spain represented almost half of the global photovoltaic market in that year (EPIA 2009a). Because of legislative changes at the end of 2008, the grid-connected capacity increased by only 166 MW in 2009, to a total of 3,624 MW. The rapid growth of the installed solar capacity in Spain gave solar electricity an increasing share in the national electricity mix. In August 2009 already four percent of the total electricity demand in Spain was provided by solar photovoltaic plants (Garcia Breva 2009). In 2010, the Spanish market grew by 661 MW (CNE 2011).

The French national solar photovoltaic market developed much more slowly than those in Germany and Spain, and started only around 2007. In the 1990s the installed solar photovoltaic capacity in France was negligible, increasing from 1 kW in 1992 to 55 kW in 1999 (ADEME 2006). The first megawatt of solar capacity was installed only in 2002. Even after the important legislative change in 2006, which guaranteed far higher tariffs than the first feed-in tariff scheme of 2002, the sector's growth was relatively modest. By the end of 2007 about 30 MW were connected to the national electricity grid (E. Wael, personal communication, 14 October 2008). This can partly be explained by long lead times for grid connection.[2] The installed capacity continued to increase in 2008 (48 MW), 2009

---

2   After 2006 the former national monopolist EDF, which is responsible granting grid connection, received more than 1,000 demands for grid connection per month. By the end of 2008 more than 20,000 owners of solar photovoltaic systems were awaiting grid connection. The total installed capacity of solar photovoltaic plants in the pipeline amounted

(269 MW) and 2010 (704 MW) (RTE 2009a, SER-SOLER 2010, 2011). However, a much larger market can be expected for the near future. At the end of 2009 there was almost 3,500 MW installed capacity waiting to be connected by the grid operator (SER-SOLER 2010).[3]

7.1.1. Germany

The first German feed-in tariff law of 1990 already included solar photovoltaics as an eligible technology (see section 4.1.1). However, solar systems did not receive a technology-specific tariff payment (see section 5.1.1). Under the value-based tariff calculation approach (see section 4.3.1), solar photovoltaic systems received the same tariff payment as wind energy producers – 90 percent of the retail electricity price. Therefore, the tariff level fluctuated between 8.45 €cent/kWh and 8.84 €cent/kWh during 1991–99. There was a lack of political will to support this technology via the national feed-in tariff scheme.

The amendment of the feed-in tariff scheme in 2000 explicitly targeted a stronger promotion of solar photovoltaics (Deutscher Bundestag 1999). Based on the positive experience of cost-based tariff payment for solar photovoltaics in several cities and regions, the German government and especially the German Green Party wanted to make use of a similar approach at national level. Therefore, the remuneration for solar photovoltaics increased more than fivefold, reaching 50.62 €cent/kWh in 2000 (BGBl 2000). When fixing the tariff level for solar photovoltaics in 2000, the legislator wanted to guarantee sufficient rates of return for power producers. As well, the remuneration was coordinated with the tariff level guaranteed under the Spanish tariff scheme. However, the German legislator also took the better resource conditions in Spain into account and therefore offered slightly higher tariffs (Deutscher Bundestag 1999). The tariff payment was subject to tariff degression of five percent, and consequently the remuneration diminished on an annual basis, down to 45.7 €cent/kWh in 2003. The law explicitly focused on the promotion of roof-mounted systems, since free-standing power plants were limited to 100 kW. Besides, tariff payment was guaranteed only for the first 350 MW that were connected to the grid. This program size cap, however, was extended to 1000 MW in 2002 (see section 4.2.1).

Contrary to all other renewable electricity technologies, the feed-in tariff for solar photovoltaics was amended in 2003, not as foreshadowed in 2004

---

to 526 MW. On average, grid connection delayed projects by six months (Systèmes Solaires 2009, Assemblée Nationale 2009).

3   In order to solve the problem of projects being delayed by grid connection problems, the legislation from the Grenelle de l'Environnement set a maximum waiting time, prescribing that small-scale power producers should wait no longer than two months for grid connection (Assemblée Nationale 2010, MEEDDAT 2010c).

(BGBl 2003).[4] This early amendment had become necessary since the German government's 100,000 roof-top systems program came to an end. The program offered additional investment incentives in the form of low-interest loans. Since these were now no longer available, the tariff level for solar installations needed to be increased; the tariff level was not sufficient to maintain the emerging market – despite the production costs reduction achieved in the previous years (BMU 2003). The amendment of 2003 for the first time differentiated between roof-mounted and free-standing solar photovoltaic systems. The law guaranteed a tariff payment of at least 45.7 €cent/kWh for all installations. Even higher tariff levels applied to roof-top solar systems – 57.4 €cent/kWh for all installations up to 30 kW capacity, 54.6 €cent/kWh for all plants up to 100 kW and 54 €cent/kWh for all larger, roof-mounted installations. Moreover, an additional tariff of 5 €cent/kWh was granted for building-integrated solar systems. The tariffs for new power plants decreased by five percent every year. A higher degression rate of 6.5 percent was implemented in the case of free-standing systems. The new photovoltaic remuneration scheme triggered another boost in the development of the national market. Between 2004 and 2008, the installed capacity increased from 1105 MW to 5,979 MW (BSW 2010).[5]

The amendment of 2009 was prepared in 2007. Because of the considerable market growth and the expected cost reduction of solar modules arising from economics of scale and technological learning, the progress report of 2007 suggested reducing tariff levels marginally by 1 €cent/kWh. As well, the scientific analyses of the solar photovoltaic market recommended increasing the degression rate over time (up to eight percent in 2011). Even though the industry association BSW had argued that the five percent tariff degression should not be altered (BSW 2008), large companies in the German photovoltaic sector indicated that they were able to reduce costs by seven to ten percent per annum. Eventually, the new degression rate varied between eight and ten percent. The concrete degression rate depended on the system type and the annual market growth (see section 5.4.1). In addition, the tariff levels for all types of photovoltaic installations were reduced. The tariff for free-standing systems was fixed at 31.94 €cent/kWh. Roof-mounted installations larger than 1 MW now received 33 €cent/kWh, power plants between

---

4   All regulations for solar photovoltaics of 2003 were included without modification in the amendment of the German feed-in tariff scheme in 2004 (BGBl 2004).

5   After the amendment of 2004, the market for free-standing systems started to develop since the 100 kW project size cap had been removed. Nonetheless, the legislator stressed the priority of promoting roof-mounted systems (Deutscher Bundestag 2004a). This is also reflected in the fact that tariff payment for free-standing plants is guaranteed only for all power plants connected before 2015 (BGBl 2000, 2004). The German legislator wanted to create only a niche market for free-standing systems in order to avoid potential ecological problems (BMU 2007b). Nonetheless, the share of free-standing solar photovoltaic installations has increased considerably over time. During 2004–06 the share was below seven percent of the total installed capacity (Staiß et al. 2007).

100 kW and 1,000 KW 39.58 €cent/kWh, systems between 30 kW and 100 kW 40.91 €cent/kWh, and all power plants below 30 kW 43.01 €cent/kWh (BGBl 2008).

The amendment of 2009 abolished a special tariff payment for building-integrated solar photovoltaic systems. The special tariff was 'sacrificed' during the final tariff negotiations of the political decision-making process in summer 2008. In 2008 building-integrated systems account for little more than one percent of the German solar market (Heup 2009). The additional tariff payment of 5 €cent/kWh which was granted until 2009 has not been sufficiently high to compensate for the extra costs related to necessary construction works. Besides, building-integrated systems generally have a lower energy yield since they often do not have the ideal angle towards the sun (BMU 2007b). For the first time, the amendment of 2009 offered a special remuneration option for direct consumption. When power producers consumed their electricity themselves, they received compensation of 25.01 €cent/kWh for not feeding the electricity into the grid (see section 6.1.1). This remuneration option was included to anticipate grid parity, that is, generation costs of solar photovoltaics equalling the retail electricity price (Morris 2009a). This way, power producers were already able to gather experience with renewable electricity direct consumption. However, this provision was not very widely taken up in the market, so the legislator aimed to strengthen the incentive with the revision of solar photovoltaic support in 2010 (see below).

After this round of amendments, the prices for solar modules continued to decrease considerably. In 2009 alone prices fell by more than 30 percent, in some cases by even 40 percent (Schlandt 2009, EUPD Research 2010). Therefore, shortly after the general elections in September 2009, the new government coalition agreed that solar tariffs needed to be reduced in the short term. The solar photovoltaic tariff was amended mainly to factor in the decreasing module costs in order to avoid windfall profits. Falling module prices had allowed for much higher returns on investment than originally anticipated and needed by the investors (BMU 2010d). Besides, the legislator had an interest in controlling the additional costs for the final consumer. During debate on the amendment of 2008, members of the CDU and FDP, the parties that formed the new governing coalition, called for more drastic cuts in remuneration for solar tariffs than were eventually agreed upon (Dagger 2009). They became more enthusiastic when drastic cost reductions in the solar photovoltaic industry became apparent. Besides, the installed capacity increased much faster than initially anticipated by the government.[6]

The amendment included significant reductions of the tariff levels, new regulations for direct consumption (see section 6.1.1) and modifications of the degression rates (see section 5.4.1). Based on a compromise agreed the mediation committee of the upper and lower houses of the German parliament, the law

---

6   In October 2008, the annually published lead study – the reference scenario of the responsible German ministry – still anticipated an annually installed solar capacity of about 1.2 GW (Nitsch 2008).

of 2010 imposed tariff reductions of 11–16 percent, depending on the type of installation.[7] Accordingly, as of October 2010 tariff levels were reduced to 24.26 €cent/kWh for free-standing systems (25.37 €cent/kWh for installations located on land converted from economic or military use),[8] 24.79 €cent/kWh for power plants larger than 1 MW, 29.73 €cent/kW for all installations between 100 kW and 1,000 kW, 31.42 €cent/kWh for all systems between 30 kW and 100 kW and 33.03 €cent/kWh for small-scale units up to 30 kW (BGBl 2010).

In July 2011, the 2012 feed-in tariff amendment included new tariff rates for the coming years (BMU 2011). Due to the German regulation of tariff degression (see section 5.4.1) free-standing solar PV systems received 21.11 €cent/kWh as of July 2011 (22.07 €cent/kWh for installations located on land converted from economic or military use). Depending on the installed capacity, tariffs for roof-mounted solar photovoltaic systems were fixed at 28.74 €cent/kWh (up to 30 kW), 27.33 €cent/kWh (up to 100 kW), 25.86 €cent/kWh (up to 1000 kW) and 21.56 €cent/kWh (above 1000 kW). If the German market should grow by more than 7.5 GW in 2011, in January 2012 tariffs will be reduced by an additional 24 percent (Fulton et al. 2011, BMU 2011).

7.1.2. Spain

The first Spanish feed-in tariff legislation of 1994 included solar energy as an eligible technology but did not provide for a technology-specific tariff. As in the case of wind energy and other renewable electricity technologies, the tariff level for solar producers was 6.9 €cent/kWh in 1995 and 1996, 6.8 €cent/kWh in 1997 and 6.5 €cent/kWh in 1998 (BOE 1994a, 1995, 1996, 1997a). The total remuneration depended also on other factors, such as the time of electricity generation and the provision of reactive power (BOE 1994a).

Conditions improved considerably under the regulation of 1998. The new electricity law of 1997 represented a great advance in the promotion of solar energy in Spain, since it allowed for an exceptionally high remuneration for this technology. While the support for all other renewable electricity technologies had to be in the range of 80–90 percent of the average electricity price that was fixed every year by the legislator, considerably higher tariff levels were allowed for solar photovoltaics. The subsequent Royal Decree 2818/1998 offered solar energy producers a choice between a fixed tariff level option and a premium feed-in tariff option (see section 6.1.2). Under the fixed tariff option, solar photovoltaic producers received a tariff of 39.7 €cent/kWh for small-scale installations (less than 5 kW) and only 21.6 €cent/kWh for larger plants (BOE 1998a). These tariffs applied until 2004 (BOE 1998a, 2002a).

The new regulation of 2004 indexed tariff payments to the average retail electricity price. The primary goal of this new approach was to increase investment

---

7   These tariff reductions took place in two steps, in July and October 2010.
8   Modules located on agricultural land are no longer eligible for tariff payment.

security, a long-standing demand of the renewable electricity industry associations (Bechberger 2009). Royal Decree 436/2004 prescribed that solar photovoltaic producers should receive 575 percent of the retail electricity price in the event that their power plants were not bigger than 100 kW. Accordingly, the highest tariff payment amounted to 41.42 €cent/kWh in 2004, 42.14 €cent/kWh in 2005 and 44.04 €cent/kWh in 2006 (BOE 2004a, 2004b, 2005b, 2006b). Larger-scale systems received only a 300 percent equivalent. Tariff payment was guaranteed for the full lifetime of a power plant. After 25 years, the tariff levels were reduced to 460 percent and 240 percent respectively.

Tariff indexation to the average retail electricity price was abolished under the regulation of 2007. The Royal Decree 661/2007 continued guaranteeing fixed tariffs for the full lifetime of power plants[9] and differentiated tariff payments according to plant size (see section 5.2.2). All installations up to 100 kW received 44.04 €cent/kWh, power plants between 100 kW and 10 MW received 41.75 €cent/kWh and larger power plants up to 50 MW were granted 22.98 €cent/kWh. The tariff payment was guaranteed for the first 400 MW of solar capacity. Because of these incentives, especially in the case of larger-scale projects up to 10 MW, this target was reached just three month after the law came into force. However, Royal Decree 661/2007 had incorporated a one-year transitional period in order to increase investment security. Investors or developers that had already started to work on a project would have the time to finish the power plant and receive the tariff payment they had initially anticipated. Accordingly, in September 2007 the responsible ministry issued a resolution which prolonged the payment period for one year (BOE 2007b). In the meantime, the ministry started working on a new tariff regime for 2008–11.[10]

A first draft of the new decree for solar photovoltaics was tabled by the responsible ministry in September 2007 (MITYC 2007a).[11] After long political negotiations and extraordinary market growth – from September 2007 to September 2008 the Spanish market grew by more than 2.5 GW – the royal decree entered into force in September 2008. In view of the market development

---

9   As under the previous regulation, solar power producers still received a slightly lower tariff after the first 25 years of operation.

10   Since the Spanish feed-in tariff scheme is reviewed every four years, the next major amendment of the support instrument is planned for 2011. In addition, the Spanish legislator has already planned to table a new national plan for the development of renewable energy sources in 2010. This explains why the new tariff regime was intended only for an interim period of three years.

11   For the first time, the Spanish legislator explicitly differentiated roof-mounted and free-standing systems. Tariffs for free-standing systems were considerably reduced and tariff degression of five percent was suggested. The draft of the new decree included a capacity cap of 200 MW for roof-mounted systems and 1,000 MW for free-standing systems. The Spanish industry association for photovoltaics, ASIF, rejected the first draft most of all because of the implied legal insecurity. Instead of an interim solution for 2008–10, ASIF wanted a long-term framework until 2020 (ASIF 2007).

of the previous years, the primary objective of the Spanish policymaker was to control the market and thus the costs for the final consumer (IFIC 2009). Therefore, the new Royal Decree 1578/2007 capped the newly installed capacity which produced the collapse of the Spanish solar photovoltaic market. The newly installed capacity in 2009 was of 166 MW – about five percent of the newly installed capacity of 2008.

The newly installed capacity was now limited by decree to 400 MW annually, whereas 233 MW are reserved for free-standing photovoltaic systems. This annual cap can be increased by 10 percent every year if tariffs decrease by the same percentage (BOE 2008a, Rutschmann 2007). The total annual capacity is subdivided into quarterly shares. To become eligible for tariff payment, power producer have to sign up to a national registry.[12] The tariff level depends on the actual growth of the solar photovoltaic sector, differentiating small-scale roof-mounted projects, larger-scale roof-mounted projects and free-standing systems (see section 5.4.2). Since the quarterly targets for roof-mounted systems were not achieved in all four applications rounds of 2009, the tariff remained at 34 €cent/kWh (for plants up to 20 kW) and 32 €cent/kWh (for larger-scale plants). Because of the high demand for free-standing projects, the tariff was reduced from 32 €cent/kWh in 2008 to 29.09 €cent at the end of 2009 (MITYC 2009a). In mid-2010, the tariffs had already fallen to 32.19 €cent/kWh for small-scale roof-mounted systems, 28.68 €cent/kWh for larger-scale, roof-mounted systems and only 25.86 €cent/kWh for free-standing systems (MITYC 2010).

In summer 2010 the Spanish government started to amend the feed-in tariffs for all renewable electricity technologies – in line with the periodic revision that is supposed to take place every four years. The government decided to reduce tariff levels for solar photovoltaics by up to 45 percent in the case of free-standing systems, 20 percent in the case of larger-scale roof-mounted systems and 5 percent for small-scale roof-mounted systems. According to RD 1565/2010, as of July 2011 small-scale roof-mounted systems received 28.88 €cent/kWh, larger-scale roof mounted systems were granted 20,37 €cent/kWh and free-standing solar photovoltaic systems were guaranteed 13.46 €cent/kWh (BOE 2010b).

7.1.3. France

The first feed-in tariff for solar photovoltaics in France was issued in March 2002 (J.O. 2002a). The support consisted of two different tariffs, a relatively low tariff of 15.25 €cent/kWh for installations on the French mainland and one twice as high (30.5 €cent/kWh) for Corsica and the French overseas department. Besides, tariff levels were subject to a five percent tariff degression and tariff indexation. Even though the French renewable industry associations and other non-governmental

---

12   The new RD 1578/2008 allowed only for an installed capacity of 58–94 MW per trimester. However, developers signed up for 15 times more capacity at each approval round (Pause 2009).

organisations such as HESPUL called for cost-based tariff levels, the responsible ministry wanted to grant only a tariff level which reflected the avoided costs of conventional power generation plus certain ecologic advantages (M. de l'Épine, personal communication, 7 August 2007). This also explains the higher tariff for the overseas department, since electricity generation in isolated locations is far more expensive. Soon after the decree of 2002 was published, it became clear that the tariff level for installations on the French mainland was not sufficiently high to allow power plants to operate profitably (ADEME 2006). Only in combination with additional investment incentives from regional authorities – solar photovoltaic producers were granted investment incentives of one to two euro per installed kilowatt (E. Wael, personal communication, 14 October 2008) – were a small number of projects realised. However, the installed capacity increased only from 1 MW in 2002 to 22 MW in 2006.

The considerable improvement of the feed-in tariff regime for solar photovoltaics in 2006 came as a complete surprise to the industry associations and other proponents of renewable energy sources. Several months earlier, the responsible ministry had announced only a doubling of the tariff payment (22.5 €cent/kWh). In response, the proponents of solar electricity generation had argued that this tariff level would not have been sufficiently high and consequently called for tariff levels in the range of 40–50 €cent/kWh (M. de l'Épine, personal communication, 7 August 2007). Eventually, the responsible ministry indeed offered tariff levels calculated on the basis of generation costs of solar photovoltaics – at least in the case of building-integrated systems (see section 7.1.3). Under the new decree of 2006, solar photovoltaics received a remuneration of 30 €cent/kWh plus 25 €cent/kWh for building integration. The tariff level for systems in the overseas department was 40 €cent/kWh plus 15 €cent/kWh building integration.

When the French legislator established the 2006 tariff levels, it had previously analysed the European solar market and the support policies in other European countries – that is, the German market – since the French legislator wanted to establish a national industry. To avoid competition with the well-established German (and Spanish) solar photovoltaic industries, the French policymakers wanted to establish a national industry in the niche market of building-integrated modules (E. Wael, personal communication, 14 October 2008; J.-L. Bal, personal communication, 24 October 2008; Assemblée Nationale 2009). This explains the high extra payment for this type of installation.[13] With 25 €cent/kWh extra payment for building-integrated solar photovoltaic modules, France has set up one of the most attractive incentives for this market segment in Europe. An analysis of the current market structure shows that the French legislator was successful in targeting the niche market for building integration. As of 2008, building integrated

---

13   Since competition with other European markets does not have to be feared in the French overseas departments, and actual power generation is a more important issue, the basic tariff is 10 €cent/kWh higher and the additional premium for building integration 10 €cent/lower.

systems accounted for 58 percent of the total French photovoltaic market (Heup 2009). Besides, the French solar market is largely dominated by small-scale systems. At the end of 2009, 91 percent of all solar photovoltaic systems were smaller than 3 kW, representing 44 percent of the total installed capacity (SER-SOLER 2010).

The tariff level of 30 €cent/kWh for non-building-integrated systems, that is, free-standing solar photovoltaic systems, was too low when the tariff was first implemented. Between 2006 and 2009 interest groups often advocated a higher tariff level for free-standing systems, but without success. The development of free-standing installations started only in 2008, when the falling costs for photovoltaic modules enabled producers to operate plant profitably despite the relatively low tariff payment (P.G. Thérond, personal communication, 22 October 2008; E. Wael, personal communication, 14 October 2008). As well, tariff levels for solar photovoltaics are currently increasing year by year as a result of tariff indexation. The tariff for free-standing systems increased from 30 €cent/kWh in 2006 to 32.8 €cent/kWh in 2009.[14] Indexation also affected the tariff for building-integrated systems, amounting to 60.18 €cent/kWh in 2009 (Thomas 2008).[15] Tariff degression was not implemented in 2006, since it was initially believed that this tariff regime would remain in place only for a maximum of two years (see section 5.4.3).

In August 2007 the French government set up an expert committee to elaborate future options for the promotion of solar photovoltaics under the feed-in tariff scheme. The committee comprises representatives of the French government, the industry associations CLER, SER and HESPUL, and ADEME. The committee suggests increasing the overall cap for solar photovoltaics to 7 GW in 2020 and implementing tariff degression again from 2010 onwards (Hirshman and Rutschmann 2008). In 2008 it became clear that the responsible ministry planned to establish an additional tariff for roof-mounted solar photovoltaic systems on commercial buildings. Up to 2011 the French feed-in tariff mechanisms targeted small-scale, building-integrated systems or, to a minor extent, larger-scale, free-standing applications. However, the market for medium-sized, roof-mounted systems had so far been neglected. The French legislator planned a remuneration of 45 €cent/kWh for roof-mounted systems on industrial building, supermarkets and farm houses. As well, the responsible ministry announced that tariff degression would start only in 2012, thus increasing investment security for a number of years (MEEDDAT 2008a). The ministry intended to leave unchanged the tariff for free-standing applications and building-integrated systems.

---

14   In addition to this relatively low tariff level, free-standing solar photovoltaic systems were supported via the national tender scheme. In April 2009 the first tender for free-standing systems was issued, auctioning a maximum capacity of 300 MW (MEEDDAT 2009b).

15   In Corsica and the overseas departments, the remuneration level of 2009 reached 43.76 €cent/kWh.

The announcement of a new tariff regime for solar photovoltaics led to a run on the French market. According to the responsible ministry, the first draft created a financial bubble which could have triggered investments of more than €50 billion over 20 years. By the end of 2009 the responsible ministry had already received hundreds of applications amounting to an overall installed solar photovoltaic capacity of 4.5 GW (Hirshman et al. 2010). This might have led to electricity price increases of 10 percent. Therefore, the new French feed-in tariff regime was issued only in January 2010; it comprised certain modifications of the first draft, including a plant-size cap. Accordingly, the highest tariff payment for building-integrated applications was accessible only for installations up to 250 kW (MEEDDAT 2010a). The high tariff for building integration of 58 €cent/kWh remained in place; but only private homes, schools and health-care buildings could profit from it.[16] For other buildings, including offices and those used for agricultural or commercial purposes, a reduced tariff of 50 €cent/kWh was paid (MEEDDAT 2010a). The aforementioned 'intermediate' tariff was eventually introduced for simplified building integration. Initially, a tariff was planned for roof-mounted systems (without building integration).[17] The new tariff category was intended to target solar installations on commercial buildings where fully building-integrated systems were difficult to realise (MEEDAAT 2010a). The tariff for free-standing solar photovoltaic systems was retained. In addition, from 2010 onwards tariff payment for free-standing solar photovoltaic plants larger than 250 kW was subject to tariff differentiation according to location (see section 5.3.3).

In August 2010 the tariff levels for solar photovoltaics were again reduced. In view of the fact that projects with more than 4.5 GW capacity were in the queue and that the French government was targeting only this amount of solar capacity for 2020, the legislator was trying to control the national market growth, as well as the additional costs for the final consumer. Tariff cuts affected mostly larger-scale solar photovoltaic systems. The tariff for simple building integration, that is, roof-mounted photovoltaic installations, was reduced to 37 €cent/kWh. For free-standing photovoltaic systems on the French mainland, tariffs were reduced to 27.6 €cent/kWh. The high tariff of 58 €cent/kWh for building-integrated systems remained only for installations up to 3 kW on private houses (J.O. 2010d).

After a three-month moratorium new solar photovoltaic tariffs were issued in March 2011 (J.O. 2011). In additional to the implementation to trimestral capacity

---

16　Even though a focus on building-integrated applications was viewed as positive, the rapporteur of the French parliament suggested reducing the tariff level of 60 €cent/kWh (Assemblée Nationale 2009).

17　The new French 'intermediate' tariff was criticised by non-governmental organisations because of its technical requirements. The legislation of January 2010 prescribes that the solar system has to replace certain elements of the building. More precisely, solar photovoltaic systems have to 'close', 'seal' or 'cover' the building, thus creating very high hurdles to becoming an eligible producer. Therefore, it is feared that many projects would not be eligible for this tariff category (CLER and Hespul 2010).

caps (see section 4.2.3) and increased tariff degression (see section 5.4.3) tariff levels were reduced considerably. For free-standing systems tariff levels were reduced to 12 €cent/kWh. In the case of roof-mounted systems, tariff levels for small scale systems (up to 36 kW) tariffs can be as high as 30.35 €cent/kWh. Building-integrated systems for individual households (up to 9 kW) can receive tariff payment of up to 46 €cent/kWh.

7.1.4. Preliminary Conclusion

At first sight, the tariff level development for solar photovoltaics in Germany, Spain and France reveals certain similarities. When one compares the tariff levels in those three countries, one first notices the steep tariff increases – in Spain in 1998, in Germany in 2000 and in France in 2006. This tariff jump can be attributed in all three countries to a change in the national tariff calculation methodology. As described in more detail in section 4.3, all three countries have moved from a value-based approach towards a cost-based approach, from calculating the avoided costs towards calculating the actual power generation costs of each technology. The effects of this change in the tariff calculation methodology are especially obvious in the case of solar photovoltaic systems since the actual generation costs and the avoided costs for conventional power generation (plus the environmental benefits of renewable electricity) differ considerably.

For a more detailed analysis, it is necessary to distinguish different system types, that is, free-standing solar photovoltaic systems, roof-mounted systems and building-integrated installations. This is necessary because, generally speaking, large-scale, free-standing solar photovoltaic systems can produce solar electricity more cheaply than smaller-scale, roof-mounted systems. This is because of the size of the power plants and the greater complexity for project installers. The installation of building-integrated systems is even more demanding and therefore even more costly.

In the case of free-standing solar photovoltaic systems, tariff convergence can be observed over time (see Figure 7.2). However, when comparing the tariff development we should keep in mind that these types of power systems started to spread only from 2004 onwards, since prohibitively low tariffs or national plant-size caps de facto prohibited the development of this market segment for a long time. The low tariff payment for solar photovoltaics in Germany and Spain in the 1990s did not allow for the development of this technology. This was also the case for the first French feed-in tariff, which offered a tariff level in the range of 15 €cent/kWh until 2006. Even the first significant tariff increase in Spain in 1998 did not lead to the development of this market segment. The tariff level of 21.64 €cent/kWh for all installations larger than 5 kW did not allow for the profitable operation of this type of power plant; besides, the tariff level for smaller-scale units was almost twice as high and therefore far more interesting for potential investors. The German feed-in tariff amendment of 2000 guaranteed tariff payment for free-standing solar photovoltaic systems only if the installations were no larger than

100 kW. Through this regulation, the development of this market segment was de facto prohibited.

Free-standing solar photovoltaic markets started to develop only with the amendment of the German feed-in tariff scheme in 2004. In Germany, the new law for the first time included a specific tariff category for this market segment; and the limit on plant size to a maximum of 100 kW was abolished. In Spain, investment in free-standing solar photovoltaic systems started in 2004 and was further accelerated by the 2007 amendment. The royal decree of 2004 differentiated between plants that were smaller and larger than 100 kW. As under the previous royal decree, the remuneration for smaller-scale power plants was almost twice as high. However, project developers now started to subdivide larger-scale power plants into many small units of 100 kW. Therefore, even large-scale free-standing power plants were de facto granted a tariff level above 40 €cent/kWh. The royal decree of 2007 even explicitly granted a tariff level of more than 40 €cent/kWh for all power plants up to 10 MW, that is, for all realised free-standing solar photovoltaic systems. This explains the short-term tariff peak depicted in Figure 7.2.

Once this market segment started to develop in Germany and Spain, tariff levels were reduced significantly, partly to control the market growth and partly to factor in the decrease in generation costs over time. In 2009 tariff levels in all three countries were very similar, with 31.94 €cent/kWh in Germany, 29.09 €cent/kWh in Spain and 32.35 €cent/kWh in France.

The latest tariff amendment in Germany has again considerably reduced tariff payment for this market segment. By offering a prohibitively low tariff level of only 24.26 €cent/kWh[18] from October 2010 onwards and 21.11 €cent/kWh as of July 2011, the German legislator is apparently trying to lower the share of this market segment. Because of flexible tariff degression and tariff cuts, the tariff in Spain also fell to 13.46 €cent/kWh in 2011. In France, the 2011 amendment led to even lower tariffs of just 12 €cent/kWh. When the French legislator implemented a tariff level of about 30 €cent/kWh in 2006, it was clear that no projects were to be developed at such a low tariff because power projects were very unlikely to be profitable. However, the tariff remained in place for a number of years and even increased as a result of tariff indexation and a lack of tariff degression. After the generation costs for solar photovoltaics fell considerably, the first projects started to be developed in 2008 and 2009. Almost coincidently, the tariff level of 32.35 €cent/kWh – which was guaranteed until summer 2010 – was higher than the tariff guaranteed in Germany and Spain. The same might be the case for the prohibitively low tariffs granted in France in 2011.

---

18   Free-standing solar photovoltaic systems located on land converted from economic or military use receive a slightly higher tariff level.

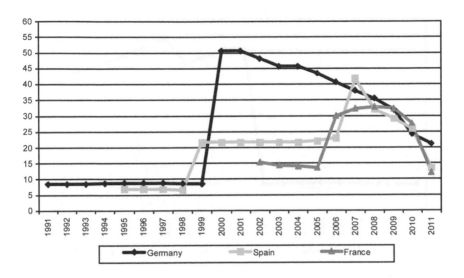

Figure 7.2 Tariff level development for solar photovoltaic systems (free-standing) in Germany, Spain and France from 1991–2011 (€cent/kWh)[19]

Source: Author based on BGBl (1990, 2000, 2003, 2009a, 2010) BMU 2011, BOE (1994a, 1998a, 2004a, 2007c, 2008a, 2010b), Bechberger (2009), J.O. (2002a, 2006f, 2010a, 2010d, 2011)

The remuneration for roof-mounted solar photovoltaic systems has also become increasingly similar over time. In 2004 the difference was greatest between feed-in tariff levels. Germany had just increased the tariff level because the 100,000 roof-top program had come to an end and tariff payment needed to be increased to 54 €cent/kW. Spain also guaranteed a tariff level in line with generation costs (41.44 €cent/kWh); and France still calculated tariffs in terms of avoided costs, and so offered a substantially lower tariff level of 13.98 €cent/kWh. The smallest tariff differences existed in 2011, with 28.74 €cent/kWh in Germany, 28.88 €cent/kWh in Spain and 30.35 €cent/kWh in France (see Figure 7.3). Now, legislators in all three countries calculated tariff levels based on the technology-specific generation costs.

---

19 Where national legislators in Germany, Spain or France did not explicitly use the term 'free-standing systems', the comparison is based on the tariff levels for large-scale installations. In the case of Spain, this refers to all installations above 100 kW during 1998–2004, and all installations between 100 kW and 10 MW during 2004–07.

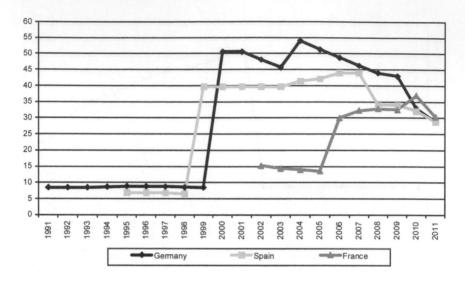

Figure 7.3   Tariff level development for solar photovoltaic systems (roof-mounted) in Germany, Spain and France, 1991–2011 (€cent/kWh)[20]

Source: Author based on BGBl (1990, 2000, 2003, 2009a, 2010) BMU 2011, BOE (1994a, 1998a, 2004a, 2007c, 2008a, 2010b), Bechberger (2009), J.O. (2002a, 2006f, 2010a, 2010d, 2011)

The tariff level development for building-integrated solar photovoltaic systems is very similar to that for roof-mounted plants. Spain never granted any additional tariff payments for this market segment, so the tariff level is the same as for roof-mounted systems. Only Germany and France tried to promote these types of installations though increased tariff payment. Germany offered an additional tariff of 5 €cent/kWh from 2004 until 2009. The amendment of 2009 dropped this special remuneration, since it was 'sacrificed' during the final tariff negotiations of the political decision-making process in summer 2008.[21] In contrast, the French legislator is still largely promoting this market segment. Starting in 2006, the

---

20   This comparison does not include additional incentives for technological innovations. Tariff degression and tariff indexation are taken into account where applicable (see section 5.4). The comparison is based on the smallest plant-size categories, in other words, the highest possible tariff for this market segment.

21   The CDU/CSU wanted to further reduce tariff payment for solar photovoltaics but the Social Democrats opposed this. During the final negotiation round, the Social Democrats managed to prevent drastic tariff cuts. However, they 'sacrificed' the additional tariff payment of 5 €cent/kWh since this market segment was literally non-existent.

national legislation offered a premium payment of 25 €cent/kWh, thus increasing the total remuneration to 55 €cent/kWh. The highest tariff difference existed in 2004 (59 €cent/kWh in Germany, 41.44 €cent/kWh in Spain and 15.25 €cent/kWh in France). Tariffs for this market segment converged in 2006 (after France had significantly increased the tariff level) but later diverged again, after Germany decided to stop additional support for this market segment. Now, only the German and the Spanish tariff levels seem to be converging because neither country is paying any additional money. As of 2011, the highest possible tariff for building integration in France is 46 €cent/kWh.[22]

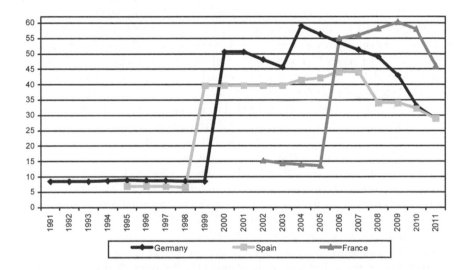

Figure 7.4   Tariff level development for solar photovoltaic systems (building integrated) in Germany, Spain and France, 1991–2011 (€cent/ kWh)[23]

Source: Author based on BGBl (1990, 2000, 2003, 2009a, 2010) BMU 2011, BOE (1994a, 1998a, 2004a, 2007c, 2008a, 2010b), Bechberger (2009), J.O. (2002a, 2006f, 2010a, 2010d, 2011)

---

22   It should be noted that the French tariff scheme for building integration was more complex than suggested by Figure 7.4, since it differentiated installations on the French mainland and in the overseas departments. Tariffs are even differentiated for different types of building. For the sake of this comparison we always refer to the highest possible tariff for building integration.

23   This comparison does not include additional incentives for technological innovations. Tariff degression and tariff indexation are taken into account where applicable (see section 5.4).

How much did transnational communication influence the tariff level development in the three countries under analysis? The analysis above shows that national legislators have frequently observed the tariff levels in other countries when determining the remuneration for power producers in their own countries. In 2000 the German legislator referred to the existing tariff level for solar photovoltaics in Spain – the first support instrument worldwide where the tariff level for solar photovoltaics was calculated on the basis of the actual generation costs.[24] Later, the German support level served as a reference point for policymakers in France and Spain. Representatives of non-governmental organisations in France argued for tariff levels similar to those in Germany, allowing for a 15-year payback period for all costs (M. de l'Épine, personal communication, 7 August 2007). Interestingly, the tariff level for building-integrated systems as issued by the French legislator in 2006 (55 €cent/kWh) was very close to the German tariff level for this market segment (53.74 €cent/kWh). In fact, the German feed-in tariff law of 2000 had become a model for a large number of countries, as it has frequently been cited as groundbreaking for the development of solar photovoltaics, even though Spain had offered similarly high tariffs in 1998. The importance of the German law can be primarily attributed to its success in bringing about newly installed capacity. Unlike with the Spanish decree of 1998 which capped the installed capacity, the German solar photovoltaic market grew (almost) without restrictions.[25]

But it was not only the German tariffs that were taken as a point of reference. National industry associations also pointed to tariff levels in European countries other than Germany. When the Spanish legislator suggested a significant tariff reduction in 2007 (ranging from 29 to 33 €cent/kWh) (MITYC 2007a), the national industry associations ASIF and APPA argued for higher tariffs of 35–37 €cent/kWh in order to remain competitive with other European feed-in tariff schemes, that is, the Italian support instrument (Gellings and Rutschmann 2008). An interviewee from France also stated that a comparison with the German tariff level would not be appropriate, since the German market was much more developed and therefore the cost of installing solar photovoltaic systems was lower (E. Wael, personal communication, 14 October 2008). However, Germany continued to be a model. Over the years, the German and the Spanish industries had developed into the leading European photovoltaic markets. In 2006 the French legislator implemented a very attractive tariff regime in order to catch up with the leading European countries in the field of solar photovoltaics. As France also wanted to become a world leader in the development of this technology,

---

24   It has to be noted, however, that this explicit reference to the Spanish support level can also be understood as a justification for this comparatively high tariff level in Germany. The European Commission was very critical of the German feed-in tariff scheme at the time and therefore the German legislator was referring to similar policies in other countries (Bechberger 2000).

25   The existing capacity limit of 350 MW was extended to 1 GW in 2002 and completely removed in 2003.

the French policymaker continued to study the framework conditions in those markets. Accordingly, a 2009 report from the French parliament included a case study on the 'photovoltaic miracle in Germany' and a very critical case study on 'how to build castles in the air in Spain' (Assemblée Nationale 2009). This shows that policy learning was also based on the negative experience in other European countries. The aforementioned report from the French parliament explicitly referred to Spain's negative experience with exponential market growth over a few years. The report suggested that a French support scheme should by all means avoid such 'anarchic growth' and implement certain design options for controlling market growth (Assemblée Nationale 2009).[26]

When determining the tariff levels for the new decree of 2010, the French legislator examined tariff levels in other European countries and carried out its own estimate of generation costs (J.-L. Bal, personal communication, 24 October 2008). However, the inspection of the support levels in other countries and the generation costs for this technology had already started in 2008. The actual decree, however, was passed only in January 2010. This was also the major point of criticism of the national regulatory commission, CRE. It stated that the large cost reductions of this technology during 2009 have not been factored in by the responsible ministry and therefore double-digit rates of return had been made possible (CRE 2009a). Tariff levels were accordingly again reduced in August 2010.

This case reveals a potential problem related to transnational communication and policy adaptation in the case of a rapidly developing technology such as solar photovoltaics. Apparently, the costs of solar photovoltaic systems decreased between the time they were analysed and the time the legislation was implemented, and the French legislator had not anticipated cost reductions.[27] This partly explains the tariff divergence amongst the three countries. While Germany and Spain are now anticipating technological learning through annual tariff reductions based on tariff degression, the tariff level in France has remained stable or even increased since 2008. When one compares the tariff level for roof-mounted systems, it becomes apparent that the French tariff established in January 2010 (42 €cent/kWh) is similar to the tariff level that the German and Spanish legislators granted in 2008 (43.99 €cent/kWh and 44.04 €cent/kWh respectively) – the year when the French legislator analysed the costs.[28]

Regulatory competition also played an important role in the development of the solar photovoltaic tariff. In line with the theoretical framework described in section 2.1.1, national industries have threatened to move their production facilities elsewhere should tariffs be reduced. This was the case in Spain in 2007, after the

---

26 The rapporteur of the French parliament explicitly referred to the system of flexible tariff degression which was later implemented (see section 5.4.3).

27 In recent decades, the European photovoltaic industry association EPIA has calculated that, with each doubling of the globally installed capacity, the module price was reduced by about 22 percent (EPIA 2009c).

28 The Spanish tariff level of 34 €cent/kWh was implemented only in September 2008.

responsible ministry had tabled the first draft law with substantially lower tariffs. However, the industry associations were only partially successful in influencing the legislator's decision on the tariff level; the final draft of the decree of 2008 included only slightly higher tariffs than those in the initial version. However, the Spanish legislator tried to meet the industry's demand for more sustainable market growth – an important prerequisite for the establishment of a national industry – and therefore implemented the aforementioned system of annual market caps that are subdivided into quarterly shares.

More importantly, however, national legislators in Germany, France and Spain were competing with each other indirectly on the international market. To understand the impact of regulatory changes in one country on political decisions (and market development) in other countries, it is necessary to grasp the importance of the German and the Spanish markets for the worldwide development of this technology. As shown by Figure 7.5, the German market has made up a large share of the total world market for solar photovoltaics since 2003. In 2004, 2005 and 2006 the German market accounted for around 50 percent of the worldwide installed capacity.[29]

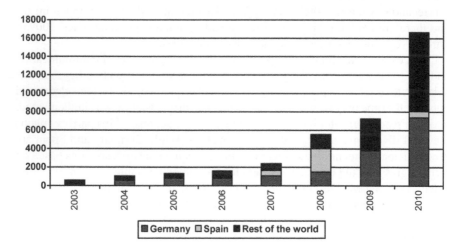

Figure 7.5   Historical development of annually global installed solar photovoltaic capacity (MW) in Germany, Spain and the rest of the world, 2003–2010

Source: Author based on EPIA (2009b, 2010, 2011)

---

29 The German progress report of 2007 explicitly referred to the lead role of the German market in the development of solar photovoltaics worldwide. It stated that the currently established support mechanisms for solar photovoltaics in other countries will absorb a substantial share of the world market only after 2010. Therefore, the report recommended that the support framework should not be drastically changed but the tariff level should be only marginally modified (BMU 2007).

In 2007 the Spanish market also started to grow significantly; and in the following year Spain overtook Germany for the first (and only) time. In 2008 more than half of the worldwide installed photovoltaic capacity was located in Spain and Germany together (EPIA 2009a).[30] Because of the unintended market growth in Spain, at the end of 2008 the Spanish legislator capped the market to 400 MW annually and reduced tariff levels by 20–30 percent. The one-year transitional period, as implemented by the Royal Decree 661/2007, was abandoned. Since the global photovoltaic market is characterised by an oversupply, module producers started to look for new markets for their products in 2009. This partly explains the run on the French solar photovoltaic market in the same year, with more than 4.5 GW in the queue. It also partly explains the amendment of the German feed-in tariff scheme for solar photovoltaics in 2010. In 2009 the installed capacity was of 3.8 GW, twice as much as in the previous year and three times more than initially expected by the responsible ministry. Apparently, a large number of the modules that were installed in Spain in 2008 were now shifted to the German market.

It is important to note that the German feed-in tariff scheme is the only support mechanism for photovoltaics worldwide which is not limited by any sort of capacity cap (see section 4.2.1). Unlike with other national support instruments, the German legislator is trying to control market growth by adjusting the tariff level to the actual generation costs – or even less. At a conference organised by the German renewable energy association BEE in 2010, Mr Pfeiffer, a parliamentarian of the governing CDU and coordinator of energy issues, explained that a tariff reduction had become necessary since the Spanish market had collapsed and the German market share had consequently increased (Pfeiffer 2010). This example shows the close interrelatedness of support mechanisms in different countries.

Unlike Germany, which tried to control and limit the number of national power plants by implementing low tariff levels, France increased its tariff level in order to be more attractive to investors. In France, the high tariffs for solar photovoltaics can clearly be linked the national legislator's intention to establish a national industry and to catch up with the leading European countries, that is, Germany and

---

30   The dramatic market increase in Spain was not intended by the Spanish legislator and can be primarily traced back to a number of policy design mistakes. First, it is surprising that the legislator offered the relatively high tariff level for all types of installation between 100 kW and 10 MW of 41.75 €cent/kWh, which was almost as high as the tariff for small roof-mounted systems (44.04 €cent/kWh). The large range of this plant size category encompasses solar photovoltaic systems with completely different cost structures, including larger-scale roof-mounted and free-standing systems. Second, the Royal Decree 661/2007 established a one-year transitional period in which tariff payment would still be guaranteed even though the national target had already been achieved. This one-year transitional period was based on the experience with wind energy, where the entire planning process normally exceeds this time period. However, the Spanish legislator did not consider that a photovoltaics project can easily be planned and realised within one year. Therefore, the royal decree triggered a race once it became clear that (high) tariff payment was guaranteed only for power plants installed within the coming year.

Spain. In November 2008 the French Minister for the Environment, Jean-Louis Borloo, declared that it was France's ambition to be one of the world's leading countries driving the technological development of solar energy (MEEDDAT 2008c). On several occasions the French legislator stressed that France was offering 'the highest tariff payment in the world' for solar photovoltaics, referring to the significantly higher tariff for building integrated solar photovoltaic systems (MEEDDAT 2010a, 2008a, Hirshman and Schug 2009).

This comparison shows the different political objectives associated with the support for solar photovoltaics in the countries under analysis. In Germany and Spain, where the market has increased rapidly since 2005, cost control of the overall support scheme has become a primary political objective. In Germany the total remuneration for solar photovoltaic plants already exceeded €2 billion in 2008. The high tariff level for solar photovoltaics significantly increases the costs of the overall system, even though the share of solar power remains relatively small. In 2008 the average remuneration for solar photovoltaics was 51 €cent/kWh, compared with an overall average of 12 €cent/kWh for all renewable energy technologies. The tariff payment for solar photovoltaics accounted for more than 20 percent of the total system costs, while photovoltaics produced only little more than five percent of all renewable electricity under the feed-in tariff scheme (IFNE 2009). This shows why cost control through tariff reduction has become such an important issue in Germany. However, it remains to be seen whether Germany will manage to control the market only by adjusting tariffs or whether – as is widely contested by the renewable energy industry – the legislator will finally have to implement some sort of capacity cap. As long as the world market for solar photovoltaic equipment is characterised by oversupply, the German market always risks being the backstop market for all modules which cannot be installed elsewhere. One way around this problem might be a stronger combination of feed-in tariff payment with targeted research and investment support. This strategy was partly applied during the last amendment round. When the national legislator decided to cut the tariff levels for solar photovoltaics in 2010, the government simultaneously increased research spending for this sector.[31]

---

31   In 2009 the German government set up a €100 million fund for research and development. Contrary to feed-in tariff mechanisms, research and development spending from the state makes it possible to support the national industry directly. In recent years, the share of German modules that has been installed under the national feed-in tariff mechanism has decreased significantly. Given that the national solar module production capacity was 1.2 GW in 2010, German modules had a share of about one third since the total installed capacity was 3.8 GW (Podewils 2010c). In view of the fact that one of the major objectives of solar photovoltaics promotion in Europe is the establishment of a strong national industry, the support under the feed-in tariff might in the future be complemented with additional national incentives.

In Spain the national legislator traditionally controls the costs of renewable electricity support by capping the installed capacity. It should be noted that a cap has always existed – including in 2006 and 2007 – but that the national legislator had problems enforcing this upper limit because of design errors in the feed-in tariff scheme. In order to control the costs – it was calculated that the installed capacity of 2008 alone would cost the electricity consumer about €800 million (Gellings and Rutschmann 2008) – the Spanish legislator also frequently adjusts the tariff level to the actual generation costs. Hence the frequent adjustment of the support instrument in 2007, 2008 and 2011. Cost control is an even more important issue in Spain because of the electricity system's existing deficit (see section 4.5.2).

In France, in contrast, the primary political objective was to become a leading European market for solar photovoltaics; hence, attractive feed-in tariff levels were established. France's relatively small share of solar photovoltaics in 2008 accounted for €85 million, which is divided amongst all national ratepayers. However, the costs are also expected to increase sharply. A report from the French parliament has calculated that by 2020 the total amount of tariff payment for solar photovoltaics will reach €2.2 billion, if the national target of 5.4 GW is taken as a reference point (Assemblée Nationale 2009). Therefore, cost control is becoming a more important political objective, which explains the current political negotiations for tariff reductions.

7.2. Conclusion

As a general tendency, solar photovoltaic feed-in tariff levels in Germany, Spain and France have become more similar over time. To a large extent, these similarities can be linked to a similar tariff calculation methodology, based on the technology-specific generation costs. This methodology was first applied in Spain (1998), later in Germany (2000) and finally in France (2005). In fact, a cost-based tariff calculation methodology is a prerequisite for international tariff convergence. Indeed, tariff levels will start to converge as soon as national legislators in different countries target a similar level of profitability when calculating tariff levels.

Critics of feed-in tariff mechanisms sometimes argue that the feed-in tariff levels are the result of political negotiations. However, the more rigorous and transparent the cost-based tariff calculation methodology that is applied, the more accurately the tariff level reflects the actual technological and market developments. It cannot be denied that national policymakers also pursue certain political objectives when setting feed-in tariffs. These are often related to specific national circumstances, as it will be shown below. However, a comparative analysis of feed-in tariff levels clearly indicates that they reflect more than just national political factors. The influence of international factors, including global technological development, regulatory competition and communication with other actors in the international arena, has clearly been shown in this chapter. However, it has to be noted that a cost-based tariff calculation methodology, transnational learning via communication and regulatory competition can sometimes not be clearly differentiated and that they often influence the policymaker's decision simultaneously.

The influence of transnational communication on national tariff levels has been demonstrated in the case of solar photovoltaics. For instance, solar photovoltaic tariffs in Spain influenced the German legislator's decision in 2000. Even though legislators in all three countries conducted their own calculations of tariff levels, they seem to have cross-checked whether their calculations of technology-specific generation costs were in line with the tariff levels in other countries. Moreover, tariff levels in other European countries were frequently analysed and taken as a point of reference for all actors involved in the national political decision-making process, including non-governmental actors (which usually do not make their own calculations). However, the necessary information also spreads indirectly among actors, though industry associations and research reports on support instruments.

Interestingly, transnational communication has mostly occurred since 2000. This is again due to the choice of the tariff calculation methodology, since tariff developments in other countries are of little importance as long as the national legislator chooses to set the tariffs on the basis of the avoided costs of conventional electricity generation in the country. However, as soon as the technology-specific generation costs needed to be analysed, international market development and tariff levels in other countries around the world became of interest to national actors. However, increasing competition with other European markets has also played an important role.

The concept and theory of regulatory competition as developed by political scientists has to be slightly modified when one analyses the support for renewable electricity. Earlier studies in the field of environmental policies – the empirical basis of this theory – primarily focused on circumstances where the legislator tried to protect the environment by imposing certain restrictions or regulations. As a response, the national industry sometimes threatened to shift production facilities elsewhere, that is, to countries where such regulations did not exist. This sometimes caused a so-called race to the bottom, meaning that there was an international competition to offer the weakest possible restrictions (see section 2.1.1). In the case of financial support mechanisms, the analysis of this chapter shows that the underlying logic of regulatory competition (almost) has to be reversed. In this case, policymakers do not make use of command-and-control instruments but instead try to create new markets with market-based policies. National governments pursue the political objective of creating and incentivising a certain market in order to drive economic transition. In the case of support mechanisms for renewable electricity, national policymakers are trying to incentivise the transition to environmental friendly power-generation technologies.

These considerations could lead one to assume that national governments are competing for the highest possible tariff in order to attract investment in their country. This outcome, however, is not confirmed by the analysis above. Only in very exceptional cases did national legislators pursue this strategy, such as when the responsible minister in France offered what he proclaimed to be 'the highest tariff in the world' for building-integrated solar photovoltaics. In this case, France wanted to catch up with the advanced solar photovoltaic markets in other European countries (Spain and Germany). At that time, the share of solar photovoltaics in France was still very low, and therefore financial constraints – strong increases in the retail electricity price – were of minor importance. However, a race to the top – that is, ever-increasing tariffs because of regulatory competition – was not observed for any of the technologies under investigation. Apparently, budgetary constraints prohibited this sort of competition.

Normally, if suffices to offer a tariff level which guarantees a certain level of profitability in order to attract investment. This was certainly one reason why all three countries eventually switched to a cost-based tariff calculation methodology, targeting internal rates of return of between four and nine percent. At least medium and large-scale companies base their investment decision on an analysis of support mechanisms in different countries. If a certain level of profitability is no longer guaranteed renewable energy developers realise their projects in other countries with more attractive support frameworks.

The case of solar photovoltaics is slightly complex because of the very peculiar framework conditions of the global market. First, national legislators in Germany and Spain (and most recently in France) are confronted with very fast increases in the annually installed capacity, which forces them to control the cost of the overall support systems by limiting growth. Second, the world market is most of the time characterised by oversupply and consequently module producers are constantly looking for new markets.[32] As a result, the international market is very responsive to changes in national support mechanisms. Third, there is a limited number of countries offering attractive support frameworks. If we bear in mind the responsiveness of the global solar photovoltaic market to legislative changes, the adjustment of support in one country can have a strong impact on market development in another country. This was the case in Germany in 2009, where a large share of the modules that had formerly been installed in Spain had to be absorbed, thus doubling the newly installed capacity. This, in turn, prompted the German legislator to modify the national support instrument by decreasing tariff levels. Because of these peculiar framework conditions, there seems to be a race to the bottom in the case of solar photovoltaic tariffs in Germany and Spain. Lower and lower tariff levels are offered in order to control costs for the final consumer. Therefore, legislators in all three countries are trying to limit market growth. This race to the bottom of tariff payment is especially obvious in Germany, the only

---

32   This has not always been the case. In the first and second quarter of 2010 modules were sold out, which led to a significant price increase.

country under analysis where the total amount of solar photovoltaics eligible for tariff payment is not subject to any sort of capacity cap.

In sum, the analysis of feed-in tariff development in all three countries shows that tariff can converge at the top, or at the bottom, or around the actual power generation costs. How far oversupply, under-supply and technological maturity have influenced the direction of tariff convergence will be analysed in more detail in section 8.2.

Interestingly, other research papers, for instance the EREF price reports, indicate a correlation between the degree of tariff convergence and the maturity and market size of the technology. The degree of tariff convergence is highest in the case of wind energy, followed by solar photovoltaics and less mature technologies such as geothermal (see EREF 2009).[33] Wind energy is a rather mature technology, and technological learning is taking place slowly. Therefore, tariff reductions over time are minor,[34] and consequently legislators have few problems calculating the tariffs on the basis of the actual generation costs. As well, the global wind market development is taking place in a large number of countries, all offering sufficiently attractive support levels. While wind plants were established in only a handful of countries in the 1990s, today this technology is applied in 82 countries worldwide (REN21 2010). The fact that cost-based feed-in tariffs are established in a large number of countries also increases the general knowledge about the generation costs of wind energy. Based on the experience in other countries, a newcomer in wind energy promotion would not establish a tariff level of either 5 €cent/kWh or 12 €cent/kWh, because an international comparison would clearly indicate that such levels would be either too low or too high. Depending on the resource intensity, the tariff level would have to be located in between.[35]

In the case of solar photovoltaics, the market is characterised by rapid market growth and rapid cost reductions. Therefore, the legislator has difficulties calculating the technology-specific generation costs. In addition, the analysis above showed that the time period between tariff calculation and the issuing of the support scheme needs to be as short as possible, because otherwise the latest technological development might not be taken into account. It is even more difficult for the policymaker to estimate the future potential for cost reduction when fixing the degression rate. Therefore, tariff setting involves a certain element

---

33   The EREF price report of 2009 offers a detailed analysis of support levels for different technologies in 2009, including small hydro, wind power (onshore and offshore), biomass, solar photovoltaics, wave and geothermal. A comparison of all EU countries reveals that tariff levels are more similar in the case of small hydro, wind power and biomass than in the case of solar photovoltaics, geothermal and wave energy (EREF 2009).

34   This is also indicated by the relatively low level of tariff degression (see section 5.4).

35   The IEA has pointed out that the minimum level of remuneration in order to encourage wind power is of 7 $cent/kWh (IEA 2008a: 100).

of trial and error, and the degree of tariff convergence is therefore lower. Besides, the number of countries offering sufficiently high tariffs for solar photovoltaic development is rather small, and so the potential to learn from the tariff levels in other countries is also limited.

Chapter 8

Powering Cross-National Policy Convergence: Communication, Competition, Technological Innovations and European Policy Steering

In this book we have shown that national feed-in tariffs in the EU have become more similar over time even though the harmonisation of the support instrument failed. More generally speaking, the design of national policies in EU member states is apparently based not merely on national considerations but also on cross-national interdependencies. Although social scientists frequently assume otherwise, domestic factors alone apparently cannot explain the outcome of the national political decision-making process. Therefore, analysing countries as interdependent rather than independent actors will be of increasing importance for political scientists (Garrett et al. 2008).

This chapter will combine our empirical findings, as presented in Chapters 4 to 7, with the theoretical framework of policy convergence as presented in Chapter 2. In each sub-chapter, we will elaborate on the causal mechanisms for policy convergence as already depicted in Chapter 2.1.1 and test them against our findings from our empirical analysis. The micro-level perspective of this research project – the detailed analysis of more than 20 design options of just one policy instrument over time – has allowed us to identify the causal mechanisms leading to cross-national similarities.

Testing the theory of policy convergence, we have identified cross-national similarities with respect design options for basic feed-in tariffs (Chapter 4), design options for tariff differentiation and adjustment (Chapter 5), and solar photovoltaics tariff level development (Chapter 7). Only in the case of design options for system and market integration has the spread of design features so far been limited (Chapter 6). The primary drivers of cross-national similarities were transnational communication, regulatory competition and independent problem solving. Interestingly, the European level also influenced national policy decisions – even though member states and the European institutions could not agree on a single harmonised European support mechanism. As the influence of the European level has been identified several times in previous chapters, here we analyse how the EU has managed to influence national policy decisions even though several attempts to harmonise European support schemes have failed.

In this chapter, we further structure the preliminary conclusions of the previous chapters to unveil the impact of the different causal mechanisms of policy convergence as presented in Chapter 2. Even though it needs to be clear that these causal mechanisms sometimes interact (Holzinger and Knill 2005b, Marsh and Sharman 2009), we nonetheless analyse the impact of each of the aforementioned causal mechanisms separately. We will start by summarising the influence of transnational communication (section 8.1), followed by regulatory competition (section 8.2) and independent problem solving (section 8.3). Finally, we will show to which extent the European level managed to influence national policy decisions (section 8.4). The conclusions are presented in section 8.5, revealing ways of how to strengthen European policy coordination.

8.1. The Power of Communication: Agents of Transfer and National Reference Models

Of the three causal mechanisms which have served as the substance of the hypotheses of this research project, transnational communication certainly has had the strongest effect in homogenising national feed-in tariffs in Germany, Spain and France. Transnational communication has been identified as a causal mechanism in the case of a large number of feed-in tariff design options, including design options for the basic feed-in tariff (Chapter 4), design options for tariff differentiation and adjustment (Chapter 5), design options for system and market integration (Chapter 6) and the evolution of the tariff level (Chapter 7).

As introduced in section 2.1.1, the term 'transnational communication' includes a variety of mechanisms leading to increasing cross-national similarity, including lesson drawing from policies in other countries, joint problem solving, emulation, and international policy promotion. However, for this research project the afore-mentioned sub-category of transnational communication that is responsible for cross-national similarity is not of primary importance. It is of minor relevance for our project whether policies have converged because policy makers from one country have analysed policies in other countries (lessons drawing) or whether they have sat down together in order to come up with similar solutions (joint problem solving). We are more interested in the types of actors that have been crucial for communicating policy-related information, as this knowledge will be of importance to European policy makers if they wish to increase the impact of transnational communication in the future.

In this section we show that an increasing number of communicators (agents of transfer) have facilitated communication between national actors in different European countries. Moreover, we underline the importance of reference models, that is, pieces of national legislation which serve as a model for other countries, leading to the spread of international best practice. Because of the limited focus of this research project – the development of feed-in tariffs in Germany, Spain and France – the analysis in Chapters 4 to 7 mostly identified circumstances of direct policy transfer between two countries. However, the theoretical concept of policy diffusion goes beyond this bilateral exchange of idea, policies and programs.

The wider communication of information in the international community is also discussed in this section.

Before we elaborate in more detail the agents of transfer and the role of reference models, it is important to note the limits of our analysis. Even though we have collected evidence from many circumstances where cross-national communication influenced national policy decisions, it is highly possible that we have not been able to describe all the events where cross-national communication actually took place. When analysing the impact of transnational communication, the researcher would ideally have to be involved in the political decision-making process, or, even better, have complete knowledge of all national and international actors and events – like an omnipresent narrator in historical novels. To overcome this problem, our archival and textual analysis included a large number of sources – primary legal documents, parliamentary minutes, position papers, press releases, and others. However, it is not difficult to imagine situations where a person involved in the policy decision-making process in one country has communicated with another important actor from another country at an international conference, leaving no record of communication or learning. This is why our analysis was extended by many interviews with governmental and non-governmental actors involved in the political decision-making process.

However, it cannot be guaranteed that interviewees clearly remember all circumstance of cross-national communication, especially when analysing the reasons for policy amendments 20 years before. This can partially explain why circumstances of transnational communication have primarily been identified since 2000 and not so much in the 1990s. Besides, it has to be stressed that establishing a direct cause-and-effect relation between policy communication and the final outcome of the political decision-making process is almost impossible. Therefore, the instances of cross-national communication that we discovered are more truly an indication of the causes of policy convergence rather than solid proof. Finally, it should also be noted that instances of transnational communication often overlap with other causal factors leading to cross-national policy similarities. In the case of tariff level convergence, for instance, we have shown that national governments often analysed tariff levels in other countries (transnational communication) while at the same time regulatory competition was influencing the outcome of the political decision-making process.

8.1.1. Agents of Transfer: the Role of Actors in the International Arena

The role of actors in increasing cross-national policy similarities cannot be overestimated as it is they who need to be informed about policy developments and policy choices in other countries in order to become active and to modify national legislation (Strang and Meyer 1993, Tews 2002b). In order to understand how far transnational communication has affected the development of national feed-in tariffs over time, we had to identify actors and institutions that communicate innovative features of support mechanisms, thus increasing the possibility of transnational convergence.

Knowledge about feed-in tariff design features has increased over time, partly because of an increasing number of actors actively communicating policy innovations in this policy field. These actors have also been called 'agents of transfer' (Dolowitz and Marsh 1996).[1] In this study we have shown that certain groups of actors have been crucial for transnational communication at different moments of time. At an early stage of market development, when feed-in tariffs were not very elaborate (Germany and Spain) or had not even implemented (France), the role of governmental actors was crucial for mutual policy learning. Therefore, governmental actors have played an important role in communicating information regarding basic feed-in tariff design options (see section 8.1.1.2). At a later stage, the role of non-governmental actors, above all national industry associations, was important for communicating certain feed-in tariff design options (8.1.1.3). However, given our focus on national support instruments we have so far not debated the importance of international organisations. This is our starting point in the next sub-section (8.1.1.1).

8.1.1.1. International Organisations

As mentioned above, the promotion of policies by international organisations is one sub-mechanism in the theory of transnational communication. International organisations are dedicated to this role as they are usually perceived as reliable sources of information in the international policy discourse (Busch and Jörgens 2007, 2009). In the case of support mechanisms for renewable electricity, two international organisations are of primary importance for communicating policy design: the International Energy Agency and the European Commission.

The European Commission played a partly ambivalent role in its communication regarding renewable electricity support mechanisms. In the 1990s the Commission positioned itself clearly, arguing for the implementation of a Europe-wide quota-based mechanism and being opposed to national feed-in tariff schemes (see Chapter 3). In fact, the policy debate at European level was almost entirely dominated by the question of which policy instrument to choose until 2001 (the moment the first directive came into force) and even until 2005 (the moment when the European Commission issued the first assessment report on European support instruments

---

1   Increasing knowledge about international best practice of feed-in tariffs is well reflected in the new feed-in tariff mechanism in the UK, as of 2010 the latest European country implementing feed-in tariffs. In 2009 the British Department on Energy and Climate Change commissioned a report on design options for the forthcoming UK feed-in tariff. The consultancy that conducted the study came up with suggestions based on best-practice evaluations of international actors and institutions such as the International Feed-in Cooperation, the International Energy Agency, the Energy Economics Group (Austria) and the Optres study (Brown et al. 2009). The British feed-in tariff mechanism includes a large number of design options which can be considered international best practice.

and the issue of harmonisation). It can be argued that this strong focus on the advantages and disadvantages of different policy instruments distracted attention from the even more important question of how to design properly each support instrument.

Only since 2005 has the question of how to design each policy instrument seemed to be of increasing importance. When assessing the European support mechanisms for renewable electricity, the European Commission refrained from implementing a harmonised European support scheme but instead wanted to facilitate coordination via cooperation. Therefore, the European Commission promoted certain policy design options in order to improve the general framework conditions for renewable electricity producers. The design options promoted, however, where always of such a nature that they could be used irrespective of the support instrument chosen by the different EU member states. In its 2005 report on the performance of national support instruments, the EU recommended the use of transparent grid connection rules and a shallow connection charging approach (EU Commission 2005a). These recommendations were frequently based on international research projects that were co-financed by the European Commission (for instance, Resch et al. 2005, Ragwitz et al. 2007a). Our analyses revealed that, in the case of shallow grid connection charging, the promotion of this design option by the European Commission probably influenced policymakers' decisions in France and Spain, as both moved towards a 'shallowish' approach in 2007 and 2008 respectively.

The recommendation of certain design options and regulation was pursued in the succeeding years. The 2008 report of the European Commission on the assessment of European support mechanisms included recommendations that were more far-reaching in administrative aspects. The Commission suggested establishing a one-stop authorisation procedure in order to avoid long lead times, planning mechanisms at regional or local level to assign locations for renewable energy units, and lighter administrative procedures for small projects (EU Commission 2008a).

Another important international organisation, the International Energy Agency (IEA), also changed its opinion on feed-in tariffs over the years. For a long time the IEA has been known for its criticism of price-based support mechanisms. This criticism was partly due to the apparent incompatibility of feed-in tariffs with the idea of competition in liberalised electricity markets. As we pointed out in section 4.3, paying renewable electricity producers a tariff based on the production costs of each technology is diametrically opposed to the philosophy of competitive, liberalised electricity markets. Therefore, it is not surprising that even in 2007 the IEA urged Germany to 'consider policies other than very high feed-in tariffs to promote solar photovoltaics' (IEA 2007: 77). In a 2005 IEA paper on liberalised electricity markets, the agency clearly showed its preference for an international, quota-based support mechanism (IEA 2005).

However, this attitude apparently changed in 2008, when the IEA evaluated global renewable energy markets and policy programs. Already the draft paper for the meeting of the governing board showed that support mechanisms were now evaluated on the basis of their performance, that is, their efficiency and effectiveness. Like the analysis of the European Commission, the draft paper concluded that 'in the case of on-shore wind, well-designed and -implemented feed-in tariffs (FIT) have proven to be an effective incentive scheme in several European countries' (IEA 2008b). Contrary to its earlier preferences of a non-technology-specific quota-based support instrument, the IEA recommends technology-specific support, thus pointing in the direction of feed-in tariffs. However, the final report remains rather unspecific with respect to the actual design of support instruments. Like the European Commission, the IEA recommends more general features of support instruments that can be applied to all types of support instrument. It suggests the removal of non-economic barriers and the establishment of predictable and transparent support. However, some references can be found to design options that have been discussed in this study. For instance, the report suggests the 'introduction of transitional incentives, decreasing over time'. This clearly points into the direction of tariff degression, as discussed in section 5.4. Nonetheless, the report does not explicitly refer to feed-in tariff design (IEA 2008a).

In sum, the influence of international organisations in the spread of feed-in tariff design options has so far been relatively limited. Even though theorists perceive international organisations as well-functioning channels for cross-national communication, by now international organisations have focused only on the general description and comparison of different policy instruments for renewable electricity support. Their recommendations for policy design are not tailored to each specific support instrument but rather focus on neutral aspects of support instruments.

Other international organisations have not had a major influence on the design of national support instruments for renewable electricity. The REN21 network publishes an annual status report on renewable energy (see, for instance, REN21 2009, 2010). Even though this report indicates how many counties and jurisdictions in the world have implemented what type of support instrument, it does not make any reference to the design of support frameworks. The role of the newly established International Renewable Energy Agency (IRENA) remains to be seen. One of the major objectives of this organisation is to develop 'methods, tools and networks for promoting experience exchange and accelerating an international learning process' (IRENA 2008: 3). Therefore, IRENA might also play an important future role communicating best practice in support instruments for renewable electricity.

8.1.1.2. Governmental Actors

Our analysis has shown that governmental actors have been crucial in communicating information concerning the basic design of feed-in tariffs (see Chapter 4). Apparently, governmental actors, that is, employees in the responsible ministries and members of ruling parties, have exchanged information on the basic pillars of feed-in tariff mechanisms as we know them today. This process of policy transfer between governmental actors has been especially pronounced in the case of Germany and France in 2000 and 2001.

In this respect, the cooperation between members of the respective Green parties was of special importance. Party members involved in the political decision-making process at the time confirmed that there was frequent exchange between renewable energy experts from the German Green Party and the responsible French minister, Yves Cochet, a member of the French Green Party. As well, the first French feed-in tariff scheme was based on a proposition of the governmental organisation ADEME. The institution's viewpoint on best-practice support was largely based on and inspired by the German feed-in tariff scheme. Policy transfer between governmental actors in Germany and France was especially pronounced in the case of the implementation of location-specific tariffs for wind energy (see section 5.3). However, we have also shown that the implementation of long tariff payment durations in France was affected by the German experience (see section 4.4). And the implementation of tariff degression in France in 2001 was clearly influenced by the German experience.

In addition, we discovered other instances of transnational communication between governmental actors in the case of the setting of tariff levels. In the case of solar photovoltaics national legislators have frequently analysed tariff levels in other countries when determining the remuneration for their domestic power producers. For instance, in 2000 the German legislator referred to the Spanish remuneration for solar photovoltaics, which had already been established in 1998 – the first country offering a cost-related tariff level for this technology (see section 7.1).

Interestingly, we have also identified instances of 'negative' policy learning, that is to say, legislators have learned from mistakes in other countries via transnational communication (Rose 1991). This was most obvious when the French legislator amended the support instrument for solar photovoltaics in 2010. The French legislator wanted to prevent by any means an unsustainable market development such as took place in Spain in 2007 and 2008. The proposal from the French parliament stated that 'anarchic growth', such as had happened in Spain, had to be avoided. Therefore, the French policymaker re-implemented tariff degression (see section 5.4.3) and considerably reduced tariff levels (see section 7.1.3). In 2011, additionally trimestral quotas have been implemented.

However, cross-national policy communication between governmental actors did not happen only bilaterally. Over the years, legislators tried to institutionalise cross-national policy learning via the establishment of several international cooperations at governmental level, namely, the International Feed-in Cooperation (IFIC) and the German–French wind energy cooperation. The establishment of these international platforms can be seen as an attempt to establish feed-in tariffs in the European arena. As pointed out by Holzinger and Knill, countries that do not comply with internationally promoted policies face international pressure to legitimise their own policy (Holzinger and Knill 2008). In view of the fact that the

European Commission was for a long time in favour of tradable green certificate scheme, the foundation of the IFIC in 2004 in particular can be seen as a reaction and an attempt to establish a counterbalance.[2]

The International Feed-in Cooperation was established as an issue-specific international platform for cooperation in 2004 at the international conference 'Renewables 2004'. In the early years only Spain and Germany were members. In 2006 Slovenia joined the organisation as the third country. Despite the limited number of official members, it should be noted that the workshops organised on an annual basis are usually attended by government representatives of a large number of European and non-European feed-in tariff countries. At each workshop, the representatives present a short overview of their national support instrument, thus improving knowledge about feed-in tariff design in general. Besides, IFIC has produced some standard literature on design options for feed-in tariffs (Klein et al. 2006, 2008). In this study we have established that cross-national communication concerning premium feed-in tariffs was clearly taking place via IFIC. The debate about this design feature in Germany was started by the German ministry for the environment, which learned from the Spanish experience via IFIC.

Cooperation between European member states via IFIC was especially strong during the negotiations preceding the European directive for renewable energies in 2008. When it became clear that the proposal of the European Commission would endanger the future existence of feed-in tariffs, collaboration was intensified. Besides frequent meetings and issue-related workshops during that period, this collaboration resulted in a jointly formulated letter to energy commissioner Piebalgs, signed by the responsible ministers for Germany and Spain. In this letter the envisaged certificate trading mechanism between companies was criticised. The government representatives expressed their 'disappointment' that their earlier concerns regarding an 'EU-wide harmonised system' had so far not been taken into account by the European Commission. Finally, the letter made clear that such a trading mechanism would not be accepted by the German and Spanish governments (Clos and Gabriel 2008). As this letter was posted shortly before the Commission issued the first public proposal in January 2008, it can be assumed that this common effort influenced the Commission's decision to eventually exclude from its final proposal this type of trading between companies.

As well, the German–French coordination body for wind energy has to be taken into consideration. This platform was established by the French and the German ministries responsible for the promotion of renewable energy sources in 2006. The body is supported by industry associations and research institutes from both countries, including BWE, SER, ADEME and DENA. The coordination body serves as a network for members of ministries, administrations, industries, and financial institutions, and for lawyers, facilitating knowledge transfer and

---

2   From an economic viewpoint, Drezner noted that '[f]or governments, any agreement to coordinate standards at a point that diverges from the domestic status quo comes with economic and political costs' (Drezner 2005: 845).

overcoming language barriers.[3] Frequent workshops are organised to foster exchanges between stakeholders in the two countries. The coordination body also provides a translation service for the most important documents related to wind energy policy in Germany and France. The exchange of information via this platform has been highly valued by government officials (T. Chrupeck, personal communication, 27 October 2008; V. Oschmann, personal communication, 1 September 2008). However, the impact of this informational exchange on concrete national policy design has so far been limited.

8.1.1.3. Non-Governmental Actors

The analysis of the development of feed-in tariff design also showed that non-governmental actors played a crucial role in communicating information regarding national policy innovations. This confirms earlier findings of convergence and diffusion studies (Kern and Kissling-Näf 2002, Weidner 2002, Schreurs et al. 2009b). In this study we have frequently shown that national industry associations have been involved in the process of communicating policy information. However, at the end of this section we focus also on the role of other actors, including researchers and the issue-specific journals.

Let us first look at the role of national industry associations. In this study we have covered several cases where the information provided by these actors has been decisive for the implementation of a certain design option. Already in 2002, the Spanish solar photovoltaic industry association ASIF pointed to the need for longer tariff payment durations in order to reach a similar degree of investment security as was guaranteed in Germany (see section 4.4.2). ASIF also promoted the 'German' concept of tariff degression, which was later introduced into the Spanish feed-in tariff mechanisms (see section 5.4.2). Besides, the German wind energy association learned about additional tariff payments for providing auxiliary grid services as they existed in Spain since 2004 and communicated this information to the German legislator. Based on the BWE's analysis of the Spanish design options, this feature was later implemented in Germany (see section 6.2.4). This case is especially interesting as it reveals that industry associations can help to communicate more complex design options cross-nationally. Industry associations have very good knowledge about the latest technological developments and therefore know what kind of technological requirements are feasible and at what cost.[4]

The success of industry associations in Germany and Spain in particular in communicating certain policy innovations internationally is not surprising. Logically, the number of industry representatives determines the extent to which

---

3   More information on the objectives of the coordination body is available at http://www.wind-eole.com/.

4   In other cases, the efforts of non-governmental actors have been less successful because of specific national circumstances. This was the case with location-specific tariffs for wind energy in Spain based on the German and French experience (see section 5.3).

these non-governmental actors can influence the national political decision-making process. As the number of employees of national industry associations largely depends on the size of the national industry and the size of the national market, it is no coincidence that the influence of the French associations has been less visible. In other words, the number of communicators in one country also depends on the size of the national renewable energy market.

According to an anonymous expert on the French renewable energy market, as of 2010 the industry association SER has about 15 employees, who have to deal with all renewable energy technologies. Up until 2004, only one person was employed as official staff. This limited number of staff normally focuses on national issues and has only limited capacity to deal with European issues or even to investigate renewable energy support mechanisms in other member states. In contrast, German industry associations collectively employ more than 70 people. In Spain, about 50 people were employed. It should be noted that in the recent past German and Spanish industry associations have employed people who are exclusively responsible for European affairs. Given the limited number of personnel in France[5] it is not surprising that French actors have proved to be less knowledgeable about different feed-in tariff design options in the case of market integration (see section 6.1).

Exchange via European platforms and industry associations is equally important for the communication potential as the size of national industry associations. Various renewable energy associations at European level serve as platforms for cross-national exchange of information. Besides the general European industry associations, EREC and EREF, which represent all renewable electricity producers, each technology is again represented by its own association. The European wind energy association EWEA, for instance, organises meetings of national representative about four to five times a year. Similarly, the solar photovoltaics association as institutionalised policy exchange via a National Actor Network, including frequent annual meetings and the establishment of several working groups.[6] In addition, the EUFORES network is organising bi-annual

---

5   The number of active people in governmental organisations is even more limited. The responsible ministry employs only two people dealing exclusively with renewable energy matters. The national agency ADEME employs two wind energy experts and one solar power expert.

6   Representatives of the European solar photovoltaic industry developed an online tool which allows policymakers around the world to calculate feed-in tariffs for solar photovoltaics based on the generation costs. The tariff calculation is based on parameters such as system costs, system size, annual efficiency decrease of solar plants, annual cost of insurance, and average energy yield in a given region, and a parameter related to capital cost. The European Photovoltaic Technology Platform provides an online excel sheet for the necessary calculation. The excel sheet can be downloaded at http://www. eupvplatform.org/index.php?id=41. The National Renewable Energy Laboratory in the US has designed as similar template which can be accessed via https://www.nrel.gov/analysis/ sam/download.html.

meetings for European policy makers and industry representatives in order to exchange policy relevant information. At this point it is also worth mentioning the benchmarking reports published by the industry association EREF (EREF 2009). Since 2002 the industry association has published an annual report comparing the tariff levels for different technologies offered by different member states (and also support mechanisms other than feed-in tariffs). This report has sometimes been cited as a source of reference by legislators when fixing their national tariff levels.

We also wish to stress the role of research and international journals in the diffusion of policy knowledge. As already shown in the introduction of this book, research projects have for a long time served the interests of the European Commission, comparing and analysing the efficiency and effectiveness of different support instruments. The analysis of best-practice design of single support mechanisms, such as feed-in tariffs, has been neglected for many years. Only since 2006 has a number of studies on best-practice feed-in tariff design been published (Klein et al. 2006, 2008, Mendonça 2007, Mendonça et al 2009b, Couture et al. 2010). The increasing number of journals in the field of renewable energy technologies and politics has also had an impact on the accelerated diffusion of feed-in tariff design options since 2000. In Germany, journals like *Neue Energie*, *Solarthemen* and *Photon* frequently report on policy developments in other (European) countries. In France, *Système Solaire* has a long record in providing policy relevant information. In Spain, journals like *Energías Renovable*, *Infopower*, and *Photon España* provide useful information for policy makers and industry representatives.

Finally, it is worth noting that certain non-governmental organisations have specifically focused on the design of national feed-in tariffs. The role of the World Future Council has to be underlined in this respect, as the promotion of feed-in tariffs has been one of its major objectives. A non-governmental organisation, it was founded in 2007 and its first major focus was the diffusion of best-practice information on feed-in tariffs. The organisation provided policy-related information via a home page, including the specific design of all major feed-in tariff design features.[7] As well, a number of publications inform policymakers about feed-in tariff design. Moreover, the organisation is engaged in campaigning for feed-in tariffs in a number of countries worldwide, including the UK.

8.1.2. National Policy Innovations: The Role of Reference Models

It is important to realise that, in the process of policy learning and information exchange, actors in the international arena often filter information. The stakeholders involved in the political decision-making process do not rely on all the information that is available on a certain policy instrument but pursue a more selective approach and focus on specific 'reference models' (Holzinger and Knill 2008: 57).

---

7 The information is available from http://www.futurepolicy.org/renewableenergy. html.

The sought-after international best practice has also been referred to as 'miracle' models or 'success stories' (Simmons and Elkins 2004: 157, Meseguer 2005: 72). In this case, national policymakers search for the most successful countries in a certain policy area in the hope of learning from this country's legislation.

This also seemed to be the case with certain feed-in tariff design features. The diffusion patterns of feed-in tariff design options indicate that certain clusters of design options have become 'popular' at a given moment in time. In the 1990s feed-

Table 8.1   Development of feed-in tariff design options over time

Development of feed-in tariff design over time		
Design options for independent power producers in monopolistic markets	Design options for increased investment security	Additional design options for system and market integration
Purchase obligation Tariff calculation based on avoided costs Support for the most cost-efficient technologies Limited tariff differentiation Short tariff payment duration Small-scale production units (plant size and system caps) Deep connection charging	Tariff calculation based on generation costs/IRR Support for a large portfolio of technologies High degree of tariff differentiation to avoid windfall profits (e.g. size, and location, specific tariffs, tariff degression) Long tariff payment duration Shallow connection charging	Optional premium feed-in tariff Implementation of additional intra-day markets Forecast obligation Demand-oriented tariff payment Remuneration for auxiliary grid services Combination of different renewable energy technologies
Primary objective – Security of supply – Diversification of energy mix – Establishment of a framework for independent power producers	Primary objective – Environmental protection – Technological development – Rapid increase in the share of renewable electricity	Primary objective – Market and system integration of large shares of renewable electricity
Model legislation – No single model law – general pricing approaches for independent power producers in monopolised markets	Model legislation – German feed-in tariff of 2000 (EEG 2000)	Model legislation – Spanish feed-in tariff of 2004 (RD 436/2004)

Source: Author

in tariffs were calculated on the basis of the avoided costs of conventional power generation, payment was not guaranteed for a long time, and tariff differentiation was limited. At the turn of the century legislators started to move towards a cost-based tariff calculation approach, tariff differentiation was expanded, and the payment duration was extended. Later, design options for system and market integration became increasingly important.

Therefore, in the following sections we look in more detail at typical feed-in tariff design options at different moments of time, identifying the model legislation which was the starting point for cross-national spread of these design features. We argue that – in line with Table 8.1 – early feed-in tariffs in the 1980s and 1990s were designed to create a niche market for independent power producers in monopolised energy markets. Later, the German feed-in tariff of 2000 became a model because it included certain design options which significantly increased investment security. Once the share of renewable electricity started to increase significantly in a number of countries, in 2004 Spain became the first jurisdiction to implement a large number of design options for system and market integration, in 2004. The detailed information contained in Table 8.1 is further explained in the following sections on feed-in tariff development over time. We describe the extent to which early feed-in tariffs of the 1990s have been established in order to create niche markets for independent power producers (section 8.1.2.1), how far the German amendment of 2000 managed to increase investment security (section 8.1.2.2) as well as the extent to which Spain managed to integrate large amounts of renewable electricity via design options for system and market integration (section 8.1.2.3).

The conclusions drawn in this section are largely based on our comparative analysis executed in Chapters 4, 5, 6 and 7 and their respective conclusion. However, it should be noted that the general policy development described in this section is also confirmed by the analysis of other feed-in tariff policies around the world. In this respect it is important to stress that the author of this research project is also co-author of *Powering the Green Economy: The Feed-in Tariff Handbook* (Mendonça et al. 2009b) and co-author of the Future Policies home page on feed-in tariffs (http://www.futurepolicy.org). In both research projects more than 40 feed-in tariff mechanisms from all over the world have been analysed in detail. Some of this information is included in the following section.

8.1.2.1. Early Feed-In Tariffs: Opening Monopolistic Markets to Independent Power Producers

In the 1990s few countries had at their disposal feed-in tariffs[8] or similar legislation for renewable electricity promotion. The analysis of the three countries under

---

8   At this stage it is important to stress that early price-based support mechanisms, including the early German and Spanish support framework as presented in this study, do not qualify for a feed-in tariff under today's standards. Feed-in tariffs should offer guaranteed tariff payment over a long period of time. The early German and Spanish legislation,

investigation did not reveal any evidence of information exchange between the Spanish and the German legislators at that time. Despite several policy learning occasions among European member states,[9] the intensity of policy-related communication has certainly been limited. Renewable energy promotion strategies – for instance, the diversification of energy supply – were driven by national policy objectives. Neither international obligations under a common EU framework nor an international climate regime existed at that time. Environmental and climate protection were not the primary drivers of the promotion of renewable energy sources in the 1980s and 1990s.[10] Contrary to today's energy policy objectives, the first wave of support for renewable electricity was fuelled by the policymakers' desire to diversify energy supply and to allow non-utility actors to enter national energy markets.

This largely reflected the experience of two oil crises in the 1970s, revealing the increasing dependency of industrialised nations on imported fossil fuels.[11] The first feed-in tariff scheme worldwide was issued in the 1970s. In the United States, the PURPA Act of 1978 established the first, price-based framework for renewable electricity generators and other independent power producers worldwide (Lipp 2007, Rickerson and Grace 2007). This policy obliged electric utilities to purchase electricity from eligible producers at prices based on the utilities' avoided costs.[12] Utilities were obliged to pay all non-utility producers, not only renewable electricity producers. Besides, only power generation plants up to 80 MW were eligible for tariff payment (Hirsh 1999). Interestingly, the early feed-in tariffs in Europe displayed similar features. Tariffs were calculated on the basis of the avoided costs of conventional power generation; consequently, technology differentiation was very limited. Besides, payment levels were generally indexed to the retail electricity price. This was the case in Portugal (1988),[13] Germany (1990), Denmark

---

however, did not guarantee tariff payment for a given number of years. Therefore, by today's standards they would qualify only for preferential tariffs (Jacobs 2010).

9   Note that there are certain indications of policy transfer from Denmark to Germany (Brand-Schock 2010).

10   In some countries these considerations had already started to matter. The German support policy of the 1990s was already based on environmental concerns and the potential threat of climate change (Jacobsson and Lauber 2006).

11   A more detailed analysis was made by Hirsh (1999) for the US and Nebreda Pérez (2007) for Spain.

12   The determination of avoided costs was left up to each single state, which explains the different interpretations across the US (Lipp 2007, Hirsh 1999).

13   In Portugal, renewable electricity producers received the same guarantees as national companies operating fuel oil-based generation facilities, because the national policy at the time aimed to support electricity generation from small producers and not exclusively renewable energy producers. The tariff payment in Portugal was non-technology-specific and based on the electricity prices charged by utilities (REACT 2004b).

(1990), Spain (1994) and Greece (1994).[14] As well, tariff payment was offered only for a short period of time and the 'deep' connection-charging approach was the predominant cost-sharing methodology for grid connection – an approach that was developed on the basis of experience with large-scale, conventional electricity power plants (see section 4.6). In short, there were not many differences between general pricing policies in monopolistic power markets and feed-in tariffs for renewable energy sources.

Given that transnational communication has apparently been limited during the early years of renewable electricity promotion, the similarities at cross-national level most likely arise from the general framework conditions. Legislators in different countries tried to establish frameworks for independent power producers in largely monopolised markets. The easiest way to do so was to oblige national utilities to purchase independently produced, non-utility power – the historical origin of the purchase obligation as we know it today – and by fixing certain prices – or tariffs – for this type of electricity. Even though this was the first important policy innovation for renewable power producers, the logic of the conventional power market remained dominant. The remuneration for renewable electricity producer was subordinated to the logic of monopolistic power markets. Tariff payment was linked to the costs of conventional power generation either by explicitly referring to the avoided cost or by indexing tariff payment to the retail electricity price.

In sum, there was not one reference model of early feed-in tariffs – or, more precisely, generation-focused, price-based support. Even though the first feed-in tariff of this kind was developed in the US in 1978, there is no clear evidence of transnational communication. One could also assume Germany's first feed-in tariff law of 1990 to be a reference model for other countries, since in the following years a number of European countries have established similar support mechanisms, calculating tariffs on the basis of the avoided costs and indexing them to the retail electricity price. However, it should be noted that Germany had already learned from Denmark, and that the first legislation with similar characteristics was implemented in the United States.[15] Apparently, the common ground of these early support schemes was the legislators' intention to establish a framework for

---

14   Some of these features, namely capacity caps for eligible power plants, are still included in the Spanish and the French legislation today (see section 4.2).

15   The first Danish feed-in tariff for wind energy was implemented in May 1990, comprising a tariff payment of 85 percent of the retail electricity price (Bechberger et al. 2008). The German legislation from December 1990 is strikingly similar, offering a tariff level of 90 percent of the retail electricity price for wind energy producers. However, already by the end of the 1980s Danish wind power producers were receiving rather attractive feed-in rates. These were based not on any legal obligation but rather on voluntary agreements between wind power producers and local power suppliers (Riepe 2004). These incentives were equally important for the policy transfer process between the two countries.

independent power producers – including renewable power producers – in largely monopolised national energy markets.

8.1.2.2. The German EEG 2000: A Reference Model for Maximum Investment Security

Despite the efforts of many governments around the world, the share of renewable electricity – excluding large hydro power – remained relatively low. Only the share of the least costly technologies such as onshore wind power increased in the 1990s. In this decade, Germany managed to spur wind power deployment, reaching an installed capacity of about 4,440 MW in 2000 (BMU 2010b), while Denmark's early feed-in tariff policy led to an increase of almost 2,000 MW of wind capacity between 1992 and 2000 (Lesser and Su 2008). However, less mature technologies, such as solar photovoltaics and geothermal power, were not developed, largely because of the tariff calculation methodology applied (see section 4.3).

In order to increase the share of renewable electricity, the German legislator introduced a number of innovative design options to increase investment security. First, the tariff calculation methodology began taking the technology-specific generation costs into account (see section 4.3). Second, tariff payment was now guaranteed for a long period of time, thus making total revenues more predictable over the full lifetime of a power plant (see section 4.4). Third, Germany moved from a deep towards a shallow grid-connection charging approach (see section 4.6). All three design options were truly innovative, since they clashed with traditional electricity market regulation. In the case of all three design options, national legislators changed the established rules of the conventional electricity market, now taking more account of the needs of renewable power producers. First, for the first time, the remuneration level was decoupled from the costs of conventionally produced electricity by taking the technology-specific generation costs into account. From a historical perspective this decision was almost revolutionary, as tariff payment above the avoided costs of conventional electricity generation clashed with the political objective of providing electricity at the lowest possible cost – one of the three main energy policy objectives of the EU and its member states. Second, the German legislator fixed tariff payment for a long period of time in order to increase investment security. And third, the national legislator established specific grid connection-charging methodologies for renewable electricity producers, now separating the cost of investment in grid infrastructure from investment in renewable power generation capacity. These fundamental changes partly explain the fundamental opposition of certain actors to feed-in tariff mechanisms (see section 4.7).

Besides the implementation of design options for increased investment security, the German legislation of 2000 also introduced a number of design options for the first time intended to reduce windfall profits through the use of tariff differentiation and adjustment. These were discussed in Chapter 5. Once again, it was shown that Spain and France copied many of these design options from Germany. The German law of 2000 for the first time implemented tariff degression and introduced technology-, size- and location-specific tariffs for a large number of technologies.

The combination of all these design options in just one law explains why the Germany feed-in tariff scheme of 2000 was important for the diffusion of design options for basic feed-in tariff and for tariff differentiation as we know them today (see section 4.7 and 5.6).[16] Over the years it became apparent that these design options were crucial for a fast uptake of the renewable electricity market at the lowest possible costs. Consequently, these design features were copied by legislators from other countries, including Spain and France. Today, these design options are considered international best practice.

8.1.2.3 The Spanish RD 436/2004: A Reference Model for System and Market Integration

The constant improvement of feed-in tariff legislations led to a rapid increase of the installed renewable electricity capacity in a number of countries. The feed-in tariff mechanisms established around the turn of the century have been very successful in bringing about large-scale deployment of renewable electricity plants. However, this success also created new challenges since the increasing share of green power needed to be integrated into the existing energy market and the electricity system. One of the first countries to face problems with system and market integration was Spain, because of a combination of factors. First, the Spanish grid is less developed than the grids in other European countries and therefore less capable of including large amounts of renewable electricity. Second, the increasing share of renewable electricity is largely due to newly installed wind energy capacity, a fluctuating source of electricity generation. Third, the Spanish electricity grid can be considered an 'electricity island' as interconnections with other European countries are scarce. Besides, the national actor coalitions allowed for the implementation of these design options, since the Spanish utilities had a vital interest in developing technological solutions (see section 6.1.3). In order to allow these incumbents to increase the share of renewable electricity in the grid, the large utilities developed technological solutions that would help overcome problems with grid integration and power transmission (Stenzel and Frenzel 2008).

The Spanish feed-in tariff scheme of 2004 can be considered a reference model for market integration design options. Even though earlier legislation has included some design options for market integration – the Royal Decree of 1998 included the premium feed-in tariff for the first time – the 2004 legislation is more comprehensive, comprising the optional premium feed-in tariff, the implementation of additional intra-day markets in order to facilitate renewable power sales, demand-oriented tariff payment, a forecast obligation and extra payment for renewable electricity producers that provide auxiliary grid services

---

16 It should be noted that the German legislator facilitated the spread of these design options by translating the German feed-in tariff law into many languages (see the BMU homepage).

(see Chapter 6). Some of these design options have already been copied by other countries, including Germany.[17]

However, the Spanish Royal Decree 436/2004 was only the beginning of the implementation of design options for system and market integration. As we have shown, the German legislator tried to invent other policy innovations, such as tariff payment for the combination of several renewable electricity technologies and storage technologies. Incentives for the development of storage capacity will be of increasing importance in the future. Besides, the German legislator introduced another remuneration option for solar photovoltaic producers that decide to consume their electricity without feeding it into the grid. Whether other design options for systems and market integration will be implemented in the future remains to be seen (see Langniss et al. 2009). The future design of national support instruments and electricity markets largely depends on political objectives. Generally, selling electricity on the market – with or with a premium feed-in tariff option – increases investment risks, and therefore the cost of green electricity and the additional costs for the final consumer also go up. Consequently, legislators have to decide whether the benefits of additional bonuses to encourage the market integration of green electricity outweigh the additional costs. Two political objectives, market integration and reducing the costs for final consumer, are apparently in conflict (Klessmann et al. 2008).

The extension of the number of intra-day markets in Spain (and later in Germany and France) was crucial to enabling renewable electricity producers to sell their power on the market more easily (see section 6.4). The implementation of this design feature reveals that measures to promote renewable electricity will have an ever stronger impact on the design of national electricity markets as a whole. Some researchers have even gone so far as to say that liberalised electricity markets and a renewable energy-based power system are not compatible at all. Bode and Groscurth have pointed out that financial incentives to invest in new generation capacity will decrease if the share of renewable electricity increases. First, an increasing share of renewable power lowers the spot market price for electricity and therefore the revenues for all power producers. Second, annual operating hours for conventional power plants will be reduced because they will be frequently replaced by renewable energy capacity (Bode and Groscurth 2008a).[18]

---

17   The premium feed-in tariff, for instance, was also implemented by a number of other European countries, namely Denmark, Slovenia, the Czech Republic, Italy, Estonia and the Netherlands. Finland also plans to implement this remuneration option for wind energy in 2011 (Couture and Gagnon 2009, Mendonça et al. 2009b, Klein et al. 2008).

18   Similarly, Hvelplund has questioned the logic of today's power markets in the case of renewable energy sources. He points out that, by increasing the share of certain renewable energy sources such as wind power and solar photovoltaics, electricity production become 'automated' as production depends on climate and weather conditions. Once the wind power plant has been installed, 'the wind turbine will not work more efficiently because of competition with other wind turbines in the electricity market' (Hvelplund 2009: 287).

In order to tackle these problems, the implementation of so-called capacity markets is currently being discussed in Germany. In this case, power producers are not only paid for producing units of electricity that are fed into the grid but also receive some compensation for providing a certain amount of capacity even though it will not necessarily always be needed for power generation (Battle and Pérez-Arriaga 2007). In the case of a renewable energy-based power system, this will be necessary in order to provide sufficient reserve capacity, and to ensure the reliability of electricity supply.[19] Whether this modification of electricity market rules will suffice to prepare the market for the inclusion of more and more renewable electricity remains to be seen.

8.2. The Power of Competition: Convergence of Remuneration Levels

As introduced in section 2.1.1, the theory of regulatory competition includes two both 'economic competition' and 'political competition'. Political competition is expected to occur when national policymakers try to adjust European policy to national policy preferences in order to reduce the transaction costs that might arise when the EU 'prescribes' a certain policy or program that is not in line with the national regulatory approach (Héritier et al. 1996, Börzel 1999, Busch and Jörgens 2009).

This research project did not identify any influence of political competition on the design of national feed-in tariff mechanisms. This partly reflects the level of our analysis, that is, the analysis of micro-level design options of just one policy instrument. Political competition requires knowledge of different policies, or, in our case, knowledge of different feed-in tariff design options. The theory suggests that if a critical mass of countries has implemented a certain policy or program, previously reluctant countries will no longer be able to ignore them and will then follow the majority along this path (Busch et al. 2005). As knowledge about all feed-in tariff design options and their spread in different member states is very limited (even among national and European policymakers), one of the main ingredients for this type of competition is missing.

Besides, the European Commission has been in favour of another support instrument for a long time. According to Héritier, the influence of member states on European legislation depends on the compatibility of national regulation with the overall objectives and strategies of the European Commission (Héritier et al. 1996). As the European Commission's policy preference is for a quota obligation

Because of the different cost structures of conventional and renewable power plants, the shift from fossil fuels to renewable energy sources will lead to a shift from the electricity market to the equipment market. Therefore, competition in the equipment market will be more important than competition in the electricity market.

19   In competitive electricity markets as they exist today, investment decisions are based only on expected profitability (Klackenberg et al. 2004).

based on certificate trading, member states will find if difficult to convince the European legislator of specific feed-in tariff design options. The implementation of a harmonised European feed-in tariff scheme, which might have spurred political competition in the specific design, was not even anticipated by national legislators. Therefore, member states operating with feed-in tariff mechanisms are not competing with each other with respect to the design of this support instrument. On the contrary, they cooperate with each other – sometimes within international platforms (see section 8.1.1.2) – in order to protect their support instrument against attempts from the European level to push for a harmonisation based on a different support mechanism.

8.2.1 Economic Regulatory Competition in the Case of Positive Incentives

However, this research project has identified circumstances of economic competition in the case of tariff level development (see Chapter 7) and other design options that have an influence on the economic feasibility of renewable electricity projects, such as tariff degression (see section 5.4). The theory or regulatory competition – as introduced in section 2.1.1 – suggests that national legislators implement or modify a policy in order to remain competitive with other jurisdictions in an economically integrated world. Accordingly, regulatory competition arises from increasing economic competition in European or global markets, which leads to mutual adjustments of national policies and programs (Holzinger et al. 2008a, Garrett et al. 2008). This type of economic competition has been identified in the case of feed-in tariffs in the EU.

In Chapter 7 on the evolution of tariff levels for solar photovoltaics, we concluded that, as a general tendency, tariff levels have become more similar over time. Like transnational communication (see section 8.1) and the influence of international technological development (see section 8.3), regulatory competition affected national decisions on the tariff level. We identified regulatory competition in the case of solar photovoltaics. In 2007 the Spanish manufacturing industry threatened to move its production facilities elsewhere if tariffs were reduced as envisaged in the first draft regulation. This forced the Spanish legislator to slightly increase the tariff in 2008 (see section 7.1.2).

The influence of regulatory competition was also enhanced by more subtle international economic interdependencies. We have clearly shown instances where policy modification in one country had an impact on market development in other countries. This type of economic interlinkage was most obvious in the case of solar photovoltaics. In contrast to wind energy, there are rather few well-functioning support mechanisms for solar photovoltaics in place worldwide. Therefore, the German market and– for a certain number of years – the Spanish market were (and still are) very crucial for the worldwide solar photovoltaic industry, absorbing a large share of the modules provided by international producers. When the Spanish market suddenly started to expand dramatically in 2007 and 2008, the national market growth diverted modules from other countries, including France and

Germany. This was partly because the Spanish support scheme of 2007 offered very attractive tariffs, unintentionally offering double-digit rates of return for producers. However, with the implementation of a stricter capacity cap in Spain in 2008, producers started to focus on other countries again, leading to increasing installations in Germany and France. The rapidly increasing market shares in France and Germany, in turn, was one reason for the tariff cuts in 2010 (see section 7.2.4).

Apparently, the size of the (international) market determines the impact of regulatory competition as a result of economic interlinkage. The smaller the number of countries with well-functioning support instruments, the greater are the economic interdependencies between policy modifications in any one country and the resulting effect on policy design in other countries. The international wind power market is larger than the solar photovoltaic market. The existence of many well-functioning support mechanisms around the world reduces the impact of tariff modifications in one country on the market development in other countries. Tariff levels are generally set in the light of technological developments, that is, the actual costs of power generation; neither raising tariff levels in order to attract investment nor reducing tariff levels in order to control the newly installed capacity is necessary.[20]

In sum, the theoretical assumptions regarding regulatory competition were largely confirmed. This is somehow surprising, since we were dealing with a policy instrument offering positive incentives – guaranteed tariff payment for each unit of renewable electricity provided. We have to remember that the state-of-the-art theory of regulatory competition has been largely developed on the basis of observations related to command-and-control regulation, mostly of the environmental kind. However, command-and-control policy instruments function in a contrary manner to market-based policy instruments offering financial incentives, such as feed-in tariffs. Instead of prohibiting certain behaviour by establishing negative incentives (such as limiting the emissions of certain production units), the policy mechanism under investigation offers positive incentives. Under such incentive-based policy

---

20   This responsiveness of international markets to national support frameworks was also stressed by several interviewees during field research. In the past, some windmill producers increased their product prices because of diverging rates of return in different countries (A. Ceña, personal communication, 12 February 2008). This does not relate only to countries with feed-in tariffs. Interviewees pointed out that higher remuneration under quota- based support mechanisms also encouraged producers to demand higher equipment prices in these countries (P.G. Thérond, personal communication, 22 October 2008). Even though the biomass sector is not analysed in detail in this study, interviewees stressed that the tariff level in different European countries influenced the decisions of biomass producers on where to sell their resources. Since biomass products can be transported – in contrast to stationary renewable electricity technologies like wind power and solar photovoltaics – Spanish biomass producers sometimes decided to sell their product to countries that offered a higher rate of return (C. Cánovas del Castillo, personal communication, 26 February 2008).

instruments, national governments generally pursue the political objective of creating a certain market in order to power economic transition. Instead of using the 'stick' to prevent certain behaviour we are dealing here with a case of government using a 'carrot' in the form of financial incentives. Instead of restricting certain privileges, certain rights are granted (Parisi et al. 2001).

In this respect, we have analysed a policy instrument that functions in a diametrically opposite manner to command-and-control policies. This explains why regulatory competition in the case of feed-in tariffs has led to increasing tariff levels, as shown in the case of wind energy in Germany in 2009 and in the case of solar photovoltaics in Spain in 2008. These and similar empirical observations have induced representatives of the conventional energy sector to argue that the existence of feed-in tariffs in many EU member states leads to ever-increasing tariffs, a race to the top (Birr-Pedersen 2009). However, as the next section shows, our analysis does not bear lout this claim.

8.2.2. The Direction of Policy Convergence in the Case of Positive Incentives

In section 2.1.1 we explained that early theoretical assessments in political science and economics have often assumed a 'race to the bottom', in which policymakers redesign national regulations in order to avoid competitive disadvantages for the national industry. However, a large number of empirical studies have challenged this assumption, showing that policy might just as well converge at the top. The logic of industry representatives in the case of feed-in tariffs is very similar to that of early theoretical assessments in policy convergence theory, but applied to a policy mechanism functioning in a diametrically opposed manner. It is argued that governments would issue higher and higher tariffs because they were competing with support frameworks in other countries in an international market. Mobile capital and products would flow into the countries offering the highest incentives.

However, just as empirical studies in the case of command and control-based policies have shown that the 'race to the bottom' hypotheses cannot be verified, the 'race to the top' hypothesis is not confirmed by our empirical analysis in the case of feed-in tariffs. Because of financial restrictions – the legislator is trying to limit additional costs for the final consumer – tariff levels mostly reflect the actual, technology-specific generation costs of renewable electricity technologies. In other words, technological development (see section 8.3) is more important for the development of tariff levels. Only in exceptional cases have we demonstrated convergence at the 'top' or at the 'bottom', depending on the national and international market sizes.

At an early stage of market development, when the share of renewable electricity supported by the national feed-in tariff scheme is relatively small, legislators sometimes offer relatively high tariffs in order to attract investment. This was shown to be true in very exceptional cases, such as solar photovoltaic tariffs in France, which the French legislator promoted as the highest tariff levels in the world (see section 7.1.3). At an early stage of market development, the pressure to establish a cost-efficient feed-in tariff scheme is relatively low, because the total amount of tariff payment is still marginal and so very attractive tariffs have only a limited impact on overall system costs.[21] However, as the share of green power increases, financial restrictions oblige national legislators to establish an efficient support scheme in order to reduce costs for the final consumer. Therefore, support for renewable electricity is driven both by the need for effectiveness – for renewable electricity to reach a certain market share – and the need for efficiency – to reach this target in the most cost-effective way.

In most cases, national legislators have calculated the technology-specific generation costs, thus offering a certain internal rate of return for investment in new renewable electricity capacity. In section 4.3.4 we showed that these vary between five percent and nine percent in the three countries we have analysed. Therefore, tariff levels have been converging alongside these generation costs, that is, alongside technological development (see section 8.3). We have also identified cases where national legislators, seeking to further reduce the costs for the final electricity consumers, have offered prohibitively low tariff levels, less than actual generation costs. The drastic tariff reduction in the solar photovoltaic industry since 2008 partly reflects the sharp cost reduction, but also shows how far national policymakers will go in trying to reduce the overall system costs for renewable energy support by controlling the market growth of this technology. This is especially true for free-standing solar photovoltaic tariffs in Germany and Spain (see section 7.1.4).

To a minor degree, the worldwide supply of power plants, that is, the oversupply or under-supply of power generation equipment can influence the direction of tariff convergence. We have seen that in markets with oversupply, as it was the case for solar photovoltaics during many quarters of 2009 and 2010, legislators seem to reduce tariff payment in order to 'protect' their markets from absorbing too much of the capacity that is produced on the world market. The Spanish and the German legislators reduced tariff levels for solar photovoltaics because the national market grew faster than expected, thus increasing the additional costs for the final consumers.

Based on this empirical observation, we can try to derive more general cause-and-effect relations between the international supply of products and the direction of feed-in tariff convergence. Theoretically, one could assume that, in markets where demand for products exceeds supply, producers take advantage of their market power and sell their products in those markets with the highest tariff payment level. As a consequence, national legislators in countries with comparatively low tariffs will have to increase remuneration levels in order to achieve national

---

21   This also explains why a number of southern European countries, including Italy and Greece, offer comparatively high tariffs for solar photovoltaic deployment (despite the favourable geographic conditions). The total share of solar photovoltaics is still relatively low and so are the total costs of solar photovoltaic remuneration.

targets.[22] Theoretically, this would lead to tariff level convergence 'at the top', that is, ever-increasing tariff levels as a result of regulatory competition. In the case of oversupply, producers will start selling their products to the most attractive markets but will soon move to markets with lower tariff levels. In this case, policymakers in countries with relatively high tariffs will be able to reduce tariff levels. Contrary to the situation characterised by oversupply, one can assume that the tariff level is going to decrease in all countries, that is, tariff level convergence at the 'bottom'. However, these speculations should be taken with a pinch of salt, as national legislators usually do not respond to short-term market developments, that is, the switch from oversupply to under-supply, and vice versa. Usually, tariff levels are amended after a pre-defined number of years and national legislators do not constantly react to the aforementioned market structures. Besides, the production facilities for renewable electricity products become more and more flexible so that oversupply, as previously observed in the case of solar photovoltaics, will no longer appear and 'cause' the aforementioned problems.

In sum, tariffs are generally decreasing (due to technological learning) and therefore empirical observations point to convergence at the bottom. However, this direction of convergence is a result not primarily of regulatory competition but rather of technological advances. In the case of regulatory competition, there are no clear signs of whether convergence will take place at the top or at the bottom. In fact, this depends on whether effectiveness (rapid increase in the installed capacity) or efficiency (support at the lowest possible cost per unit) is the guiding principle of policy design.

8.3. The Power of Technological Innovations: Independent but Similar Policy Responses

As introduced in section 2.1.1, the concept of independent problem solving is based on the observation that as a result of parallel problem pressure various national governments implement similar policies independently. In other words, governments find similar solutions to common problems without learning from policies or programs in other jurisdictions. This theoretical framework of 'independent problem solving' can be extended on the basis of the findings of this research project. In this section, we try to identify causal mechanisms that have led to similar policy responses even without transnational interaction.

When introducing the theory of policy convergence we showed that other causal mechanisms, for instance transnational communication, also include a number of sub-mechanisms. Over time, researchers have identified several causal factors that can all be linked to transnational communication. However,

---

22 This, of course, is only true for countries with technology-specific targets. Otherwise, the reduced deployment rate for one technology might be substituted by a stronger increase in other technologies.

the causal mechanism of 'independent problem solving' has not yet been further differentiated by scholars. In fact, this mechanism has often been used as a catch-all explanatory factor for cross-national similarities that could not be explained by communication or competition. Here, we will make an attempt to identify different factors that can lead to independent but similar policy solutions. As shown below, extending and distinguishing this causal mechanism will also require a new, more encompassing term to describe these cause-and-effect relations.

The state-of-the-art theory of policy convergence operates with the term 'independent problem solving', since similar policy responses have usually been caused by similar problem pressure (Bennett 1991: 231). However, our analysis of feed-in tariff evolution has shown that, in the case of incentive-driven, market-based policy instruments, independent but similar policy innovations usually arise from new 'opportunities' created by technological innovations rather than from common 'problems' and similar problem pressure. Therefore, similar policy responses are 'opportunity-driven' rather than 'problem-driven'. Recent studies of policy convergence have not sufficiently taken into account the influence of technological innovations on national policy decisions. Only a few earlier research studies have considered technology-driven policy innovations (see Arrow 1962). For instance, Jänicke (2000) analysed the interplay between innovative environmental policies and technological innovations. However, the analysis is linked more closely to regulatory competition ('lead markets') than to independent problem solving. Bennett (1988) focused on the technological dimension of independent policy responses when analysing data protection policies in European and the United States. However, in this case technological advances in telecommunication technologies have been perceived as a threat rather than an opportunity. Therefore, his analysis is substantially different from the present research project.

According to our findings, a more comprehensive term might be more appropriate for describing the phenomenon that we called 'independent problem solving' in section 2.1.1. We suggest the term '*independent but similar policy responses*', as this would encompass both problem-driven policy responses and opportunity-driven policy similarities arising from technological advances. The need for a more encompassing term is also reflected in the use by previous researchers of a wide variety of expressions, including 'technocratic determinism' (Rose 1991), 'technological determinism' (Bennett 1988), 'clustering' (Simmons and Elkins 2004), and 'parallel domestic pressure' (Hoberg 2001).

The importance of opportunity-driven policy similarities is not surprising in the case of feed-in tariffs. In fact, one of the major objectives of renewable electricity support is to further develop these technologies, to reduce the generation costs and, finally, to make them compatible with conventional power-generation technologies. From this point of view, it is interesting to observe the mutual influence between policy developments and technological development. Inasmuch as feed-in tariff policies have been shown to be successful in reducing the generation costs of renewable electricity technologies, this study also shows that this technological development reacted on national policy design., Legislators first establish a certain policy in order to develop a certain technology, and later the support instrument needs to be modified because technological development is actually taking place.

8.3.1. Opportunity-Driven Policy Convergence Arising from Technological Advances

The implementation of certain design options in all three countries under analysis can clearly be related to the overall technological development of renewable energy sources and can therefore be termed 'opportunity-driven'. In section 4.1.4 we showed that new technologies became eligible under national feed-in tariff mechanisms partly because of the development of less mature technologies which, over time, left behind the stage of research and development funding and therefore became eligible for tariff payment. This was also the reason for more technology-specific support, as described in section 5.1.4. Similarly, capacity caps in all three countries could be removed because of the achieved reduction of generation costs over time. Because of the decreasing costs of renewable power generation, the national legislators had more political space to implement more ambitious renewable energy targets and to remove caps (see section 4.2.4). At the same time, more technology-specific support became possible because of the technological advances, especially in the case of less mature technologies such as geothermal, offshore wind and alternative hydro-power technologies (see section 5.1.4). The increasing size of renewable energy power plants also prompted national legislators to implement plant-size specific tariffs, as they had to accommodate the wide range of plant sizes in national legislation (see section 5.2.4). In sum, reduced power generation costs and large plant sizes have given national legislators the opportunity to expand support programs.

However, technological advances not only enabled policymakers to improve the framework conditions for renewable power producers. At the same time, technological innovations triggered policy innovations that obliged power producers to cope with certain regulatory requirements. The obligation on wind-power producers to provide certain auxiliary grid services – first in exchange for financial compensation and later for free – is a good example of this (see section 6.2.4). Besides, in Chapter 6 we showed that the special regulations for market sales of green power were also technology-driven. The decreasing costs of certain renewable electricity technologies reduces the price gap with conventional power generation, thus making it more and more interesting for power producers to start selling (parts) of the electricity on the market. This also explains the establishment of direct sales in the case of solar photovoltaics in Germany (section 6.1).

The influence of the technological development was also shown in the case of the evolution of tariff levels (Chapter 7). In the case of solar photovoltaics calculating the right tariff level is more difficult than, for instance, wind energy because of continuous and rapid technology learning (see section 7.1). This also explains why legislators in Germany, Spain and France have frequently amended the support scheme for solar photovoltaics since 2008. Even though this research project compares tariff development in only three European countries (Germany, Spain and France), a comparison of support levels in all EU 27 countries in 2009 confirms these findings (see EREF 2009). However, the evolution of solar photovoltaic tariffs in Germany, Spain and France most clearly reveals how far technological advances influenced policymakers' decisions.

We have also shown that in some circumstances national legislators have still not adjusted their policies to the latest technological development. In the case of general plant-size caps (see section 4.2) we showed that the 50 MW limit in Spain and the 12 MW limit in France are largely based on the (outdated) idea that renewable electricity plants are by definition small-scale. In the new millennium, however, projects from all over the world have demonstrated that renewable power plants can often be bigger than 100 MW. The capacity limits remain because they have been enshrined in national electricity laws, which are more resistant to changes. In both France and Spain, the actual feed-in tariff designs were implemented through subordinate decrees and ordinances. In fact, the limit of 12 MW as set in France dates back to a national electricity law from 1946 (amended in 2000, 2005 and 2010); and the Spanish legislator established the so-called special regime for small-scale, independent power producers in the 1980s.

8.3.2. Problem-Driven Policy Convergence Arising from Similar Problem Pressure

Despite the importance of technology-driven policy innovations in the case of feed-in tariffs, we have identified a number of cases where similar problem pressure has led to similar policy responses.[23] This type of problem-driven policy convergence was observed in the case of tariff calculation methodologies (see section 4.3) and – even more interestingly – in the case of flexible degression in Germany and Spain. As shown in section 5.4, flexible tariff degression was developed independently in Germany and Spain. In both countries, it was intended to reduce the overall costs of the support mechanism by adjusting tariff levels to national market growth. In other words, rapidly increasing costs for the support of solar photovoltaics was the common problem. Besides, in Germany and Spain legislators were confronted with the similar difficulty of fixing the degression rate in rapidly developing markets in order to anticipate

---

23 If we leave our micro-level perspective for a moment and look at the broader picture for renewable energy support, it becomes clear that the implementation of support instruments in general was certainly driven by similar problem pressure as well. It should be noted that the motivation for and objectives of establishing support instruments for renewable electricity are very similar across most European countries. The major reasons for supporting renewable energy sources are concerns regarding the environment and national security of supply (Lipp 2007). However, analysing the reasons for renewable energy support in general is not the focus of our research project.

cost reductions. Interestingly, the schemes were developed at more or less the same time but without the relevant actors knowing about each other's effort.

In February 2008 the author of this study was undertaking field research and interviews in Spain. During an interview, the representative of the Spanish solar photovoltaic association, ASIF, presented an unfinished document as containing an innovative approach; it elaborated the concept of flexible tariff degression as it was later published by ASIF (2008). The interviewee even stressed that the new Spanish proposal for flexible tariff degression would go beyond the general concept of tariff degression as applied in Germany (T. Días, personal communication, 25 February 2008). At the same time, the German renewable energy sector was debating the amendment of the German feed-in tariff scheme. Carsten Pfeiffer, research assistant of the parliamentarian Hans-Josef Fell, had independently developed a similar concept, which he called the 'integrated market degression model'. Mr Pfeiffer was surprised when he became aware that in Spain a similar concept has been proposed (C. Pfeiffer, personal communication, 4 September 2008). Without any interaction between the actors in the two countries, flexible degression was introduced into the German and the Spanish feed-in tariff schemes in 2008 (respectively, June and September). Therefore, the decision to implement this design option can be identified as genuinely independent problem solving.

This case reveals the difficulty for researchers in the field of policy convergence in distinguishing different causal mechanisms. It shows how researchers of policy convergence could fall into the trap of assuming transnational communication simply because of striking similarities in national policy design. Even though the detailed designs of the first German and Spanish proposals were utterly alike and the implementation took place almost simultaneously, the two policymakers did not know of each other's efforts. The case also shows the general difficulty in identifying problem-driven independent problem solving. While it is already difficult to show causal dependencies in the case of similar policy design options in different countries, it is even more difficult to guarantee the absence of mutual influence at cross-national level, be it through transnational communication or regulatory competition. Even though it is possible to investigate any influence from the international level via expert interview, it still cannot be guaranteed that one specific design option was developed independently in order to solve a specific problem, as interviewees might have forgotten about information transfer. As a matter of fact, independent problem solving can be identified more easily in the case of recent policy design implementation, as the 'intervening variable' of oblivion can be minimised.

8.4. The Soft Power of European Policy Steering – the Open Method of Coordination

This research project analyses development of feed-in tariff mechanisms in the EU. In line with our hypotheses, we have shown that feed-in tariffs in Germany, Spain and France are converging despite the lack of coercive measures based on European harmonisation. We have revealed that the increasing similarities of feed-in tariffs in the EU arise from several voluntary mechanisms, namely, transnational communication, regulatory competition and independent problem solving. However, surprisingly and contrary to our initial belief that the European level did not have any influence on national policy design because of the failure of harmonisation, the detailed analysis of the evolution of feed-in tariff design options in Chapters 4 to 7 has shown that the implementation or modification of certain design features was nonetheless influenced by the European level – without legally binding measures.

8.4.1. Targets, Reporting and the Threat of Harmonisation

The European directive from 2001, as described in Chapter 3, apparently had an effect on national policy design despite its non-binding nature. First of all, indicative national targets had a large influence on national policy designs in Germany, Spain and France. Generally speaking, indicative targets can be seen to accelerate cross-national learning in general. Suddenly, national legislators' efforts in support of renewable electricity were 'measured', and therefore they started looking for policy solutions in other countries. Besides, the European Commission fostered policy learning by publishing benchmarking reports and by threatening to harmonise member states' support instruments, in case their voluntary efforts were not be fruitful. These ways of influencing national policy decisions are analysed in the present section.

In section 4.2 we showed that the targets enshrined in the European legislation were usually adopted in national development plans for renewable energy sources. The European targets forced national legislators to implement more ambitious renewable energy targets then one could have expected under the auspices of any European regulation. This is especially true for countries with rather weak political support for renewable energies, like France. In those countries, international renewable energy targets can help to place the development of renewable energy sources on the political agenda. Proponents of renewable electricity in France – for instance, renewable energy industry associations – have frequently pointed out the European 'obligation' in order to force the national legislator to take more ambitious action. In countries where support for renewable energy sources has traditionally been strong, such as Germany and Spain, the European level can be considered as complementary to the national policy objectives.

In Chapter 4 it was shown that the eligibility of new technologies under national feed-in tariff schemes was, and will be, necessary to reach the indicative and binding national targets fixed by the European directives. At the same time, capacity caps at national level were often removed or extended in order to comply with European legislation. Proponents of renewable energy sources, such as industry associations, often managed to persuade national governments to extend and improve national support frameworks by pointing out international obligations at European level (see section 4.2.4). Equally, the switch to a cost-based tariff calculation methodology – especially in France – can be partly attributed to the need to comply with European targets. In section 4.3.4 we showed that all countries have switched from a value-based towards a cost-based tariff calculation methodology, primarily because the latter allows for more successful and more targeted support. By fixing the tariff level in the light of technology-specific generation costs, the legislator can support renewable electricity in the most effective and efficient way. The need to support renewable energies in the most cost-effective way – as incentivised by the European directive 2001/77/EC – also prompted national policymakers to increase tariff differentiation. This is especially obvious in the case of size-specific tariff payment (see section 5.2).

However, the indicative targets established in 2001 were not the only mechanism that empowered the EU to direct national policy decisions. Under Article 3 of the directive of 2001, European member states were also obliged to submit a number of reports to the European Commission (EU 2001, Sötebier 2003). In 2002 and 2007, member states were required to set national indicative targets for the following ten years, taking into account the indicative targets included in the directive. This explains why targets have frequently been adopted in national legislation (see section 4.2). The reports were also to state the measures that national governments would take to reach these indicative targets. Besides, in 2003, 2005 and 2007 member states were urged to submit progress reports to the European Commission. Based on these reports from all member states, the European Commission assessed the progress made in the different member states and also evaluated the likelihood of reaching the overall target of 21 percent by 2010. These report were published in 2004 (EU Commission 2004a), 2006 (EU Commission 2006) and 2009 (EU Commission 2009a).[24] This sort of reporting empowered the European Commission to centralise the communication process within the EU, and thus to take indirect control of national measures in support of renewable energy. The directive even gave the European Commission the power to propose mandatory targets in case 'national indicative targets are likely to be inconsistent [...] with the global indicative targets' (EU 2001). In other words,

---

24   The 2006 report concluded that, despite some progress in reaching the indicative national targets, significant renewable energy deployment took place only in a limited number of countries and that growth was primarily based on the installation of wind power plants (EU Commission 2006). In the report of 2009, it became apparent that the target of 21 percent renewable electricity by 2010 would most likely not be reached, but that a target of 19 percent might be reached (EU Commission 2009a).

the EU threatened to take stricter control of national policy measures via binding targets in the event of insufficient national effort.[25]

This threat of more hierarchical steering was also used in the case of support instruments. Even though in 2001 no consensus was reached regarding a harmonised European support instrument, the electricity directive included the possibility of harmonisation at a latter stage. Article 4 of directive 2001/77/EC prescribed that the European Commission was to assess the coexistence of different support instruments in the EU. This report was to make reference to the cost efficiency of support instruments, that is, the generation one unit of renewable electricity at the lowest possible costs, and the effectiveness of support, that is, the ability of national support instruments to reach the indicative national targets. 'If necessary, [the report might] be accompanied by a proposal for a Community framework with regards to support schemes' (EU 2001). In other word, the EU threatened to impose a harmonised support mechanism on member state governments.

This threat of harmonisation triggered a 'race' between member states searching for the most efficient and effective way to support renewable electricity. Even though mostly different support instruments were competing against each other, namely, feed-in tariffs and quota-based support instruments (Ringel 2006), we have shown that this also fostered the exchange of policy information between member states with the same support instruments in place (see section 8.1.1.2). This race is normally decided by coalition building among several member states and the strategic interest of the European Commission (Héritier et al. 1996). Research in the field of Europeanisation has established that the threat of legislation implemented at the European level is often used to increase the willingness of actors to cooperate on a 'voluntary' basis (Héritier 2002b). This partly explains the establishment of transnational cooperation platforms such as the International Feed-in Cooperation and the German–French coordination body for wind energy (see section 8.1.1.2). In the same line of thought, Héritier and Eckert (2008) argue that voluntary policy processes require a 'shadow of hierarchy' (Scharpf 1993: 70), that is, the treat of policy intervention in the event that voluntary agreements are not arrived at. This was clearly the case for renewable energy promotion in the EU.[26]

---

25 Since in the early years of the twenty-first century many member states were not on track to reach the 2010 target, the European Parliament recommended the implementation of binding national targets (EWEA 2004). However, the European Commission did not respond to this suggestion (Hinsch 2001).

26 Apart from the European framework directive, the EU managed to influence national policy design through other provisions, such as European state-aid rules. As described in detail in section 4.5, all European support mechanisms for renewable energy sources have to be in line with European state-aid regulations. In order to avoid conflicts with these regulations, most policymakers in the EU, including legislators in Germany, Spain and France, have opted for a ratepayer-based financing mechanism and equal burden sharing. The modification of the national financing mechanism as a result of pressure from the European level is especially obvious in Germany in 2000. In fact, the intention to comply

8.4.2. Moving from Harmonisation to Open Coordination

This analysis reveals that the European policy level has been able to influence national policy decisions even without harmonising European support instruments. Apparently, the EU switched to a different steering mode for policy coordination after the first harmonisation attempt failed in 2000 (Lauber 2007). Researchers have termed this less hierarchical, 'soft' mode of policy steering in the EU the 'open method of coordination'. This 'new' mode of governance was developed at the end of the 1990s in order to strike a balance between the need for diversity among European member states and the unity established by the European institutions.[27] In this sense, this method of policy making 'supports and radicalizes the principle of subsidiarity' (Hodson and Maher 2001: 719). In addition, this new mode of governance was developed against a background in which, despite the formal completion of a single market, economic disfunctionalities continued to exist in partly integrated or non-integrated areas. Lastly, the open method of coordination was established as a response to a legitimacy crisis of the EU in the 1990s and the early 2000s – the 'large gap existing between the European political elites and the publics' (Borrás and Jacobsson 2004: 186). The use of the open method of coordination was first officially referred to in the Lisbon summit conclusions in March 2000 (EU Council 2000), thus initiating a political and scholarly debate on new modes of governance.

The open method of coordination was established 'as the means of spreading best practice and achieving greater convergence towards the main EU goals'; and it should 'help member states to progressively develop their own policies' (EU Council 2000: 12). According to the Council's conclusions of March 2000, this new open method of coordination would be based on:

• the fixing of guidelines (and their potential translation into national and regional policies);
• the setting of specific timetables for achieving goals in the short, mid- or long term;
• the establishment of qualitative or quantitative indicators or benchmarks as a way of comparing best practice; and
• periodic evaluation and monitoring of policies organised as mutual learning processes (EU Council 2000, Hodson and Maher 2001, Scott and Trubek 2002, Bisopoulos 2003, Borrás and Jacobsson 2004, Laffan and Shaw 2005, Bruno et al. 2006).

---

with the rules of the European trade regime not only influenced the design of the financing mechanism, but also led to the implementation of tariff degression (see section 5.4.1).

27   In fact, coordination is a well-established governance mechanism in European politics. Since the Treaty of Rome (1957), member states have claimed to coordinate their national economic policies. In the 1970s coordination also started to encompass employment issues. Later, coordination was extended to other policies fields (Hodson and Maher 2001).

Almost all of these elements can be found in the first directive for renewable electricity of 2001 (EU 2001). As shown above, the directive included an (indicative) European target which was translated into (indicative) national renewable electricity consumption targets. As well, the long-term targets for 2010 were translated into shorter-term targets so that the European Commission could periodically evaluated which member states were 'on track' for reaching their national targets and which were not. Indicators were established by measuring the effectiveness and the efficiency of support mechanisms within the two assessment reports published in 2005 and 2009. In its 2005 report on the assessment of national policy instruments, the European Commission described this sort of policymaking as 'coordination', 'co-operation', and 'optimisation' (EU Commission 2005a).

What then have we implicitly done in this research project? While identifying the reasons for cross-national policy similarities in EU member states – initially we expected transnational communication, regulatory competition and independent problem solving to be the major drivers – we 'stumbled' across cases where, surprisingly, the European level influenced national policy decisions – even though the attempted harmonisation of all European support instruments failed twice. However, through our detailed qualitative analysis at a micro level – the analysis of more than 20 design options of just one policy instrument – we have unintentionally measured the success of the open method of coordination that was initiated by the EU in this policy field in 2001. In other words, all the circumstances where we have identified the influence of the European level are indications of the success of this steering mode, which is based on voluntarism, subsidiarity, flexibility and participation (Jacobsson 2001, Scott and Trubek 2002).

Our analysis also indicates possible ways of combining research on alternative governance modes in the EU, such as the open method of coordination, with the theory of policy convergence. The combination of policy convergence theory with Europeanisation studies is nothing new. In fact, Europeanisation in political science centres on the question of whether European member states are becoming more alike. Earlier research projects have already shown how European integration led to the convergence of national policies or institutions (see, for instance, Knill 2001). However, researchers analysing the open method of coordination have pointed out that describing this new community method is relatively easy, while actually measuring its effects is difficult because of its 'essentially cognitive nature' (Goetschy 2005: 64). However, the theory of policy convergence, that is, the mechanism of 'transnational communication', provides a framework for measuring such cognitive processes. This is not surprising, since both the empirically rooted open method of coordination and the concept of transnational communication focus on one and the same issue but approach it from different angles: voluntary mechanisms leading to cross-national policy convergence.

As shown in section 2.1.1, transnational communication covers a number of mechanisms leading to cross-national policy similarities, including lesson drawing from experience in other countries, emulation (the desire for conformity), and the promotion of certain policies by international organisations. In the case of all these sub-mechanisms, the link with the concept of the open method of coordination is more than obvious. In a nutshell, the open method of coordination is a policy-oriented tool trying to encourage the spread of certain policies and programs, whereas theorists of policy convergence have identified causal mechanisms leading to cross-national policy similarities (Nedergaard 2007). Since the scope of the present research project is limited, further research will be needed to explore the suitability of policy convergence theory for measuring the effect of the open method of coordination. However, further research in this area is also needed.

8.5. Conclusion: Strengthening European Policy Convergence

In this chapter we have summarised the causal mechanisms for cross-national policy similarities as they have been identified throughout this research project. We have depicted how communication, competition, technological innovations and European policy steering have been responsible for increasing similarities of national feed-in tariffs. However, our analysis has also shown that the potential for enabling cross-national policy similarities has so far not been fully tapped. There are still opportunities to strengthen cross-national policy communication, to foster regulatory competition between member states and to enhance the influence of the European level. These potential for strengthening European policy convergence in the future is presented in the following three sections.

8.5.1. Strengthening Informational Exchange

Even though transnational communication already exerted a strong influence on the convergence of feed-in tariffs in the three countries under analysis, there are also other ways to improve international information flows and the benefits of mutual learning among EU member states. First of all, it can be expected that the potential for communication will increase in coming years because the industry is expected to grow – especially in countries with currently relatively small markets. Our analysis of non-governmental actors has shown that larger industry associations in countries with large markets have more often been able to influence the national political decision-making process than have industry representatives in smaller-scale markets. This will also increase the knowledge of actors in those countries about more exotic design options, such as for system and market integration.

We have also revealed how transnational communication can be problematic in the case of rapidly developing technologies. The case of solar photovoltaics in France showed that between the analysis of tariff levels in other European countries that took place in 2008 and the actual implementation of the decree in January 2010, power generation costs decreased considerably. Because of the slow political decision-making process – from analysing the market to the actual issuing of the law – the new decree of January 2010 offered relatively high tariffs, not taking the technological advance of the previous two years into account.

Consequently, our comparison in section 7.1.4 shows that French tariff levels from early 2008 resembled the remuneration levels that were granted in Germany and Spain in 2008. This shows the need for up-to-date information.

The European institutions could provide such information to policymakers all over Europe. By facilitating information exchange in a more centralised way, the aforementioned problems could be partly overcome. Interestingly, the latest large-scale European research project (FUTURES-E), which was co-financed by the European Commission, focused on the optimisation and coordination of European support mechanisms. In order to increase policy learning from international best practice, the researchers recommended the establishment of 'a website with details of RES policy framework in all EU Member States, including experience on how the policy framework functions in practice' (Resch et al. 2009). Interestingly, in June 2011, the European Commission issued an invitation to tender for a research project called 'renewable energy policy database and support'. Based on this research study the Commission wants to establish 'a publically accessible (web-based) database of key national policy measures promoting renewable energy, including a detailed and exhaustive of the support provided for all forms of renewable energy ...' (EU Commission 2011: 3).

To go one step further, it would be equally useful for policymakers to share information about the technology-specific generation costs of renewable energy sources. Our analysis revealed the importance of a common tariff calculation methodology for tariff convergence. However, no evidence exists that the governments in the three countries have exchanged information on their tariff calculations methods. The central position of European institutions could be very useful in gathering and providing this industry-related information.

8.5.2. Strengthening Competition

Regulatory competition between legislators in EU member states can be further strengthened. In section 8.2 we showed that the impact of competition on tariff level convergence increases with the size of the international market. The more countries establish well-functioning support mechanisms for a given technology, the more options manufacturers have for selling their products. This way, the pressure on national legislators to establish tariff levels in line with technology-specific generation costs will increase. A wide range of markets already exists in the case of wind power. This reduces the negative effects of policy modifications in one country on market growth in other countries, as observed in the case of solar photovoltaics. As the world market for less mature renewable electricity technologies increases, the impact of regulatory competition can be expected to increase automatically.

However, our analysis and the interviews conducted with national actors in Germany, Spain, France and Belgium have also revealed barriers to regulatory competition. As set out in section 2.1.1 and 3.3.2, the influence of regulatory competition depends on the degree of economic interlinkage and integration. In the European context, the four freedoms – the free movement of people, goods, services and money – are decisive. However, these freedoms do not show full their effect in the case of renewable energy products. Although we analysed three neighbouring countries within a common European market, the price for renewable energy equipment can differ between them. This country-specific price differentiation for one and the same product has been identified in the case of wind energy and solar photovoltaics. As one anonymous interviewee pointed out, wind power plant manufacturers were asking for a different price for the same product in each of the countries where the wind park was to laid out. A large European utility, which wanted to install wind power plants both in the UK and on the European continent, was not able to purchase all wind power at a single price. Instead, it had to pay higher prices for the windmills that were planned to be installed in the UK, because the quota obligation mechanism in the UK allows for higher profit margins than feed-in tariffs on the European continent. Similarly, an anonymous interviewee active in the solar photovoltaic business confirmed that large-scale manufacturers differentiate their prices according to the country of instalment. The purchaser of solar photovoltaic modules has to provide information about the country where the modules will be installed, and the producers adjust the prices accordingly. This type of price differentiation undermines the effects of regulatory competition, which is based on the mobility of capital and the free movement of goods in international markets. If these price differentiations could be limited or even prohibited, the effects of regulatory competition on tariff level convergence would most likely increase.

8.5.3. Strengthening European Policy Coordination

One of the most important findings of this research project is that the European institutions have been able to exercise influence on national policy decisions even though two attempts to establish a harmonised European support instrument have failed. We have shown that, simply by establishing a common framework based on voluntary targets and other mechanisms in line with the open method of coordination, the European Commission was able to influence national policy decisions, thus strengthening European coordination. The open method of coordination was already applied by the European institutions in a variety of issue areas, including education and training, health care and immigration policies. Our research project has shown that it was also successfully applied in the case of European energy policy.

The influence of the European policy framework can be expected to increase in the future. As shown above (section 8.4.2), the European Commission started to apply the open method of coordination with directive 2001/77/EC. We have shown that that the directive included a number of mechanisms that were intended to foster policy coordination, including target setting, reporting and the threat of harmonising support instruments. However, no obligatory mechanisms were included in 2001. In fact, the implementation of binding national targets was debated in 2000 but rejected by the majority of member states (Rowlands 2005). With the new directive of 2009 – the European framework for 2010–20 – the European level gained further powers for steering cross-national policy similarities. In particular, the establishment of mandatory national targets and punishment for non-compliance through general infringement procedures will most likely increase the pressure on member states to learn from other countries on the basis of international best practice.

As well, the information exchange will be managed in a more centralised way by the European Commission. Like the first directive of 2001, the directive of 2009 obliged member states to submit frequent reports to the European Commission. Along with others, these reports will provide information about the shares of renewable energy sources in the different sectors (electricity, heating and cooling, and transport), a description of support mechanisms implemented and details of progress made towards eliminating non-economic barriers. These reports have to be submitted for the first time at the end of 2011 and thereafter every two years. The European Commission will also play a key role in monitoring the progress made by member states.

An important innovation of the new directive is that national member states have to draft national renewable energy action plans in order to demonstrate how the national targets will be reached within the different sectors (electricity, heating and cooling, and transport). The action plans have to be elaborated according to a detailed template that the European Commission published in 2009 (EU Commission 2009b). This template includes a section in which national governments will have to provide detailed information about the specific design of support schemes. The specific questions regarding feed-in tariffs reflect the design options as described in our research project, including eligibility, capacity caps, technology-specific support, duration of support, periodic revision, and premium feed-in tariff design (EU Commission 2009b). Even though the use to which the information will be put in the end is not fully clear, it could serve as a basis for future evaluations of best practice at European level.[28] With these action plans the European Commission will create a centralised pool gathering all information regarding design options for support instruments in the EU.

In sum, the degree of hierarchical steering of renewable energy support has constantly increased over recent decades – despite the failure of harmonisation. The EU has managed to increase its competences in renewable energy support policies over time. In the 1990s, Europe had no influence on national policy decisions regarding renewable energy support. With the first directive of 2001, the EU managed to influence national policy decisions via the open method of coordination. Lately, the directive of 2009 even includes binding elements (binding

---

28   The European Commission will have the opportunity to review the action plans. Director General Matthias Ruete of DG Energy and Transport claimed that the European Commission will be analysing 'whether each Member State is actually describing a credible pathway to its share of the target' (Ruete 2008a).

targets), which leads us to the conclusion that the European steering capacity will increase over the coming years. Finally, the new directive of 2009 offers the possibility of establishing joint support mechanisms and joint projects bilaterally or multilaterally. The implementation of these flexibility mechanisms shows that the European institutions have not given up on including more and more elements that could finally lead to better coordination or even (partial) harmonisation of support instruments.

Chapter 9

Conclusion

In this book we have analysed the evolution of feed-in tariffs in Germany, Spain and France in order to understand the reasons for policy convergence in this area between the three countries. We have shown that national policies in the EU are becoming more similar over time – even without the legally binding harmonisation of support instruments. We have tested and largely confirmed the state-of-the-art theory of policy convergence. Even though we applied a micro-level focus (analysing more than 20 design options of a single policy instrument) the qualitative research approach allowed us to identify three mechanisms – transnational communication, regulatory competition and independent problem solving – which operate on a voluntary basis.

However, with two of the causal mechanisms – regulatory competition and independent problem solving – it was necessary to extend the theory of policy convergence. In the case of regulatory competition this was because our analysis is one of the first to apply the theory of policy convergence to a policy instrument that creates positive (financial) incentives. Until now, the convergence theory elaborated by political scientists has been developed primarily from empirical observations of the operation of command-and-control policies that use negative incentives to prohibit certain actions. In contrast, feed-in tariffs use positive incentives in the form of a fixed tariff payment for each unit of renewable electricity: 'carrot' rather than 'stick'. The theoretical assumptions of the theory of regulatory competition were largely confirmed even though feed-in tariff mechanisms function in a manner that is diametrically opposed to the policies and programs that were analysed previously by political scientists. However, interestingly, regulatory competition in the case of feed-in tariffs has often led to improved framework conditions for renewable power producers, because such producers have been able to threaten to move their activities to other countries. Yet no general 'race to the top' could be observed with tariff levels. Financial restrictions and the need to design policy instruments in the most cost-effective way counterbalanced national policymakers' ambition to create investment conditions for renewable electricity producers that were more attractive than those in other countries.

In addition, in the case of independent problem solving it became clear that independent policy responses from national legislators have not primarily been problem-driven – as generally assumed by policy convergence theory – but rather opportunity-driven. This, too, is indirectly liked to the fact that our analysis focused on a policy instrument that creates certain financial incentives in order to spur technological learning. Feed-in tariff mechanisms have been successful in developing renewable energy technologies by making them less expensive and larger in size. Consequently, policymakers have been able to expand national support frameworks by including more technologies, extending capacity caps, and so on. As a result of these observations we suggest a more comprehensive term for describing this phenomenon in political science policy convergence theory, namely, 'independent but similar policy responses'. Within this category, we distinguish between opportunity-driven policy convergence arising from technological advances and problem-driven policy convergence arising from similar problem pressure. Since this research project has been one of the first to analyse the convergence of an incentive-based policy instrument, it should be viewed only as a first step towards the consolidation of policy convergence theory.

Our analysis has also shown that – despite the failure of harmonisation – the EU has not been a powerless observer of national policy decisions. In fact, the so-called open method of coordination has given the European Commission a potent tool for influencing national policymakers' design decisions, even without the implementation of legally binding measures. This research project is among the first to describe and analyse in detail how successful the EU has been in triggering cross-national policy coordination through the open method of coordination, which relies on voluntary steering mechanisms. Theoretically, this is interesting because critiques of this soft mode of governance have challenged its effectiveness in view of its non-binding nature and the lack of sanctions (Héritier 2002b, Chalmers and Lodge 2003). However, despite the lack of a binding legal framework at European level, the European institutions have managed to influence policy decisions and policy design at national level through indicative policy targets, benchmarking reports and the threat of harmonisation. Even though other research projects have already shown that the EU has managed to coordinate national policies through the open method of coordination, for instance in the case of European social policies (de la Porte and Pochet 2002, Zeitlin et al. 2005), the micro-level analysis of this research project has allowed us to differentiate the influence exercised at European level from that of other causal factors, such as competition and communication, and therefore strengthens our argument.

List of References

ADEME 2003. Les énergies et matières premières renouvelables en France. Situation et perspectives de développement dans le cadre de la lutte contre le changement climatique, March 2003. Paris: Agence de l'Environnement et de la Maîtrise de l'Energie. Available from http://www.industrie.gouv.fr/debat_ energie/site/pdf/enr-1.pdf [accessed 31 July 2010].

ADEME 2006. Le marché photovoltaïque en France, Etat des lieux, mise en perspective, rentabilité financière des systèmes, vision du développement de la filière, October 2006, Paris: Agence de l'Environnement et de la Maîtrise de l'Energie.

Advocate General Jacobs, F.G. 2000. Opinion of Advocate General Jacobs delivered on 26 October 2000. Case C-379-98, PreussenElektra AG vs. Schleswag AG, European Court of Justice. Available from http://eur-lex. europa.eu/LexUriserv/LexUriserv.do?uri=CELEX:61998C0379:EN:PDF [accessed 31 July 2010].

AEE 2004. *Wind energy in Spain – Overview 2004*. Madrid: Asociación Empresarial Eólica.

AEE 2006. *Eólica 2006 – Anuario del Sector: Análisis y datos*. Madrid: Asociación Empresarial Eólica.

AEE 2008a. *La retribución de la eólica descendió en 2007 a los niveles de los años 2003 y 2004*, press release, 10 January 2008. Madrid: Asociación Empresarial Eólica.

AEE 2008b. *Eólica 2008, Anuario del sector: Análisis y datos*, Madrid: Asociación Empresarial Eólica.

AEE 2009. *La eólica ha cubierto más del 50% de la demanda eléctrica*, press release, 9 November 2009, Madrid: Asociación Empresarial Eólica.

AEE 2010. *Eólica 2010*. Madrid: Asociación Empresarial Eólica.

Agentur für Erneuerbare Energien 2010. *Strompreise: Experten bringen Klarheit über Kostenfaktor Erneuerbare Energien*, press release, 25 February 2010. Berlin: Agentur für Erneuerbare Energien.

Agnolucci, P. 2008. 'Factors influencing the likelihood of regulatory changes in renewable electricity policies', *Renewable and Sustainable Energy Reviews*, 12:1, 141–61.

Alt, F, Claus, J., Scheer, H. (eds) 1998. *Windiger Protest, Konflikte um das Zukunftspotenzial der Windkraft*. Bochum: Ponte Press.

APPA 2007. *APPA valora la nueva regulación de las energías renovables*, press release, 7 June 2007. Madrid: Asociación de Productores de Energías Renovables.

APPA 2009. *Seis asociaciones empresariales critican duramente el Real Decreto-Ley 6/2009 y solicitan al Gobierno que retire del mismo el Artículo 4 y las referencias negativas a las energías renovables*, press release, 17 May 2009. Madrid: Asociación de Productores de Energías Renovables.

APPA 2010. *APPA exige al Gobierno estabilidad regulatoria y seguridad juridical*, press release, 17 June 2010. Madrid: Asociación de Productores de Energías Renovables.

APPA and Greenpeace 2009. *Anteproyecto de Ley para el Fomento de las Energías Renovables*, Madrid: Asociación de Productores de Energías Renovables. Available from http://www.appa.es/descargas/LeyEnergiasRenovables_v11B. pdf [accessed 31 July 2010].

Arentsen, M., Bechberger, M., Di Nucci, M.-R. 2007a. *Realise-Forum, Final Report*, CSTM Studies and Reports (No. 318), University of Twente, Enschede, March 2007.

Arrow, K. 1962. 'The economic implications of learning by doing', *Review of Economic Studies*, 29:3, 155–73.

ASIF 2002. *Desarrollo de la electricidad solar photovoltaica en España – Hacia un sistema estructurado de generación limpia*, Informe ASIF, October 2002. Madrid: Asociación de la Industria Fotovoltaica.

ASIF 2007. *El sector fotovoltaico quiere crecer un 20% anual e igualar su coste de generación con el precio doméstico de la luz antes de 2020*, press release, 27 November 2007. Madrid: Asociación de la Industria Fotovoltaica.

ASIF 2008. *La Tarifa Fotovoltaica Flexible reduciría la retribución entre un 2% y un 10% cada año*, press release, 13 March 2008. Madrid: Asociación de la Industria Fotovoltaica.

ASIF 2009. *ASIF propone modificar la regulación para que los consumidores puedan producir y consumir su propia electricidad fotovoltaica*, press release, 4 November 2009. Madrid: Asociación de la Industria Fotovoltaica.

ASIF and APPA 2007. *El sector fotovoltaico quiere crecer un 20% anual e igualar su coste de generación con el precio doméstico de la luz antes de 2020*, press release, 27 November 2007. Madrid: Asociación de la Industria Fotovoltaica.

Assemblée Nationale 2009. *Rapport d'information déposé en application de l'article 145 du règlement par la commission des affaires économiques sur l'énergie photovoltaïque*, Rapporteur: Serge Poignant, 16 July 2009. Available from http://www.assemblee-nationale.fr/13/rap-info/i1846.asp [accessed 31 July 2010].

Assemblée Nationale 2010. Projet de loi portant engagement national pour l'environnement, texte définitif, 29 June 2010, Paris. Available from http://www.assemblee-nationale.fr/13/dossiers/engagement_environnement.asp [accessed 1 August 2010].

Auer, H., Obersteiner, C., Prüggler, C., Weissensteiner, L., Faber, T., Resch, G. 2007. *Action Plan – Guiding a least cost grid integration of RES-electricity in an extended Europe, GreenNet-Europe*, May 2007. Available from http://

greennet.i-generation.at/files/Action%20Plan%20GreenNet-Europe.pdf [accessed 31 July 2010].

Battle, C. and Pérez-Arriaga, I. 2007. 'Design criteria for implementing a capacity mechanism in deregulated electricity markets', *Utilities Policy*, 16:3, 184–93.

Baur, J.F., Blask, H. 2002. 'Regelungszuständigkeiten der EG im Bereich Energie', *Energiewirtschaftliche Tagesfragen*, 9/2002, 636–41.

BDEW 2009a. *BDEW zum Kabinettsbeschluss Erneuerbare-Energien-Gesetz (EEG): Verordnung sorgt für mehr Effizienz und Transparenz*, press release, 27 May.2009. Berlin: Bundesverband der Energie- und Wasserwirtschaft.

BDEW 2009b. *BDEW zum Ausgleichsmechanismus im Erneuerbare-Energien-Gesetz: Mehr Durchblick für Stromkunden*, press release, 2 June 2009. Berlin: Bundesverband der Energie- und Wasserwirtschaft.

BDEW 2009c. *Informationen zur Direktvermarktung nach § 17 EEG, Letzte Aktualisierung: 10.08.2009*, Berlin: Bundesverband der Energie- und Wasserwirtschaft.

BDEW 2009d. *BDEW zur heutigen Bekanntgabe der EEG-Umlage 2010: Kosten für Strom aus erneuerbaren Energien steigen*, press release, 15 October 2009. Berlin: Bundesverband der Energie- und Wasserwirtschaft.

BDEW 2010. *Informationen zur Direktvermarktung nach § 17 EEG, Letzte Aktualisierung: 23 March 2010*, Berlin: Bundesverband der Energie- und Wasserwirtschaft.

Bechberger, M. 2000. *Das Erneuerbare-Energien-Gesetz (EEG): Eine Analyse des Politikformulierungsprozesses*. FFU Report 06/2000, Freie Universität Berlin.

Bechberger, M. 2009. *Erneuerbare Energien in Spanien, Erfolgsbedingungen und Restiktionen*. Stuttgart: ibidem-Verlag.

Bechberger, M., Körner, S., Reiche, D. 2003. *Erfolgsbedingungen von Instrumenten zur Förderung erneuerbarer Energien im Strommarkt*, FFU-Report 01-2003, Freie Universität Berlin.

Bechberger, M. and Reiche, D. 2004. 'Renewable energy policy in Germany: Pioneering and examplary regulation', *Energy for Sustainable Development*, 8:1, 47–57.

Bechberger, M. and Reiche, D. 2007. 'The spread of renewable energy feed-in tariffs (REFITs) in the EU-25', in Mez (ed.), pp. 31–50.

Bechberger, M. and Reiche, D. (eds) 2006. Ökologische Transformation der Energiewirtschaft – Erfolgsbedingungen und Restriktionen. Berlin: Erich-Schmidt Verlag.

Bechberger, M., Mez, L., Sohre, Λ. (eds) 2008. *Windenergie im Ländervergleich. Steuerungsimpulse, Akteure und technische Entwicklungen in Deutschland, Dänemark, Spanien und Großbritannien*. Frankfurt: Peter Lang.

Beck, F. and Martinot, E. 2004. 'Renewable energy policies and barriers', *Encyclopedia of Energy*, 5/2004, 365–83.

BEE 2008a. *'EU FIT'– Europaweites Mindestpreismodell zur Stromerzeugung aus Erneuerbaren Energien*, January 2008. Berlin: Bundesverband Erneuerbare Energien.

BEE 2008b. *Vorschlag zur Ergänzung des §29 EEG – Kabinettsentwurf um einen 'Integrationsbonus' mit dem Ziel einer verbesserten Netz- bzw. Systemintegration Erneuerbarer Energien ohne volkswirtschaftliche Mehrkosten*, position paper, 8 May 2008. Berlin: Bundesverband Erneuerbare Energien.

BEE 2010a. *BEE kritisiert interessengeleitete Debatte um vermeintliche Kostensteigerung durch Erneuerbare Energien*, press release, 22 March 2010. Berlin: Bundesverband Erneuerbare Energien.

BEE 2010b. *Erneuerbare Energien Jahreszahlen 2009*, Stand: 18 February 2010. Berlin: Bundesverband Erneuerbare Energien.

Bennett, C. 1988. 'Different processes, one result. The convergence of data protection policy in Europe and the United States', *Governance*, 1:4: 162–83.

Bennett, C. 1991. 'Review article: What is policy convergence and what causes it?', *British Journal of Political Science*, 21:2, 215–33.

Berry, T. and Jaccard, M. 2000. 'The renewable portfolio standard: design considerations and an implementation survey', *Energy Policy*, 29:4, 263–77.

BGBl 1990. *Gesetz über die Einspeisung vom Strom aus erneuerbaren Energien in das öffentliche Netz (Stromeinspeisegesetz)*, Bundesgesetzblatt I, 7 December 1990, p. 2633.

BGBl 1994. *Gesetz zur Sicherung des Einsatzes von Steinkohle in der Verstromung und zur Änderung des Atomgesetzes und des Stromeinspeisegesetzes*, Bundesgesetzblatt I, 28 July 1994, p. 1618.

BGBl 1998. *Gesetz zur Änderung des Energiewirtschaftsrechts*, Bundesgesetzblatt I, 28 April 1998, p. 730

BGBl 2000. *Gesetz für den Vorrang Erneuerbarer Energien*, Bundesgesetzblatt I, 31 March 2000, p. 305.

BGBl 2001. *Verordnung über die Erzeugung von Strom aus Biomasse (Biomasseverordnung – BiomasseV)*, Bundesgesetzblatt I, 27 June 2001, p. 1234.

BGBl 2003. *Zweites Gesetz zur Änderung des Erneuerbare-Energien-Gesetzes*, Bundesgesetzblatt I, 22 December 2003, p. 3074.

BGB 2004. *Gesetz zur Neuregelung des Rechts der erneuerbaren Energien im Strombereich*, Bundesgesetzblatt I, 31 July 2004, p. 1918.

BGB 2005. *Erste Verordnung zur Änderung der Biomasseverordnung*, Bundesgesetzblatt I, 17 August 2005, p. 2419.

BGBl 2006. *Erstes Gesetz zur Änderung des Erneuerbare-Energien-Gesetzes*, Bundesgesetzblatt I, 15 November 2006, p. 2550.

BGBl 2008. *Gesetz zur Neuregelung des Rechts der erneuerbaren Energien im Strombereich und zur Änderung damit zusammenhängender Vorschriften*, Bundesgesetzblatt I, 31 October 2008, p. 2074.

BGBl 2009a. *Verordnung zur Weiterentwicklung des bundesweiten Ausgleichsmechanismus (AusglMechV)*, Bundesgesetzblatt I, 24 July 2009, p. 2101.

BGBl 2009b. *Verordnung zu Systemdienstleistungen durch Windenergieanlagen (Systemdienstleistungsverordnung – SDLWindV)*, Bundesgesetzblatt I, 10 July 2009, p. 1734.

BGBl 2010. *Erstes Gesetz zur Änderung des Erneuerbare-Energien-Gesetzes vom 11 August 2010*, Bundesgesetzblatt I, 17 August 2010, p. 1170.

Biermann, F., Brohm, R., Dingwerth, K. (eds). 2001. *Berlin Conference on the Human Dimensions of Global Change – Global environmental change and the nation state*, Potsdam: Potsdam Institute for Climate Impact Research.

Bird, L., Wüstenhagen, R., Aabakken, J. 2002. 'A review of international green power markets: recent experience, trends, and market drivers', *Renewable and Sustainable Energy Review*, 6:6, 513–36.

Birr-Pedersen, K. 2009. *Podium: Förderung alternativer Energien – teuer oder innovativ?*, presentation from Katja Birr-Pedersen (DONG Energy) at the Humboldt Forum für Wirtschaft, 10 June 2009, Berlin.

Bisopoulos, A. 2003. 'Die Europäische Beschäftigungsstrategie: Innovative Steuerung durch Peer Review', in Holzinger, Knill and Lehmkuhl (eds), pp. 151–65.

Bleuel, M. and Hillebrand, B. 2005. *Das Integrationsmodell für Erneuerbare Energiequellen – vom Staat zum Markt* Executive Summary, Energy Environment Forecast Analysis GmbH, Untersuchung im Auftrag des Verbandes der Elektrizitätswirtschaft – VDEW, Münster, October 2005.

Blok, K. 2006. 'Renewable energy policies in the European Union', *Energy Policy*, 34:3, 251–5.

BMU 2001. EU-Richtlinie zur Förderung der Erneuerbaren Energien ist in Kraft Getreten (Stand: Dezember 2001). Berlin: Federal Ministry for the Environment, Nature Conservation and Nuclear Safety.

BMU 2003. Aktualisierung des Berichtes über den Stand der Markteinführung und der Kostenentwicklung von Anlagen zur Erzeugung von Strom aus erneuerbaren Energien (Erfahrungsbericht zum EEG), Ausschussdrucksache 15(15)121, Berlin, June 2003. Berlin: Federal Ministry for the Environment, Nature Conservation and Nuclear Safety.

BMU 2007a. *Gabriel entlastet besonders stromintensive Unternehmen*, Press Release No. 358/07, 21 December 2007. Berlin: Federal Ministry for the Environment, Nature Conservation and Nuclear Safety.

BMU 2007b. *Erfahrungsbericht 2007 zum Erneubaren-Energien-Gesetz (EEG) gemäß § 20 EEG*, 7 November 2007. Berlin: Federal Ministry for the Environment, Nature Conservation and Nuclear Safety.

BMU 2007c. Entwurf eines Gesetzes zur Neuregelung des Rechts der Erneuerbaren Energien im Strombereich, Stand: 9 October 2007. Berlin: Federal Ministry for the Environment, Nature Conservation and Nuclear Safety.

BMU 2007d. *Entwurf der Begrüdung zum Erneuerbare-Energien-Gesetz*, Stand: 10.10.2007. Berlin: Federal Ministry for the Environment, Nature Conservation and Nuclear Safety.

BMU 2008a. *Depiction of the methodological approaches to calculate the costs of electricity generation used in the scientific background reports serving as the basis for the Renewable Energy Source Act (EEG) Progress Report 2007*, Extract from Renewable Energy Source Act (EEG), Progress Report 2007, Chapter 15.1. Berlin: Federal Ministry for the Environment, Nature Conservation and Nuclear Safety.

BMU 2008b. *Verbesserung der Systemintegration der Erneuerbaren Energien im Strombereich, Handlungsoptionen für einen Modernisierung des Energiesystems*, 9 May 2008. Berlin: Federal Ministry for the Environment, Nature Conservation and Nuclear Safety.

BMU 2009a. *Verordnung zu Systemdienstleistungen durch Windenergieanlagen (Systemdienstleistungsverordnung SDLWindV)*, draft version (KI III 4), Stand: 2 März 2009. Berlin: Federal Ministry for the Environment, Nature Conservation and Nuclear Safety.

BMU 2009b. *Bericht der Bundesregierung über ein Kozept zur Förderung, Entwicklung und Markteinführung von geothermischer Stromerzeugung und Wärmenutzung*. Berlin: Federal Ministry for the Environment, Nature Conservation and Nuclear Safety.

BMU 2009c. *Gabriel: Wichtiger Schritt zur Modernisierung des Stromnetzes, Bundestag beschließt Energieleitungsausbaugesetz*, Press realease, No. 129/09, 7 May 2009. Berlin: Federal Ministry for the Environment, Nature Conservation and Nuclear Safety.

BMU 2010a. *Stabile Rahmenbedingungen für einen dynamischen Ausbau der Photovoltaik, Stellungnahme des Bundesumweltministeriums*, Stand: 24.02.2010. Berlin: Federal Ministry for the Environment, Nature Conservation and Nuclear Safety.

BMU 2010b. *Entwicklung der erneuerbaren Energien in Deutschland im Jahr 2009*, Stand: 18 März 2010. Berlin: Federal Ministry for the Environment, Nature Conservation and Nuclear Safety.

BMU 2010c. *Informationen zur Anwendung von § 40 ff. EEG (Besondere Ausgleichsregelung) für das Jahr 2010*, Stand: 3.5.2010. Berlin: Federal Ministry for the Environment, Nature Conservation and Nuclear Safety.

BMU 2010d. *Röttgen welcomes Mediation Committee conclusions on support for solar energy*, press release No. 104/10, 6 July 2010. Berlin: Federal Ministry for the Environment, Nature Conservation and Nuclear Safety.

BMU 2010e. *Einfluss der Förderung erneuerbarer Energien auf den Haushaltsstrompreis in den Jahren 2009 und 2010 – einschl. erster Ausblick auf das Jahr 2011*, 13 July 2010. Berlin: Federal Ministry for the Environment, Nature Conservation and Nuclear Safety.

BMU 2011. Gesetz für den Vorrang Erneuerbarer Energien, Konsolidierte (unverbindliche) Fassung des Gesetzestextes in der ab 1. Januar 2012 geltenden Fassung, Federal Ministry for the Environment, July 2011.

Bode, S. 2006. *On the impact of renewable energy support schemes on power prices*, HWWI Research Paper (ISSN 1861-504X), September 2006.

Bode, S. and Groscurth, H.M. 2008a. 'Liberalisierter Strommarkt: Naht das Ende?', *Wirtschaftsdienst*, 89:4, 274–280.

Bode, S. and Groscurth, H.M. 2008b. *Incentives to invest in electricity production from renewable energy under different support schemes*, Arrhenius, Institute for energy and climate policy, discussion paper 1E, Hamburg, March 2008.

BOE 1981a. *Real Decreto 1217/1981, de 10 de abril, para el fomento de la producción hidroeléctrica en pequeñas centrales*, BOE No. 150, 28 June 1981, p. 14435.

BOE 1981b. *Ley 82/1980, de 30 de diciembre, sobre conservación de la Energía*, BOE No. 23, 27 January 1981, p. 1863.

BOE 1994a. *Real Decreto 2366/1994, de 9 de diciembre, sobre Produccion de energía eléctrica por instalaciones hidráulicas, de cogeneración y otras abastecidas por recursos o fuentes de energía renovables*, BOE No. 313, 31 December 1994, p. 39595.

BOE 1994b. *Ley 40/1994, de 30 de diciembre, de ordenación del sistema eléctrico nacional*, BOE No. 313, 31 December 1994, p. 39362.

BOE 1995. *Real Decreto 2204/1995, de 28 de diciembre, por el que se establece la tarifa eléctrica para 1996*, BOE No. 311, 29 December 1995, p. 37452.

BOE 1996. *Real Decreto 2657/1996, de 27 de diciembre, por el que se establece la tarifa eléctrica para 1997*, BOE No. 313, 28 December 1996, p. 38658.

BOE 1997a. *Real Decreto 2016/1997, de 26 de diciembre, por el que se establece la tarifa eléctrica para 1998*, BOE No 310, 27 December 1997, p. 38031.

BOE 1997b. *Ley 54/1997, de 27 de noviembre, del sector eléctrico*, BOE No. 285, 28 November 1997, p. 35097.

BOE 1997c. *Real decreoto 2017/1997, de 26 de diciembre, por el que se organiza y regula el procedimiento de liquidación de los costes de transporte, distribución y comercialización a tarifa, de los costes permanentes del sistema y de los costes de diversificación y seguridad de abastecimiento*, BOE No. 310, 27 December 1997, p. 28037.

BOE 1998a. *Real Decreto 2818/1998, de 23 de diciembre, sobre producción de energía eléctrica por instalaciones abastecidas por recursos o fuentes de energía renovables, residuos y cogeneración*, BOE No. 313, 30 December 1998, p. 44077.

BOE 1998b. *Real Decreto 2821/1998, de 23 de diciembre, por el que se establece la tarifa eléctrica para 1999*, BOE No. 312, 30 December 1998, p. 44106.

BOE 1999. *Real Decreto 2066/1999, de 30 de diciembre, por el que se establece la tarifa eléctrica para 2000*, BOE No. 313, 31 December 1999, p. 46482.

BOE 2000. *Real Decreto 3490/2000, de 29 de diciembre, por el que se establece la tarifa eléctrica para 2001*, BOE No. 313, 30 December 2000, p. 46764.

BOE 2001. *Real Decreto 1483/2001, de 27 de diciembre, por el que se establece la tarifa eléctrica para el 2002*, BOE No. 311, 28 December 2001, p. 50044.

BOE 2002a. *Real Decreto 841/2002, de 2 de agosto, por el que se regula para las instalaciones de producción de energía eléctrica en régimen especial su incentivación en la participación en el mercado de producción, determinadas obligaciones de información de sus previsiones de producción, y la adquisición por los comercializadores de su energía eléctrica producida*, BOE No. 210, 2 September 2002, p. 31968.

BOE 2002b. *Real Decreto 1436/2002, de 27 de diciembre, por el que se establece la tarifa eléctrica para 2003*, BOE No. 313, 31 December 2002, p. 46389.

BOE 2003. *Real Decreto 1802/2003, de 26 de diciembre, por el que se establece la tarifa eléctrica para 2004*, BOE No. 310, 27 December 2003, p. 46239.

BOE 2004a. *Real Decreto 436/2004, de 12 de marzo, por el que se establece la metodología para la actualización y sistematización del régimen jurídico y económico de la actividad de producción de energía eléctrica en régimen especial*, BOE No. 75, 27 March 2004, p. 13217.

BOE 2004b. *Real Decreto 2392/2004, de 30 de diciembre, por el que se establece la tarifa eléctrica para 2005*, BOE No. 315, 31 December 2004, p. 42766.

BOE 2004c. *Real Decreto 2351/2004, de 23 de diciembre, por el que se modifica el procedimiento de resolución de restricciones técnicas y otras normas reglamentarias del mercado eléctrico*, BOE No. 309, 24 December 2004, p. 41686.

BOE 2005a. *Ley 24/2005, de 18 de noviembre, de reformas para el impulso a la productividad*, BOE No. 277, 19 November 2005, p. 37846.

BOE 2005b. *Real Decreto 1556/2005, de 23 de diciembre, por el que se establece la tarifa eléctrica para 2006*, BOE No. 310, 28 December 2005, p. 42576.

BOE 2006a. *Real Decreto-Ley 7/2006 de 23 de junio, por el que se adoptan medidas urgentes en el sector energético*, BOE No. 150, 24 June 2006, p. 23979.

BOE 2006b. *Real Decreto 1634/2006, de 29 de diciembre, por el que se establece la tarifa eléctrica a partir de 1 de enero de 2007*, BOE No. 321, 30 December 2006, p. 46656.

BOE 2006c. *Resolución de 24 de mayo de 2006, de la Secretaria General de Energía, por la que se aprueban diversos procedimientos de operación para su adaptación a la nueva normativa eléctrica (P.O. 14.4: Derechos de cobro y obligaciones de pago por los servicios de ajuste del sistema)*, BOE No. 128, 30 May 2006, p. 20399.

BOE 2007a. *Real Decreto 1028/2007, de 20 de julio, por el que se establece el procedimiento administrativo para la tramitación de las solicitudes de autorización de instalaciones de generación eléctrica en el mar territorial*, BOE No. 183, 1 August 2007, p. 33171.

BOE 2007b. *Resolución de 27 de septiembre de 2007, de la Secretaría General de Energía, por la que se establece el plazo de mantenimiento de la tarifa regulada para la tecnología fotovoltaica, en virtud de lo establecido en el artículo 22 del Real Decreto 661/2007, de 25 de mayo*, BOE No. 234, 29 September 2007, p. 39745.

BOE 2007c. *Real Decreto 661/2007, de 25 de mayo, por el que se regula la actividad de producción de energía eléctrica en régimen especial*, BOE No. 126, 26 May 2007, p. 22846.

BOE 2007d. *Correción de errores del Real Decreto 661/2007, de 25 de mayo, por el que se regula la actividad de producción de energía eléctrica en régimen especial*, BOE No. 177, 25 July 2007, p. 32248.

BOE 2008a. *Real Decreto 1578/2008, de 26 de septiembre, de retribución de la actividad de producción de energía eléctrica mediante tecnología solar fotovoltaica para instalaciones posteriores a la fecha límite de mantenimiento de la retribución del Real Decreto 661/2007, de 25 de mayo, para dicha tecnología*, BOE No. 234, 27 November 2008, p. 39117.

BOE 2008b. *Real Decreto 222/2008, de 15 de febrero, por el que se establece el régimen retributivo de la actividad de distribución de energía eléctrica*, BOE No. 67, 18 March 2008, p. 16067.

BOE 2009. *Real Decreto-ley 6/2009, de 30 de abril, por el que se adoptan determinadas medidas en el sector energético y se aprueba el bono social*, BOE No. 111, 7 May 2009, p. 39404.

BOE 2010a. *Real Decreto 437/2010, de 9 de abril, por el que se desarrolla la regulación del proceso de titulización del déficit del sistema eléctrico*, BOE No. 96, 21 April 2010, p. 34930.

BOE 2010b. *Real Decreto 1565/2010, de 19 de noviembre, por el que se regulan y modifican determinados aspectos relativos a la actividad de producción de energía eléctrica en régimen especial*, BOE No. 283, 23 November 2010, p. 97428.

BOE 2010c. *Real Decreto 1614/2010, de 7 de diciembre, por el que se regulan y modifican determinados aspectos relativos a la actividad de producción de energía eléctrica a partir de tecnologías solar termoeléctrica y eólica*, BOE 298, 8 December 2010, p. 101852.

BOE 2010d. *Real Decreto-ley 14/2010, de 23 de diciembre, por el que se establecen medidas urgentes para la corrección del déficit tarifario del sector electrico*, BOE 312, 23 December 2010, p. 106386.

Borrás, S. and Jacobsson, K. 2004. 'The open method of co-ordination and new governance patterns in the EU', *Journal of European Public Policy*, 11:2, 185–208.

Börzel, T. 1999. 'Towards Convergence in Europe? Institutional Adaptation to Europeanization in Germany and Spain', *Journal of Common Market Studies*, 39:4, 573–596.

Boston Consulting 2004. *Donner un nouveau soufflé à l'éolien terrestre – Développement de l'éolien terrestre en France*, June 2004. Paris: Boston Consulting.

Botcheva, L. and Martin, L. 2001. 'Institutional effects or state behaviour: convergence or divergence', *International Studies Quarterly*, 45:1, 1–26.

Brand-Schock, R. 2010. *Vergleichende Analyse der Policy-Netzwerke in den Politikfeldern Strom aus erneuerbaren Energien und Biokraftstoffe in Deutschland und Frankreich*, Dissertation, FFU, Freie Universität Berlin.

Brand, E. and Witthohn, A. 2002. 'Die energiepolitischen Handlungsmöglichkeiten der EG', *Energiewirtschaftliche Tagesfragen*, 4/2002, 253–7.

Bräuer, W. 2002. 'Ordnungspolitischer Vergleich von Instrumenten zur Förderung erneuerbarer Energien im deutschen Stromsektor', *Zeitschrift für Umweltpolitik*, 1/2002, 61–103.

Braun, M. 2008. *Reactive Power Supplied by Wind Energy Converters – Cost-Benefit-Analysis*, paper presented at the EWEC 2008 conference, 3 April 2008, Brussels.

Braun, D., Gilardi, F., Füglister, K., Luyet, S. 2007. 'Ex Pluribus Unum: Integrating different strands of policy diffusion theory', in Holzinger, Jörgens and Knill (eds), pp. 39–55.

Braun, D. and Gilardi, F. 2006. 'Taking "Galton's Problem" Seriously. Towards a Theory of Policy Diffusion', *Journal of Theoretical Politics*, 18:3, 298–322.

Braun, D. and Giraud, O. 2003. 'Steuerungsinstrumente', in Schubert and Bandelow (eds), pp. 147–74.

Bröer, G. 2007. 'EEG-Strom für die Börse?', *Solarthemen 268*, 8–9.

Bröer, G. 2009. 'Klarheit beim PV-Eigenverbrauch', *Solarthemen 301*, 1–2.

Brown, M., Omom, D., Madden, B. 2009. *Qualitative issues in the designing of GB feed-in tariffs, A report to the Department of Energy and Climate Change (DECC)*. Element Energy and Pöyry Energy Consulting, June 2009.

Bruno, I., Jacquot, S., Mandin, L. 2006. 'Europeanization through its instrumentation: benchmarking, mainstreaming and the open method of co-ordination … toolbox or Pandora's box?', *Journal of European Public Policy*, 13:4, 519–36.

BSW 2008. *Tausende Solar-Jobs in Handwerk und Mittelstand in Gefahr*, press release, 5 March 2008. Berlin: Bundesverband Solarwirtschaft.

BSW 2010. *Statistische Zahlen der deutschen Solarstrombranche (Photovoltaik)*, June 2010. Berlin: Bundesverband Solarwirtschaft.

Bundesnetzagentur 2007. *Eckpunkte für die Direktvermarktung von EEG-Strom, BK6-07-003*, Bonn: Bundesnetzagentur.

Bundesnetzagentur 2009. *Bundesnetzagentur veröffentlicht Vergütungssätze für Photovoltaikanlagen*, press release, 30 October 2009. Bonn: Bundesnetzagentur.

Bürer, M. and Wüstenhagen, R. 2009. 'Which renewable energy policy is a venture capitalist's best friend? Empirical evidence from a survey of international cleantech investors', *Energy Policy*, 37:12, 4997–5006.

Busch, P.-O. 2003. *Die Diffusion von Einspeisevergütungen und Quotenmodellen: Konkurrenz der Modelle in Europa*. FFU-report, 03-2003, Freie Universität Berlin.

Busch, P.-O. and Jörgens, H. 2005a. 'The international sources of policy convergence – explaining the spread of environmental policy innovations', *Journal of European Public Policy*, 12:5, 860–84.

Busch, P.-O. and Jörgens, H. 2005b. 'Globale Ausbreitungsmuster umweltpolitischer Institutionen', in Tews and Jänicke (eds), pp. 55–194.

Busch, P.-O. and Jörgens, H. 2005c. 'International patterns of environmental policy change and convergence', *European Environment*, 15:2, 80–101.

Busch, P.-O. and Jörgens, H. 2007. 'Dezentrale Politikkoordinierung im internationalen System – Ursachen, Mechanismen und Wirkungen der internationalen Diffusion politischern Innovationen', in Holzinger, Jörgens and Knill (eds), pp. 56–84.

Busch, P.-O. and Jörgens, H. 2009. *Governance by diffusion. International environmental policy coordination in the era of globalization*, Dissertation, Environmental Policy Reseach Centre (FFU), Freie Universität Berlin.

Busch, P.-O., Jörgens, H., Tews, K. 2005. 'The global diffusion of regulatory instruments: The making of a new international environmental regime', *The Annals of the American Academy of Political and Social Science*, 589:1, 146–67.

Büsgen, U. and Dürrschmidt, W. 2009. 'The expansion of electricity generation from renewable energies in Germany. A review based on the Renewable Energy Sources Act Progress Report 2007 and the new German feed-in legislation', *Energy Policy*, 37:7, 2536–45.

Business Europe 2007. *Promote a secure, competitive and climate-friendly energy system*, Briefing, November 2007. Brussels: Business Europe.

Bustos, M. 2005. *Los sistemas de apoyo a la electricidad renovable en la Unión Europea – La visión de los productores españoles*, Documentos APPA 1/05. Madrid: Asociación de Productores de Energías Renovables.

Butler, L. and Neuhoff, K. 2008. 'Comparison of feed-in tariff, quota and auction mechanisms to support wind power development', *Energy Policy*, 33:8, 1854–67.

BWE 2007. *Stellungnahme des Bundesverband WindEnergie (BWE) zum Antrag des Landes Schleswig-Holstein im Bundesrat vom 9. Oktober 2007 (Drucksache 702/07), Entschließung des Bundesrates zur Änderung des § 10 des Gesetzes für den Vorrang Erneuerbarer Energien (Erneuerbare-Energien-Gesetz – EEG)*, position paper, 19 October 2007. Berlin: Bundesverband Wind Energie.

BWE 2008. *Stellungnahme des Bundesverbands WindEnergie (BWE) zur Vergütung für Windenergieanlagen an Land im Regierungsentwurf des Erneuerbare-Energien-Gesetzes (EEG)*, March 2008. Berlin: Bundesverband Wind Energie.

Calzada Álvarez, G., Merino Jara, D., Rallo Julián, J., García Bielsa, J. 2009. *Study of the effects on employment of public aid to renewable energy sources*, March 2009. Universidad Rey Juan Carlos, Madrid.

Campoccia, A., Dusonchet, L., Telaretti, E., Zizzo, G. 2009. 'Comparative analysis of different supporting measures for the production of electrical energy by solar PV and Wind systems: Four representative European cases', *Solar Energy* 83:3, 287–97.

CDU/CSU 2009. *Pfeiffer/Flachsbarth: SPD gegen Weiterentwicklung Erneuerbarer Energien*, press release, CDU/CSU Bundestagsfraktion, 7 May 2009, Berlin.

Ceña, A. 2005. 'Grid Penetration. Study Looks at Stability and Forecasting', *Wind Directions (EWEA)*, July/August 2005, 9–10.

Chabot, B. 2008. *Why and how to protect wind feed-in tariffs systems from potential negative effects of inflation?*, presentation at EWEC 2008 conference, Brussels, April 2008.

Chabot, B, Kellet, P., Saulnier, B. 2002. *Defining advanced wind energy tariffs systems to specific locations and applications: lessons from the French tariff system and examples*, paper for the Global Wind Power Conference, April 2002, Paris.

Chalmers, D. and Lodge, M. 2003. *The Open Method of Co-ordination and the European Welfare State*, Discussion Paper No. 11, June 2003, ESRC Centre for Analysis of Risk and Regulation, London School of Economics and Political Science.

CLER 2008. *Coût des énergies renouvelables sur la facture électrique*, Info, 15 April 2008. Paris: Comité de Liaison Energies Renouvelables.

CLER and Hespul 2010. *La 'bombe photovoltaïque': un alibi bancal pour une politique boiteuse*, press release, 14 January 2010. Paris: Comité de Liaison Energies Renouvelables.

Clos, J. and Gabriel, S. 2008. *Letter to Commissioner Piebalgs on behalf of the International Feed-in Cooperation*, 10 January 2008. Ministerio de Industria, Turismo y Comercio, Madrid.

CNE 2004. *Informe 4/2004 sobre la propuesta de Real Decreto por el que se establece la metodología para actualiziación y sistemantización del régimen jurídico y económico de la acticidad de producción de energía eléctrica en réginmen especial*, 22 January 2004. Madrid: Consejo Nacional de Energía.

CNE 2007a. *Vota particular que formulan la Presidenta María Teresa Costa Campi y los Consejeros Sebastià Ruscallada i Gallart, Jaime González Gonzáles y Jorge Fabra Utray al acuerdo del Consejo de Administración de la CNE de fecha 14 de febrero de 2007 por el que se aprueba el informe a la Propuesta de Real Decreto por el que se regula la actividad de producción de Energía Eléctrica de Régimen Especial y de determinadas instalaciones de tecnologías asimibles del Régimen Ordinario, Nota interior, voto particular, Ref.: VP0307*, 8 March 2007. Madrid: Consejo Nacional de Energía.

CNE 2007b. *Informe 3/2007 de la CNE relativo a la propuesta de Real Decreto por el que se regula la actividad de producción de energía eléctrica en el régimen especial y de determinadas instalaciones de tecnología asimilables del régimen ordinario*, 14 February 2007. Madrid: Consejo Nacional de Energía.

CNE 2011. *Información estadística sobre las ventas de energía del* régimen especial, Dirección de Energía Eléctrica, February 2011, available at http://www.cne.es/.

Cochet, Y. 2000. Stratégie et moyens de développement de l›efficacité énergétique et des sources d›énergie renouvelables en France: Rapport au Premier ministre, Paris. Available from http://lesrapports.ladocumentationfrancaise.fr/BRP/014000086/0000.pdf [accessed 3 November 2007].

Coenraads, R., Reece, G., Kleßmann, C., Ragwitz, M., Held, A., Resch, G., Panzer, C., Konstantinaviciute, I., Chadim, T. 2008a. *Renewable energy country profiles*, final version (Progress), February 2008. Utrecht (Netherlands).

Coenraads, R., Reece, G., Voogt, M., Ragwitz, M., Resch, G., Faber, T., Haas, R., Konstantinaviciute, I., Krivosik, J., Chadim, T. 2008b. *Progress: Promotion and Growth of Renewable Energy Sources and Systems*, Final report, March 2008, Utrecht (Netherlands).

Connor, P. and Mitchell, C. 2004. *Identification of barriers to convergence of support for renewable energy within the European Union*, FavoRES Project, August 2004.

Cory, K., Couture, T., Kreycik, C. 2009. *Feed-in Tariff Policy: Design, Implementation, and RPS Policy Interactions*, NREL Technical report, March 2009. Golden (CO): National Renewable Energy Laboratory.

Couture, T. and Gagnon, Y. 2009. 'An analysis of feed-in tariff remuneration models: implications for renewable energy investment', *Energy Policy*, 28:2, 955–65.

Couture, T., Cory, K., Kreycik, C., Williams, E., 2010. *Policymakers' Guide to Feed-in Tariff Policy Design*. NREL, Technical Report, July 2010. Golden (CO): National Renewable Energy Laboratory.

CRE 2001a. *Rapport d'activité*, June 2001. Paris: Commission de Régulation de l'Énergie.

CRE 2001b. *Avis sur l'arrêté fixant les conditions d'achat de l'électricité produite par les installations utilisant l'énergie mécanique du vent*, 5 June 2001. Paris: Commission de Régulation de l'Énergie.

CRE 2002. *Rapport d'activité*, June 2002. Paris: Commission de Régulation de l'Énergie.

CRE 2006. *Rapport d'activité*, June 2006. Paris: Commission de Régulation de l'Énergie.

CRE 2009a. *Avis de la Commission de régulation de l'énergie du 3 décembre 2009 sur le projet d'arrêté fixant les conditions d'achat de l'électricité produite par les installations utilisant l'énergie radiative du soleil telles que visées au 3° de l'article 2 du décret n° 2000-1196 du 6 décembre 2000*, Délibération, 3 December 2009. Paris: Commission de Régulation de l'Énergie.

CRE 2009c. *Communication de la Commission de régulation de l'énergie du 11 février 2009 relative aux charges de service public de l'électricité et à la contribution unitaire pour 2009*, Délibération, 11 February 2009. Paris: Commission de Régulation de l'Énergie.

Darflot, I. 2004. *Analysis of the barriers to the convergence of support to renewables in Europe*. The Norwegian University of Science and Technology, June 2004, Madrid.

DBFZ 2009. *Monitoring zur Entwicklung der Stromerzeugung aus Biomasse*, Zwischenbericht, March 2009. Available from http://www.erneuerbare-energien.de/files/pdfs/allgemein/application/pdf/zwischenber_mon_bio.pdf [accessed 6 September 2010].

Dagger, S. 2009. En*ergiepolitik & Lobbying – Die Novellierung des Erneuerbare-Energien-Gesetzes (EEG) 2009*. Ibidem: Stuttgart.

DB Climate Change Advisors 2009. *Paying for renewable energy: TLC at the right price – Achieving scale through efficient policy design*. New York, NY: The Deutsche Bank Group.

de Jager, D. 2007. *Characteristics of a future European internal market for RES-e, based upon the Gurantee of Origin*, Ecofys, November 2007, Utrecht.

de la Porte, C. and Pochet, P. 2002. *Building social Europe through the open method of co-ordination*. Brussels: Peter Lang.

de Miera, S., del Río, G., Vizcaíno, I. 2008. 'Analysing the impact of renewable electricity support schemes on power prices: The case of wind electricity in Spain', *Energy Policy*, 36:9, 3345–59.

del Río, P. 2005. 'A European-wide harmonised tradable green certificate scheme for renewable electricity – is it really so beneficial?', *Energy Policy*, 33:10, 1239–1250.

del Río, P. 2008. 'Ten years of renewable electricity policy in Spain. An analysis of successive feed-in tariff reforms', *Energy Policy*, 36:8, 2917–29.

del Río, P. and Gual, M 2007. 'An integrated assessment of the feed-in tariff system in Spain', *Energy Policy*, 35:2, 994–1012.

Deutscher Bundestag 1990a. *Gesetzentwurf der Fraktionen der CDU/CSU und FDP, Entwurf eines Gesetzes über die Einspeisung von Strom aus erneuerbaren Energien in das öffentliche Netz (Stromeinspeisegesetz)*, Drucksache 11/7816, 7 September 1990.

Deutscher Bundestag 1999. *Gesetzentwurf der Fraktionen SPD und BÜNDNIS 90/DIE GRÜNEN Entwurf eines Gesetzes zur Förderung der Stromerzeugung aus erneuerbaren Energien (Erneuerbare-Energien-Gesetz – EEG) sowie zur Änderung des Mineralölsteuergesetzes*, Drucksache 14/2341, 13 December 1999.

Deutscher Bundestag 2000. *Beschlussempfehlung und Bericht des Ausschusses für Wirtschaft und Technologie (9. Ausschuss) zu dem Gesetzentwurf der Fraktionen SPD und BÜNDNIS 90/DIE GRÜNEN – Drucksache 14/2341 – Entwurf eines Gesetzes zur Förderung der Stromerzeugung aus erneuerbaren Energien (Erneuerbare-Energien-Gesetz – EEG) sowie zur Änderung des Mineralölsteuergesetzes*, Drucksache 14/2776, 23 February 2000.

Deutscher Bundestag 2002. *Bericht über den Stand der Markteinführung und der Kostenentwicklung von Anlagen zur Erzeugung von Strom aus erneuerbaren Energien (Erfahrungsbericht zum EEG)*, Drucksache 14/9807, 16 July.2002.

Deutscher Bundestag 2003. *Entwurf eines Ersten Gesetzes zur Änderung des Erneuerbare-Energien-Gesetzes*, Drucksache 15/810, 8 April 2003.

Deutscher Bundestag 2004a. *Bericht des Ausschusses für Umwelt, Naturschutz und Reaktorsicherheit (15. Ausschuss) zu dem Gesetzentwurf der Fraktionen SPD und BÜNDNIS 90/DIE GRÜNEN – Drucksache 15/2327 – Entwurf eines Gesetzes zur Neuregelung des Rechts der Erneuerbaren Energien im Strombereich und zu dem Gesetzentwurf der Bundesregierung – Drucksachen*

15/2539, 15/2593 – Entwurf eines Gesetzes zur Neuregelung des Rechts der Erneuerbaren Energien im Strombereich, Drucksache 15/2864, 1 April 2004.

Deutscher Bundestag 2004b. *Antwort der Bundesregierung auf die Kleine Anfrage der Abgeordneten Kurt-Dieter Grill, Dagmar Wöhrl, Karl-Josef Laumann, weiterer Abgeordneter und der Fraktion der CDU/CSU – Drucksache 15/2036 – Umfang und Auswirkungen der aktuellen und künftigen EEG-Härtefallregelung für den Wirtschaftsstandort Deutschland,* Drucksache 15/2370, 16 January 2004.

Deutscher Bundestag 2008a. *Beschlussempfehlung und Bericht des Ausschusses für Umwelt, Naturschutz und Reaktorsicherheit (16. Ausschuss) zu dem Gesetzentwurf der Bundesregierung – Drucksachen 16/8148, 16/8393 – Entwurf eines Gesetzes zur Neuregelung des Rechts der Erneuerbaren Energien im Strombereich und zur Änderung damit zusammenhängender Vorschriften,* Drucksache 16/9477, 4 June 2008.

Deutscher Bundestag 2008b. *Gesetzentwurf der Bundesregierung, Entwurf eines Gesetzes zur Neuregelung des Rechts der Erneuerbaren Energien im Strombereich und zur Änderung damit zusammenhängender Vorschriften,* Drucksache 16/8148, 18 February 2008.

Deutscher Bundestag 2008c. *Stellungnahme der Zellstoff- und Papierindustrie zum Entwurf eines Gesetzes zur Neuregelung des Rechts der Erneuerbaren Energien im Strombereich (Bundestags-Drucksache 16/8148 vom 18.02.08),* Ausschussdrucksache 16(16)393(A), 19 March 2008.

DiMaggio, P. and Powell, W. 1991. 'The iron cage revisited. Institutionalized isomorphism and collective rationality in organisational fields', in Powell and DiMaggio (eds), pp. 63–82.

Dinica, V. 2006. 'Support systems for the diffusion of renewable energy technologies – an investors perspective', *Energy Policy,* 34:4, 461–480.

Di Nucci, R. and Mez, L. 2007. 'Actors' expectations and viewpoints on the RES-e support system in the EU: The experience of the German desk of Realise Forum', in: Mez (ed.), pp. 71–93.

Di Nucci, R., Mez, L., Reiche, D., Bechberger, M. 2007. *Country Report Germany, Second, updated version,* Workpackage 3 (Realise Forum), January 2007, Berlin.

Dolowitz, D. and Marsh, D. 1996. 'Who learns from whom? A review of policy transfer literature', *Political Studies,* 44:2, 343–57.

Dolowitz, D. and Marsh, D. 2000. 'Learning from abroad: The role of policy transfer in contemporary policy making', *Governance: An International Journal of Policy and Administration,* 13:1, 5–24.

Drezner, D.W. 2001. 'Globalisation and policy convergence', *International Studies Review,* 3:1, 53–78.

Drezner, D.W. 2005. 'Globalization, harmonization, and competition – the different pathways to policy convergence', *Journal of European Public Policy,* 12:5, 841–59.

Drillisch, J. 2001. *Quotenmodell für regenerative Stromerzeugung: Ein umweltpolitisches Instrument auf liberalisierten Elektrizitätsmärkten*, Essen: Oldenbourg Industrieverlag.

Droege, P. (ed.) 2009. *100 per cent renewable*, Earthscan: London.

Dupuis, P., Turenne, J., Bal, J.-L. 2008. *Plan de développement des énergies renouvelables à haute qualité environnemental – 2008–2012–2020, Grenelle de l'environnement, Comité opérationnel numéro 10*. Available from http://www.energies-renouvelables.org/portail/pdf/COMOP_10_Final.pdf [accessed 6 October 2008].

EDF 2009. *EDF Energy's Response to the 2008 Renewable Energy Strategy Consultation*, Executive Summary. Available from http://www.scotland.gov.uk/Resource/Doc/917/0076748.pdf [accessed 7 September 2010].

EFET 2007. *Towards an EU target of 20% renewable power production by 2020 – Ideas for the reform and harmonisation of renewable energy support schemes in EU States*, position paper, 22 November 2007. Brussels: European Federation of Energy Traders.

EFET 2008. *Electricity market participants challenge EU on closed national support schemes for renewable energy*, press release 45/08, 10 October 2008. Brussels: European Federation of Energy Traders.

Eikeland, O. and Saeverud, I.A. 2007. 'Market diffusion of renewable energy in Europe: Explaining front-runner and laggard positions', *Energy and Environment*, 18:1, 13–36.

Elkins, Z. and Simmons, B. 2005. 'On waves, clusters, and diffusion: a conceptual framework', *The Annals of the American Academy of Political and Social Sciences*, 598:1, 33–51.

E.On 2009. *UK renewable energy strategy consultation, Response by E.On.* Available from http://renewableconsultation.berr.gov.uk/offline_org_e-m_(113).pdf [accessed 11 April 2009].

Energelia 2007. Clos quiere un sector fotovoltaico menos 'turbo', 20 December 2007. Available from http://energelia.com/solar-fotovoltaica/clos-quiere-un-sector-fotovoltaico-menos-turbo.html [accessed 31 January 2008].

Energías Renovables 2009. *Los Ingenieros Industriales de Madrid alertan sobre las pérdidas de proyectos termoeléctricos por el RD 6/2009*. Available from http://www.energias-renovables.com/PAGINAS/Contenidosecciones.asp?ID=42&Cod=17286&Tipo=&Nombre=Solar%20termoel%C3%83%C2%A9ctrica [accessed on 10 August 2009].

EPIA 2005. *European PV Associations' position paper on a feed-in tariff for photovoltaic solar electricity*, position paper. Brussels: European Photovoltaic Industry Association.

EPIA 2007. *The announced European Framework Directive on renewable energy sources: a unique occasion to shape a sustainable European Energy Policy*, position paper, 21 May 2007. Brussels: European Photovoltaic Industry Association.

EPIA 2008. *European Commission Proposal for a European Directive on the Promotion of Renewable Sources: A Good Starting Point for the deployment of Solar Photovoltaic Electricity by 2020*, press release, 23 January 2008. Brussels: European Photovoltaic Industry Association.

EPIA 2009a. 2008: *An exceptional year for the Photovoltaic market*, press release, 24 March 2009. Brussels: European Photovoltaic Industry Association.

EPIA 2009b. *Global market outlook for photovoltaics until 2013*, Brussels: European Photovoltaic Industry Association. Available from http://www.epia.org/fileadmin/EPIA_docs/publications/epia/Global_Market_Outlook_Until_2013.pdf [accessed 12 December 2009].

EPIA 2009c. *Set for 2020 – Solar photovoltaic electricity: A mainstream power source in Europe by 2020*, Executive summary. Brussels: European Photovoltaic Industry Association.

EPIA 2010. *Global market outlook for photovoltaics until 2014*, May 2010 update. Brussels: European Photovoltaic Industry Association.

EPIA 2011. *Global market outlook for photovoltaics until 2015*. Brussels: European Photovoltaic Industry Association.

EPIA, EREF, WFC 2007. *Joint letter to EU Commissioner Piebalgs against mandatory certificate*, 10 October 2007. Brussels: European Photovoltaic Industry Association.

EREC 2004. *EREC's Position paper on the future of support systems for the promotion of electricity from renewable energy sources*. Brussels: European Renewable Energy Council.

EREC 2005a. *EREC's position paper on the future of support systems for the promotion of electricity from renewable energy sources*, position paper, January 2005. Brussels: European Renewable Energy Council.

EREC 2005b. *EREC dinner debate: 'Too early for the harmonisation of support mechanisms for Renewable Electricity'*, press release, 25 January 2005. Brussels: European Renewable Energy Council.

EREC 2007. *EREC's position on the Framework Directive for Renewable Energy Sources*, 1 October 2007. Brussels: European Renewable Energy Council.

EREF 2009. *Price for renewable energies in Europe: Report 2009*. Brussels: European Renewable Energy Federation.

Espey, S. 2001a. *Internationaler Vergleich energiepolitischer Instrumente zur Förderung regenerativer Energien in ausgewählten Industrieländern*, Norderstedt: Books on Demand.

Espey, S. 2001b. 'Renewable Portfolio Standard – a means for trade with electricity from renewable energy sources?', *Energy Policy*, 29:7, 557–66.

ETSO 2003. *Report on renewable energy sources*, 18 December 2003. Brussels: European Transmission System Operators.

EU 1986. *Single European Act*. Official Journal of the European Communities, 25 ILM. 506.

EU 1996. *Directive 96/92/EC of the European Parliament and of the Council of 19 December 1996 concerning common rules for the internal market in electricity*, Official Journal of the European Communities, L 027/XX.

EU 2001. *Directive 2001/77/EC of the European Parliament and of the Council of 27 September 2001 on the promotion of electricity produced from renewable energy sources in the internal electricity market*, Official Journal of the European Communities, L 283/33.

EU 2003. *Directive 2003/54/EC of the European Parliament and the Council of 26 June 2003 concerning common rules for the internal market in electricity and repealing Directive 96/92/EC*, Official Journal of the European Union, L 176/37.

EU 2006. *Decision No 1364/2006/EC of the European Parliament and of the Council of 6 September 2006 laying down guidelines for trans-European energy networks and repealing Decision 96/391/EC and Decision No 1229/2003/EC*, Official Journal of the European Union, L 262/1.

EU 2007. *Treaty of Lisbon, Amending the treaty on European Union and the treaty establishing the European Community*. Official Journal of the European Union (2007/C 306/01).

EU 2008. *Consolidated version of the Treaty on the functioning of the European Union*, Official Journal of the European Union (C 115/49).

EU 2009. *Directive 2009/28/EC of the European Parliament and the Council of 23 April 2009 on the promotion of the use of energy from renewable sources and amending and subsequently repealing Directives 2001/77/EC and 2003/30/EC*, Official Journal of the European Union, L 140/16.

EU Commission 1985. *Completing the internal market: White paper from the Commission to the European Council*, COM(85) 310.

EU Commission 1988. *The internal energy market*. Commission Working Document, COM(88) 238 final.

EU Commission 1995. *White paper: An energy policy for the European Union*, COM(95) 682 final.

EU Commission 1996. *Energy for the future – Renewable sources of energy. Greenpaper for a community strategy*, COM(96)576 final, 19 November 1996, Brussels.

EU Commission 1997. *Energy for the future – Renewable sources of energy. White Paper for a community strategy and action plan*, COM(97)599 final, 26 November 1997, Brussels.

EU Commission 1998. *Report to the Council and the European Parliament on harmonization requirements: Directive 96/92/EC concerning common rules for the internal market in electricity*, COM(1998) 167 final.

EU Commission 1999. *Commission working document, Electricity from renewable energy sources and the internal electricity market*, SEC(1999) 470 final.

EU Commission 2000a. *Amended proposal for a Directive of the European Parliament and of the Council on the promotion of electricity from renewable energy sources in the internal electricity market*, COM/2001/0884 final – COD 2000/0116, issued December 29, published to the Official Journal C154E.

EU Commission 2000b. *Proposal for a Directive of the European Parliament and of the Council on the promotion of electricity from renewable energy sources in the internal electricity market*, COM/2000/0279 final – COD 2000/0116, issued May 2000, published in the Official Journal C311E.

EU Commission (2001): *European governance: A white paper*, COM(2001), 428 final.

EU Commission 2004a. *Communication from the Commission to the Council and the European Parliament – The share of renewable energy in the EU*, COM(2004) 366 final.

EU Commission 2004b. *Commission staff working document, The share of renewable energy in the EU, Country Profiles, Overview of renewable energy sources in the enlarged European Union*, SEC(2004) 547.

EU Commission 2004c. *Medium term vision for the internal electricity market*, strategy paper (DG Energy and Transport). 1 March 2004, Brussels.

EU Commission 2005a. *The support of electricity from renewable energy sources, Communication from the Commission*, COM(2005) 627 final.

EU Commission 2005b. *Annex to the support for electricity from renewable energy sources, Impact assessment*, COM(2005) 627 final.

EU Commission 2005c. *Communication from the Commission to the Council and the European Parliament, Report on progress in creating the internal gas and electricity market*, COM(2005) 568 final.

EU Commission 2006. *Green paper follow-up action report on progress in renewable electricity, Communication from the Commission to the Council and the European Parliament*, COM(2006) 849 final.

EU Commission 2007a. *DG Competition report on energy sector inquiry*, 10 January 2007, SEC(2006) 1724 final.

EU Commission 2007b. *Renewable Energy Roadmap, Renewable energies in the 21st century: Building a more sustainable future, Communication from the Commission to the Council and the European Parliament*, COM(2006) 848 final.

EU Commission 2007c. *Commission staff working document, Accompanying document to the Communication from the Commission to the Council and the European Parliament, Renewable Energy Roadmap, Renewable energies in the 21st century: Building a more sustainable future*, COM(2006) 848 final.

EU Commission 2007d. *Directive of the European Parliament and of the Council on the promotion of the use of renewable energy sources*, Version 6.3.3. (clean), unpublished (leaked), internal draft proposal, November 2007.

EU Commission 2007e. *Communication for the Commission to the Council and the European Parliament – Priority Interconnection Plan*, COM(2006) 846 final.

EU Commission 2008a. *The support of electricity from renewable energy sources, Commission staff working document, accompanying document to the proposal for directive of the European Parliament and of the Council on the promotion of the use energy from renewable sources*, SEC(2008) 57, 23 January 2008, Brussels.

EU Commission 2008b. *Proposal for a Directive of the European Parliament an of the Council on the promotion of the use of energy from renewable sources*, COM(2008) 19 final.

EU Commission 2008c. *Impact assessment, Document accompanying the Package of Implementation measures for the EU's objective on climate change and renewable energy for 2020, Commission staff working document*, SEC(2008) 85.

EU Commission 2008d. *Annex to the impact assessment, Document accompanying the Package of Implementation measures for the EU's objective on climate change and renewable energy for 2020, Commission staff working document*, SEC(2008) 85 VOL. II.

EU Commission 2009a. *The Renewable Energy Progress Report: Commission Report in accordance with Article 3 of Directive 2001/77/EC, Article 4(2) of Directive 2003/30/EC and on the implementation of the EU Biomass Action Plan*, COM(2005)628, Communication from the Commission to the Council and the European Parliament, COM(2009) 192 final.

EU Commission 2009b. *Commission decision of 30.6.2009 establishing a template for National Renewable Energy Action Plans under Directive 2009/28/EC*, C(2009) 5174–1.

EU Commission 2011. *Invitation to tender No. ENER/C1/436-2010 concerning 'renewable energy policy database and support'*, Brussels, 17 June 2011.

EU Council 1986. Council Resolution of 16 September 1986 concerning new Community energy policy objectives for 1995 and convergence of the policies of the Member States, Official Journal C 241.

EU Council 2000. *Presidency Conclusions*, Lisbon European Council, 23 and 24 March 2000.

EU Council 2007. *Presidency Conclusion, 8/9 March 2007*, 7224/07, Brussels.

EU Council 2008a. *Presidency Conclusions of the Brussels European Council (13/14 March 2008)*, 7652/08, Brussels, 14 March 2008.

EU Council 2008b. DE/PL/UK joint proposal for the flexibility regime, meeting document from the General Secretariat of the Council to the delegations, interinstitutional file 2008/0016 (COD), Council of the European Union, Brussels, 18 June 2008.

EU Parliament 1997. *Report on the communication from the Commission on Energy for the Future: Renewable Sources of Energy – Green Paper for a Community Strategy (COM(96)0576 – C4-0623/96)*, Committee on Research, Technological Development and Energy, Rapporteur: Mrs Mechtild Rothe (A4-0168/97).

EU Parliament 2007. *Report on the roadmap for renewable energy in Europe (2007/2090(INI))*, Committee on Industry, Research and Energy, Rapporteur: Britta Thomsen, A6-0287/2007 final, 20 July 2007, Brussels.

EU Parliament 2008. *Draft European Parliament Legislative Resolution on the proposal for a directive of the European Parliament and of the Council on the promotion of the use of energy from renewable sources (COM(2008)0019 – C6-0046/2008 – 2008/0016(COD))*, Committee on Industry, Research and Energy, Rapporteur: Claude Turmes, 2008/0016(COD).

EUPD Research 2008. *The booming Spanish market is not yet ready for private rooftop systems*, press release, 31 October 2008. Bonn: EUPD Research.

EUPD Research 2010. *The German PV market*, July 2010. Bonn: EUPD Research.

EurActiv 2010. *Oettinger back EU-wide renewables feed-in tariff*, published 9 July 2010. Available from http://www.euractiv.com/en/energy/oettinger-backs-eu-renewables-feed-tariff-news-496148 [accessed 3 August 2010].

Eurelectric 1999. *Promotion of Renewables in the EU and possible Harmonisation: Eurelectric View*, October 1999. Eurolectric: Brussels.

Eurelectric 2007. *RES Directive: Eurelectric Again Calls For Market Approach*, web news article, 19 December 2007. Available from http://www2.eurelectric.org/Content/Default.asp [accessed 7 January 2010].

Eurelectrics and RECs 2004. Integrating renewable energy sources into the competitive electricity market – A shared vision, Position paper, November 2004, Eurolectric: Brussels.

Eurelectric, RECs, EFET 2007. *Incentives for renewable energy must fit within the European market framework*, press release, 7 November 2007, Eurolectric: Brussels.

Evans, M. and Davies, J. 1999. 'Understanding Policy Transfer: A Multi-level, Multidisciplinary Perspective', *Public Administration*, 77:2, 361–385.

Evrard, A. 2005. Renouveler la politique énergétique? L'intégration des énergies renouvelable aux politique de l'électricité: une comparaison France-Allemagne, Mémoire pour l'obtention du diplôme du Master de recherche, Institut d'Études Politique à Paris, Septembre 2005.

Eyestone, R. 1977. 'Confusion, Diffusion, and Innovation', *The American Political Science Review*, 71:2, 441–447.

EWEA 2001. *Response to the European Commission's Green Paper: Towards a European strategy for the security of energy supply*, position paper, November 2001. Brussels: European Wind Energy Association.

EWEA 2003. *European wind industry; Another record year*, press release, 6 February 2003. Brussels: European Wind Energy Association.

EWEA 2004. *On the future of EU support systems for the promotion of electricity from renewable energy sources*, position paper, November 2004. Brussels: European Wind Energy Association.

EWEA 2005. *Support schemes for renewable energy, A comparative analysis of payment mechanisms in the EU*, Overview. Brussels: European Wind Energy Association.

EWEA 2010. *Wind in power, 2009 European statistics*, February 2010. Brussels: European Wind Energy Association.

EWIS 2010. *European wind intregration study*, EWIS final report, March 2010, Brussels.

Eyestone, R. 1977. 'Confusion, Diffusion, and Innovation', *The American Political Science Review*, 71:2, 441–447.

Falkner, G., Treib, O., Hartlapp, M., Leiber, S. 2005. *Complying with Europe: EU harmonisation and soft law in the member states*. Cambridge: Cambridge University Press.

FEE 2009. *Etat du parc éolien, Février 2009*, Branche éolienne du Syndicat des Energies Renouvelables. Paris: France Energie Eolienne.

Fell, H.-J. 2009a. *Feed-in tariffs for renewable energies: An effective stimulus package without new public borrowing*. Deutscher Bundestag, March 2009.

Fell, H.-J. 2009b. *Mögliche Gesetzentwicklung zur Realisierung erster Leitungen und Kraftwerke in Nordafrika*, presentation at the Official Launch of the Desertec Foundation, 17 March 2009, Berlin.

Finon, D. and Menanteau, P. 2004. 'Static and Dynamic Efficiency Criteria of Instruments of Promotion of Renewables', *Energy Studies Review*, 12:1, 53–81.

Finon, D. and Perez, Y. 2007. 'The social efficiency of instruments of promotion of renewable energies: A transaction-cost perspective', *Ecological Economics*, 62:1, 77–92.

Fouquet, D. 2006. *Feed-in tariff systems and state-aid*, presentation at the third workshop of the International Feed-in Cooperation, 23 and 24 November 2006, Madrid.

Fouquet, D. and Johansson, T.B. 2008. 'European renewable energy policy at cross-roads – Focus on electricity support mechanisms', *Energy Policy*, 36:11, 4079–92.

Fouquet, D., Grotz, C., Sawin, J., Vassilakos, N. 2005. *Reflections on a possible unified EU financial support scheme for renewable energy systems (RES): A comparison of minimum-price and quota systems and an analysis of market conditions*, Brussels and Washington DC, January 2005.

Fulton, M., Kahn, B., Mellquist, N., Soong, E., Baker, J., Cotter, L., Kreibiehl, S., Rickerson, W., Meister, H.-P. 2010. *GET FiT Program – Global Energy Transfer Feed-in Tariffs for Developing Countries*, Greenpaper, April 2010. New York, NY: The Deutsche Bank Group.

Fulton, M., Mellquist, N., Rickerson, W., Jacobs, D. 2011. *The German feed-in tariff for PV:Managing volume success with price response*, May 2011, DB Research: New York.

Fürsch, M., Golling, C., Nicolosi, M., Wissen, R., Lindenberger, D. 2010. *European RES-e policy analysis*, Zusammenfassung, EWI, April 2010.

Garcia Breva, J. 2009. *The Spanish experience*, Presentation at the conference: Climate change, global threats & low carbon prosperity towards Copenhagen, 12 November 2009, Istanbul.

Garrett, G., Dobbin, F., Simmons, B. 2008. 'Conclusion', in Simmons, Dobbin and Garrett (eds), pp. 344–60.

Gellings, R. and Rutschmann, I. 2008. 'Drying up', *Photon International*, 8/2008, 70.

Genschel, P. 2002. *Steuerwettbewerb und Steuerharmonisierung in der Europäischen Union*. Frankfurt am Main: Campus.

Genschel, P. and Ganghof, S. 2008. 'Taxation and democracy in the EU', *Journal of European Public Policy*, 15:1, 58–77.

George, A. and Bennett, A. 2005. *Case studies and theory development in the social sciences*. Cambridge (MA): MIT Press.

Gerring, J. 2008. 'Review article: The mechanismic worldview: Thinking inside the box', *British Journal of Political Science*, 38:1, 161–79.

Gipe, P. 2010a. *Grading North American feed-in tariffs*, May 2010. Hamburg: World Future Council.

Gipe, P. 2010b. Renewable energy tariffs in Malta, press release, 22 July 2010. Available from http://www.wind-works.org/FeedLaws/Malta/MaltaList.html [accessed 20 October 2010].

Gipe, P. 2010c. Finland moving closer to feed-in tariffs in 2011, press release, 22 July 2010. Available from http://www.wind-works.org/FeedLaws/Finland/FinlandMovingClosertoFeed-inTariffsin2010.html [accessed 20 October 2010].

Glachant, J.-M. and Finon, D. 2005. 'A competitive fringe in the shadow of a state owned incumbent – The case of France', *Energy Journal, European Electricity Liberalisation Special*, 26:1, 181–204.

Goetschy, J. 2005. 'The open method of coordination and the Lisbon strategy: the difficult road from potential to results', *Transfer: European Review of Labour and Research*, 1/2005, 64–80.

Goggin, M., Bowman, A., Lester, J., O'Toole, L. 1990. *Implementation theory and practice: Toward a third generation*. Glenview: Scott Foresman & Co.

Grace, R., Rickerson, W., Porter, K., DeCesaro, J., Corfee, K., Wingate, M., et al. 2008. *Exploring feed-in tariffs for California: Feed-in tariff design and implementation issues and options (CEC-300-2008-003-F)*. Sacramento, CA: California Energy Commission.

Greenpeace 2008. *Trading must not hamper renewables target, press release*. 28 February 2008. Brussels: Greenpeace Europe.

Greenpeace, WWF, COGEN, EWEA, Climate Network, Friend of the Earth, E5, FGW, BWE 1999. *Principles for a renewable energy directive in the European Union*. April 1999. Available from http://archive.greenpeace.org/climate/naturalpower/principles/index.html [accessed 30 July 2010].

Grotz, C. and Fouquet, D. 2005. 'Fixed prices work better – A comparison of renewable energy promotion gives thumbs up to feed-in schemes and opposes premature EU harmonisation', *New Energy*, 2/2005, 18–21.

Gual, M., Hernández, F., del Río, P. 2001. 'Certificados verdes negociables (CVNs). Una alternative coherente para promocionar la electricidad a partir de fuentes de energía removable (E-FER)', *Energía*, 27:157, 75–82.

Haas, P.M. 1992. 'Introduction: epistemic communities and international policy coordination', *International Organization*, 46:1, 1–36.

Haas, R., Eichhammer, W., Huber, C., Langniss, O., Lorenzoni, A., Madlener, R., Menanteau, P., Morthorst, P.-E.,Martins, A., Oniszk, A., Schleich, J., Smith, A., Vass, Z., Verbruggen, A. 2004. 'How to promote renewable energy systems successfully and effectively', *Energy Policy*, 32:6, 833–9.

Haas, R., Meyer, N.I., Held, A., Finon, D., Lorenzoni, A., Wiser, R.,Nishio, K.-I. 2008. 'Promoting electricity from renewable energy sources – lessons learned from the EU, United States, and Japan', in Sioshansi (ed.), pp. 419–68.

Haase, N. 2008. *European gas market liberalisation: Are regulatory regimes moving towards convergence?*, Oxford Institute for Energy Studies, NG 24, May 2008.

Häder, M. 2005. 'Einspeisevergütungs- und Quotenmodelle zur Förderung der regenerativen Stromerzeugung – Ökonomische Analyse der Fördersysteme in den EU-Mitgliedstaaten', *Energiewirtschaftliche Tagesfragen*, 55:9, 610–15.

Hanf, K. and Jansen, A.I. (eds) 1998. *Governance and environment in western Europe*. Politics, policy and administration. Harlow (UK): Longman.

Harmelink, M., Voogt, M., Cremer, C. 2006. 'Analysing the effectiveness of renewable energy supporting policies in the European Union', *Energy Policy*, 34:3, 343–51.

Heichel, S., Pape, J., Sommerer, T. 2005. 'Is there convergence in convergence research? An overview of empirical studies on policy convergence', *Journal of European Public Policy*, 12:5, 817–40.

Heichel, S. and Sommerer, T. 2007. 'Unterschiedliche Pfade, ein Ziel? – Spezifikationen im Forschungsdesign und Vergleichbarkeit der Ergebnisse bei der Suche nach der Konvergenz nationalstaatlicher Politiken', in Holzinger; Jörgens and Knill (eds), pp. 107–30.

Heichel, S., Holzinger, K., Sommerer, T., Liefferink, D., Pape, J., Veenman, S. 2008. 'Research design, variables and data', in Holzinger, Knill and Arts (eds), pp. 64–97.

Held, A., Haas, R., Ragwitz, M. 2006. 'On the success of policy strategies for the promotion of electricity from renewable energy sources in the EU', *Energy & Environment*, 17:6, 849–68.

Hemmelskamp, J. 1999. *Umweltpolitik und technischer Fortschritt*. Heidelberg: Physica-Verlag Heidelberg.

Héritier, A. 1987. *Policy-Analyse. Eine Einführung*. Frankfurt am Main: Campus.

Héritier, A. (ed.) 1993. *Policy-Analyse, Kritik und Neuorientierung*. PVS Sonderheft 24, Opladen: Westdeutscher Verlag.

Héritier, A. 1996. 'The accommodation of diversity in European policy-making and its out-comes: regulatory policy as a patchwork', *Journal of European Public Policy* 3(2), pp. 149–67.

Héritier, A. (ed.) 2002a. *Common Goods: Reinventing European and International Governance*. Lanham: Rowman & Littlefield.

Héritier, A. 2002b. 'New modes of governance in Europe: policy making without legislating?' in Héritier (ed.), pp. 185–206.

Héritier, A., Mingers, S., Knill, C., Becka, M. (eds) 1994. *Die Veränderung von Staatlichkeit in Europa.* Opladen: VS Verlag.

Héretier, A., Knill, C., Mingers, S. 1996. *Ringing the change in Europe. Regulatory competition and the transformation of the state.* Berlin: De Gruyter.

Héritier, A., Kerwer, D., Knill, C., Lehmkuhl, D., Teutsch, M.; Douillet, A.-C. 2001. *Differential Europe. The European Union impact on national policymaking.* Lanham (Maryland): Rowman & Littlefield.

Héritier, A. and Knill, C. 2001. 'Differential responses to European policies: a comparison', in Héritier, A., Kerwer, D., Knill, C., Lehmkuhl, D., Teutsch, M., Douillet, A.-C. (eds), pp. 257–94.

Héritier, A. and Eckert, S. 2008. 'New modes of governance in the shadow of hierarchy: Self-regulation by industry in Europe', *Journal of Public Policy*, 28:1, 113–38.

Heup, J. 2009. 'Solar im Bau', *Neue Energie*, 7/2009, 44–56.

Hidalgo, R. 2009. 'El reto de la geotermia profunda en España', *Energética XXI*, January/February 2009, 114–15.

Hinrichs-Rahlwes, R. 2008. 'A harmonized feed-in system for the EU: Are we ready to seize the opportunity?', *Renewable Energy World*, 2 June 2008. Available from http://www.renewableenergyworld.com/rea/news/article/2008/06/a-harmonized-feed-in-system-for-the-eu-are-we-ready-to-seize-the-opportunity-52505 [accessed 3 November 2010].

Hinsch, C. 2000. 'Brüssel bewegt sich. EU-Kommission verabschiedet Vorschlag für eine Richtlinie zur Förderung erneuerbarer Energien', *Neue Energie*, 6/2000: 10–12.

Hinsch, C. 2001. 'Überraschende Wende in Brüssel', *Neue Energie*, 1/2001, 8–10.

Hirsh, R. 1999. 'PURPA: The spur to competition and utility restructuring', *The Electricity Journal*, 12:7, 60–72.

Hirschl, B. 2008. *Erneuerbare Energie-Politik, Eine Multi-Level Policy-Analyse mit Fokus auf den deutschen Strommarkt.* Wiesbaden: VS Research (Verlag für Sozialwissenschaften).

Hirschl, B., Hoffmann, E., Zapfel, E., Hopper-Kilpper, M., Durstewitz, M., Bard, J. 2002. *Markt- und Kostenentwicklung erneuerbarer Energien, 2 Jahre EEG-Bilanz und Ausblick.* Berlin: Erich Schmidt Verlag.

Hirshman, W.P. und Rutschmann, I. 2008. 'Kleiner Markt mit großem Potenzial', *Photon Germany*, 5/2008, 26.

Hirshman, W. and Schug, A. 2009. 'Geschmackvolle Reform, Frankreichs neue Einspeiseregelungen sind bizarre, aber dennoch äußerst attraktiv', *Photon Germany*, 12/2009, 24–7.

Hirshman, W., Espinasse, M., Schug, A. 2010. 'Spekulationsblase geplatzt?', Photon Deutschland, Feburary 2010, 22–4.

Hoberg, G. 2001. 'Globalization and policy convergence: symposium overview', *Journal of Comparative Policy Analysis: Research and Practise*, 3:2, 127–32.

Hodson, D. and Maher, I. 2001. 'The open method as a new mode of governance', *Journal of Common Market Studies*, 39:4, 719–46.

Holzinger, K.; Knill, C., Lehmkuhl, D. (eds) 2003. *Politische Steuerung im Wandel: Der Einfluss von Ideen und Problemstrukturen.* Opladen: Leske + Budrich.

Holzinger, K. and Knill, C. 2004. 'Regulatory competition and regulatory cooperation in environmental policy: individual and interaction effects', *Journal of Public Policy*, 24:1, 25–47.

Holzinger, K. and Knill, C. 2005a. 'Causes and conditions of cross-national policy convergence', *Journal of European Public Policy*, 12:5, 775–96.

Holzinger, K. and Knill, C. 2005b. *Competition, cooperation and communication. A theroretical analysis of different sources of environemental policy convergence and their interaction*, Working paper, Political sciences series 102, Vienna: Institute for Advanced Studies.

Holzinger, K. and Knill, C. 2008. 'Theoretical framework: causal factors and convergene exceptions', in: Holzinger, Knill and Arts (eds), pp. 30–63.

Holzinger, K., Jörgens, H., Knill, C. (eds) 2007a. *Transfer, Diffussion und Konvergenz von Politiken.* Politische Vierteljahresschriften, Sonderheft 38/2007, Wiesbaden: VS Verlag für Sozialwissenschaften.

Holzinger, K., Jörgens, H., Knill, C. 2007b. 'Transfer, Diffusion und Konvergenz: Konzepte und Kausalmechanismen'. Holzinger, Jörgens and Knill (eds), pp. 11–35.

Holzinger, K. and Sommerer, T. 2008. *'Race to bottom' or 'Race to Brussels'? Environmental competition in Europe*, Working Paper 01/2008, Chair of International Relations and Conflict Management, Universität Konstanz.

Holzinger, K., Jörgens, H., Knill, C. 2008a. 'State of the art – conceptualising environmental policy convergence', in Holzinger, Knill and Arts (eds), pp. 7–29.

Holzinger, K., Knill, C., Arts, B. (eds) 2008b. *Environmental policy convergence in Europe*, Cambridge: Cambridge University Press.

Holzinger, K., Knill, C., Sommerer, T. 2010. 'Umweltpolitik zwischen Annäherung und Aufholjagd: Eine Analyse umweltpolitischer Konvergenz in 24 OECD-Ländern', *Zeitschrift für Umweltpolitik & Umweltrecht*, 33:1, 1–32.

Höpner, F., 1999: 'Der Handelsplatz für regenerativen Strom – Soviel Markt wie möglich – Sowenig Staat wie nötig', *Elektrizitätswirtschaft*, 24/1999, 54–6.

Huber, C., Faber, T., Haas, R., Resch, G., Green, J., Ölz, S., White, S., Cleijne, H., Ruijgrok, W., Morthorst, P.E., Skytte, K., Gual, M., del Río, P., Hernández, F., Tacsir, A., Ragwitz, M., Schleich, J., Orasch, W., Bokemann, M., Lins, C. 2004. *Green-X – Deriving optimal promotion strategies for increasing the share of RES-E in a dynamic European electricity market*, Final Report, Wien.

Humphreys, P. 2002. 'Europeanisation, globalisation and telecommunications governance: a neo-Gramscian analysis', *Convergence: The Journal of Research into New Media Technologies*, 8:2, 52–79.

Hustedt, M. 1998. 'Windkraft – made in Germany', in Alt, Claus, and Scheer (eds), pp. 163–8.

Hvelplund, F. 2005. 'Renewable Energy: Political prices or political quantities', in Lauber (ed.), pp. 228–45.

Hvelplund, F. 2009. *Renewable energy governance systems. Political prices or political quantities. An overview*, Wind Energy International 2009/2010, World Wind Energy Association, pp. 285–93.

Iberdrola 2008. IBERDROLA valora muy positivamente el 'Green package' presentado por la Comisión Europea, press release, 23 January 2008. Madrid: Iberdrola.

IDAE 2005. *Plan de Energías Renovables en España 2005–2010*, August 2005. Madrid: Instituto para la Diversifición y Ahorro de Energía.

IEA 2004. *Renewable energy – Market and policy trends in IEA countries*. Paris: International Energy Agency.

IEA 2005. *Lessons from liberalised electricity markets*. Paris: International Energy Agency.

IEA 2007. *Energy policies of IEA countries. Germany, 2007 Review*. Paris: International Energy Agency.

IEA 2008a. *Deploying renewables: Principles for effective policies*. Paris: International Energy Agency.

IEA 2008b. *Draft paper for the next meeting of the governing board, global renewable energy markets and policies programme (GREMPP)*, 20–21 February 2008, IEA/CERT(2008)2, 07 February 2008, Paris: International Energy Agency.

IEA 2008c. *FITs, Making CHP and DHC viable – Portugal case study*, Paris: International Energy Agency. Available from www.iea.org/g8/chp/docs/portugal.pdf [accessed 10 September 2010].

IFIC 2009. *7th Workshop of the International Feed-In Cooperation, Conclusions*, 28–29 May 2009. Ljubljana: International Feed-in Cooperation.

IFNE 2009. *Beschaffungsmehrkosten der Stromlieferanten durch das Erneuerbare-Energien-Gesetz 2008 (Differenzkosten nach § 15 EEG)*, Ingenieurbüro für Neue Energie (IfnE), Gutachten im Auftrag des Bundesministeriums für Umwelt, Naturschutz und Reaktorsicherheit, Teltow, März 2009.

InterMoney Energía 2006. *Análisis y diagnóstico de la generación eólica en España*, documento preparado para AEE, 12 June 2008, Madrid.

IRENA 2008. *The role of IRENA in the context of other international organisations and initiatives*, December 2008. Abu Dhabi: International Renewable Energy Agency.

Jacobs, D. 2008. *Analyse des spanischen Fördermodells für erneuerbare Energien unter besonderer Berücksichtigung der Windenergie*, Wissenschaft und Praxis, Studie im Auftrag des Bundesverbands WindEnergie e. V., August 2008, Berlin.

Jacobs, D. 2009. The Spanish support mechanism for renewable electricity with special focus on wind energy, Wind Energy Market 2009 (BWE), 19th edition, April 2009, pp. 42–5.

Jacobs, D. 2010. 'Fabulous feed-in tariffs', Renewable Energy Focus, July/August 2010, 28–30.

Jacobs, D. and Weinhold, N. 2007. 'Ziemlich zufrieden – Spaniens Okobranche kann mit den neuen Einspeisetarifen leben', Neue Energie, 7/2007, 82–5.

Jacobs, D. and Kiene A. 2009. Renewable energy policies for sustainable African development, World Future Council, April 2009.

Jacobs, D.; Pfeiffer, C. 2009. 'Flexible degression of feed-in tariffs for photovoltaics: A new design option combining tariff payment and market growth', *PV Magazine*, May 2009, 20–24.

Jacobs, D., Butzengeiger, S., Schurig, S., Stephan, B. 2009. *Renewable energy in the post-Kyoto era – A global renewable energy policy (REP) fund for developing countries*, October 2009, Hamburg: World Future Council.

Jansen, J. and Uyterlinde, M. 2004. 'A fragmented market on the way to harmonisation? EU policy-making on renewable energy promotion', *Energy for Sustainable Development*, 8:1, 93–107.

Jacobsson, K. 2001. 'Employment and Social Policy Coordination, A New System of EU Governance', Paper for the scancor workshop on transnational regulation and the transformation of the states, Stanford, 22 to 23 June 2001.

Jacobsson, S. and Lauber, V. 2006. 'The politics and policy of energy system transformation – explaining the diffusion of renewable energy technology', *Energy Policy*, 34:3, 256–76.

Jacobsson, S., Bergek, A., Finon, D., Lauber, V., Mitchell, C., Toke, D., Verbruggen, A. 2009. 'EU renewable energy support policy: Faith or facts?', *Energy Policy*, 37:6, 2143–6.

Jansen, J.C., Gialoglou, K., Egenhofer, C. 2005. *Market stimulation of renewable electricity in the EU – What degree of harmonisation of support mechanisms is required?*, CEPS Task Force Report No. 56, October 2005.

Jänicke, M. 1992. 'Conditions for environmental policy success: an international comparison', *The Environmentalist*, 12:1, 47–58.

Jänicke, M. 1996a. 'Erfolgsbedingungen von Umweltpolitik', in Jänicke, M. (ed.), pp. 9–28.

Jänicke, M. (ed.) 1996b. *Umweltpolitik der Industrielände: Entwicklung – Bilanz – Erfolgsbedingungen*. Berlin: Edition Sigma.

Jänicke, M. 2000. *Ecological modernization: innovation and diffusion of policy and technology*, FFU-report 2000/08, Freie Universität Berlin.

Jänicke, M. 2005. 'Trend-setters in environmental policy: the character and role of pioneer countries', *European Environment*, 15:2, 129–42.

Jänicke, M. and Weidner, H. (eds). 1997. *National environmental policies. A comparative study of capacity-building*. Berlin, Heidelberg, New York: Springer.

Jänicke, M., Binder, M., Mönch, H. 1997. '"Dirty Industries": Patterns of Change in Industrial Countries', *Environmental and Resource Economics*, 9:1997, 467–91.

Jansen, A.-I., Osland, O., Hanf, K. 1998. 'Environmental challenges and institutional changes. An interpretation of the development of environmental policy in western Europe', in Hanf and Jansen (eds), pp. 277–325.

J.O. 1946. *Loi no 46-628 du 8 avril 1946 sur la nationalisation de l'électricité et du gaz*, Journal Officiel de la République Française, 9 April 1946, p. 2651.

J.O. 2000a. *Loi no 2000-108 du 10 février 2000 relative à la modernisation et au développement du service public de l'électricité*, Journal Officiel de la République Française n° 35 du 11 février 2000, p. 2143.

J.O. 2000b. *Décret no 2000-1196 du 6 décembre 2000 fixant par catégorie d'installations les limites de puissance des installations pouvant bénéficier de l'obligation d'achat d'électricité*, Journal Officiel de la République Française n° 285, 9 December 2000, p. 19550.

J.O. 2001a. Décret no 2001-365 du 26 avril 2001 relatif aux tarifs d'utilisation des réseaux publics de transport et de distribution d'électricité, Journal Officiel de la République Française n° 100, 28 April 2001 p. 6729.

J.O. 2001b. *Arrêté du 8 juin 2001 fixant les conditions d'achat de l'électricité produite par les installations utilisant l'énergie mécanique du vent, telles que visées à l'article 2-2° du décret n° 2000-1196 du 6 décembre 2000*, Journal Officiel de la République Française n° 143, 22 June 2001, p. 9889.

J.O. 2001c. *Décret no 2001-410 du 10 mai 2001 relatif aux conditions d'achat de l'électricité produite par des producteurs bénéficiant de l'obligation d'achat*, Journal Officiel de la République Française n° 110, 12 May 2001, p. 7543.

J.O. 2001d. *Arrêté du 25 juin 2001 fixant les conditions d'achat de l'électricité produite par les installations utilisant l'énergie hydraulique des lacs, cours d'eau et mers, telles que visées au 1° de l'article 2 du décret n° 2000-1196 du 6 décembre 2000*, Journal Officiel de la République Française n° 270, 21 November 2001, p. 18473.

J.O. 2001e. *Arrêté du 3 octobre 2001 fixant les conditions d'achat de l'électricité produite par les installations qui valorisent des déchets ménagers ou assimilés en utilisant le biogaz de décharge*, Journal Officiel de la République Française n° 270, 21 November 2001, p. 18479.

J.O. 2002a. *Arrêté du 13 mars 2002 fixant les conditions d'achat de l'électricité produite par les installations utilisant l'énergie radiative du soleil telles que visées au 3o de l'article 2 du décret no 2000-1196 du 6 décembre 2000*, Journal Officiel de la République Française n° 62, 14 March 2002, p. 4624.

J.O. 2002b. *Arrêté du 16 avril 2002 fixant les conditions d'achat de l'électricité produite par methanisation*, Journal Officiel de la République Française n° 105, 5 May 2002, p. 8628.

J.O. 2002c. *Arrêté du 16 avril 2002 fixant les conditions d'achat de l'électricité produite par les installations utilisant à titre principal l'énergie dégagée par la combustion de matières non fossiles d'origine végétale telles que visées au*

4° de l'article 2 du décret n° 2000-1196 du 6 décembre 2000, Journal Officiel de la République Française, n° 105, 5 May 2002, p. 8627.

J.O. 2002d. *Arrêté du 13 mars 2002 fixant les conditions d'achat de l'électricité produite par les installations utilisant l'énergie des nappes aquifères ou des roches souterraines telles que visées au 6° de l'article 2 du décret n° 2000-1196 du 6 décembre 2000*, Journal Officiel de la République Française, n° 82, 7 April 2002, p. 6186.

J.O. 2002e. *Avis de la Commission de régulation de l'électricité en date du 14 février 2002 sur l'arrêté fixant les conditions d'achat de l'électricité produite par les installations utilisant l'énergie des nappes aquifers ou des roches souterraines*, Journal Officiel de la République Française, n° 82, 7 April 2002, p. 6199.

J.O. 2003a. *Arrêté du 7 mars 2003 relatif à la programmation pluriannuelle des investissements de production d'électricité*, Journal Officiel de la République Française n° 65 du 18 mars 2003, p. 4692.

J.O. 2003b. *Décret n° 2003-229 du 13 mars 2003 relatif aux prescriptions techniques générales de conception et de fonctionnement auxquelles doivent satisfaire les installations en vue de leur raccordement aux réseaux publics de distribution*, Journal Officiel de la République Française n° 64, 16 March 2003, p. 4589.

J.O. 2003c. *Décret n° 2003-588 du 27 juin 2003 relatif aux prescriptions techniques générales de conception et de fonctionnement auxquelles doivent satisfaire les installations en vue de leur raccordement au réseau public de transport d'électricité*, Journal Officiel de la République Française n° 151, 2 July 2003, p. 11110.

J.O. 2003d. *Arrêté du 17 mars 2003 relatif aux prescriptions techniques de conception et de fonctionnement pour le raccordement à un réseau public de distribution d'une installation de production d'énergie électrique*, Journal Officiel de la République Française n° 32, 19 April 2003, p. 7008.

J.O. 2004a. *Arrêté du 23 décembre 2004 modifiant les conditions d'achat de l'électricité produite par les installations de cogénération et les installations utilisant des énergies renouvelables ou des déchets ménagers*, Journal Officiel de la République Française n° 302, 29 December 2004, p. 22214.

J.O. 2004b. *Décret n° 2004-90 du 28 janvier 2004 relatif à la compensation des charges de service public de l'électricité*, Journal Officiel de la République Française n° 24, 29 January 2004 p. 2093.

J.O. 2005a. *Décret n° 2005-1750 du 30 décembre 2005 modifiant le décret n° 2001-365 du 26 avril 2001 relatif aux tarifs d'utilisation des réseaux publics de transport et de distribution d'électricité*, Journal Officiel de la République Française n° 304, 31 December 2005, p. 20807.

J.O. 2005b. *Loi no 2005-781 du 13 juillet 2005 de programme fixant les orientations de la politique énergétique*, Journal Officiel de la République Française n° 163, 14 July 2005, p. 11570.

J.O. 2006a. *Arrêté du 10 juillet 2006 fixant les conditions d'achat de l'électricité produite par les installations utilisant l'énergie mécanique du vent telles que visées au 2° de l'article 2 du décret n° 2000-1196 du 6 décembre 2000*, Journal Officiel de la République Française n° 171, 26 July 2006, p. 11131.

J.O. 2006b. *Arrêté du 7 juillet 2006 relatif à la programmation pluriannuelle des investissements de production d'électricité*, Journal Officiel de la République Française n° 158, 9 July 2006, p. 10317.

J.O. 2006c. *Arrêté du 10 juillet 2006 fixant les conditions d'achat de l'électricité produite par les installations qui valorisent le biogaz*, Journal Officiel de la République Française n° 171, 26 July 2006, p. 11134.

J.O. 2006e. *Arrêté du 10 juillet 2006 fixant les conditions d'achat de l'électricité produite par les installations utilisant l'énergie des nappes aquifères ou des roches souterraines telles que visées au 6° de l'article 2 du décret n° 2000-1196 du 6 décembre 2000*, Journal Officiel de la République Française, n° 171, 26 July 2006, p. 11135.

J.O. 2006f. *Arrêté du 10 juillet 2006 fixant les conditions d'achat de l'électricité produite par les installations utilisant l'énergie radiative du soleil telles que visées au 3o de l'article 2 du décret no 2000-1196 du 6 décembre 2000*, Journal Officiel de la République Française, n° 171, 26 July 2006, p. 11133.

J.O. 2007a. *Arrêté du 1er mars 2007 fixant les conditions d'achat de l'électricité produite par les installations utilisant l'énergie hydraulique des lacs, cours d'eau et mers, telles que visées au 1° de l'article 2 du décret n° 2000-1196 du 6 décembre 2000*, Journal Officiel de la République Française n° 95, 22 April 2007, p.7146.

J.O. 2007b. *Arrêté du 28 août 2007 fixant les principes de calcul de la contribution mentionnée aux articles 4 et 18 de la loi n° 2000-108 du 10 février 2000 relative à la modernisation et au développement du service public de l'électricité*, Journal Officiel de la République Française, n° 200, 30 August 2007, p. 14314.

J.O. 2008a. *Avis du 30 octobre 2008 relatif au projet d'arrête fixant les conditions d'achat de l'électricité produite par les installations utilisant l'énergie mécanique du vent, Avis*, Journal Officiel de la République Française n° 121, 13 December 2008.

J.O. 2008b. *Arrêté du 23 avril 2008 relatif aux prescriptions techniques de conception et de fonctionnement pour le raccordement à un réseau public de distribution d'électricité en basse tension ou en moyenne tension d'une installation de production d'énergie électrique*, Journal Officiel de la République Française n° 0098, 25 April 2008 p. 6920.

J.O. 2008c. *Arrêté du 23 avril 2008 relatif aux prescriptions techniques de conception et de fonctionnement pour le raccordement au réseau public de transport d'électricité d'une installation de production d'énergie électrique*, Journal Officiel de la République Française n° 0098, 25 April 2008, p. 6920.

J.O. 2008d. *Décret n° 2008-386 du 23 avril 2008 relatif aux prescriptions techniques générales de conception et de fonctionnement pour le raccordement d'installations de production aux réseaux publics d'électricité*, Journal Officiel de la République Française n° 0098, 25 April 2008, p. 6908.

J.O. 2008e. *Arrêté du 17 juillet 2008 fixant les taux de réfaction mentionnés dans l'arrêté du 28 août 2007 fixant les principes de calcul de la contribution mentionnée aux articles 4 et 18 de la loi no 2000-108 du 10 février 2000 relative à la modernisation et au développement du service public de l'électricité*, Journal Officiel de la République Française, n° 200, 30 August 2007, p. 14314.

J.O. 2008f. *Arrêté du 17 novembre 2008 fixant les conditions d'achat de l'électricité produite par les installations utilisant l'énergie mécanique du vent*, Journal Officiel de la République Française, n° 0290, 13 December 2008, p. 19032.

J.O. 2009a. *Décision du 14 janvier 2010 approuvant les tarifs des prestations annexes à destination des responsables d'équilibre réalisées sous le monopole des gestionnaires de réseaux publics de distribution d'électricité*, Journal Officiel de la République Française n° 0018, 22 January 2010, p. 1387.

J.O. 2009b. *Arrêté du 28 décembre 2009 fixant les conditions d'achat de l'électricité produite par les installations utilisant, à titre principal, l'énergie dégagée par la combustion de matières non fossiles d'origine végétale ou animale telles que visées au 4° de l'article 2 du décret n° 2000-1196 du 6 décembre 2000*, Journal Officiel de la République Française, n° 0303, 31 December 2009, p. 23082.

J.O. 2010a. *Arrêté du 12 janvier 2010 fixant les conditions d'achat de l'électricité produite par les installations utilisant l'énergie radiative du soleil telles que visées au 3o de l'article 2 du décret no 2000-1196 du 6 décembre 2000*, Journal Officiel de la République Française n° 171, 26 July 2006, p. 11133.

J.O. 2010b. *Arrêté du 23 juillet 2010 fixant les conditions d'achat de l'électricité produite par les installations utilisant l'énergie des nappes aquifères ou des roches souterraines telles que visées au 6° de l'article 2 du décret n° 2000-1196 du 6 décembre 2000*, Journal Officiel de la République Française n° 169, 24 July 2010, p. 13676.

J.O. 2010c. *Loi n° 2010-788 du 12 juillet 2010 portant engagement national pour l'environnement*, Journal Officiel de la République Française n° 160, 13 July 2010, p. 12905.

J.O. 2010d. *Arrêté du 31 août 2010 fixant les conditions d'achat de l'électricité produite par les installations utilisant l'énergie radiative du soleil telles que visées au 3o de l'article 2 du décret no 2000-1196 du 6 décembre 2000*, Journal Officiel de la République Française, N° 0202, 1 September 2010, p. 15919.

J.O. 2011. *Arrêté du 4 mars 2011 fixant les conditions d'achat de l'électricité produite par les installations utilisant l'énergie radiative du soleil telles que visées au 3o de l'article 2 du décret no 2000-1196 du 6 décembre 2000*, Journal Officiel de la République Française, N° 9, 5 March 2011, p. 4218.

Johansson, T.B. and Turkenburg, W, 2004. 'Policies for renewable energy in the European Union and its member states'. *Energy for Sustainable Development*, 8:1, 5–24.

Johnston, A., Neuhoff, K., Fouquet, D., Ragwitz, M., Resch, G. 2008. 'The proposed New EU renewables directive: interpretation, problems and prospects', *European Energy and Environmental Law Review*, 17:3, 126–45.

Jordan, A. 2005. 'Policy convergence: a passing fad or a new integration focus on European Union studies', *Journal of European Public Policy*, 12:5, 944–53.

Jordan, A. and Liefferink, D. 2005. 'An "ever closer union" of national policy? The convergence of national environmental policy in the European Union', *European Environment*, 15:2, 102–13.

Jörgens, H. 2004. 'Governance by diffusion – implementing global norms through cross national imitation and learning', in Lafferty (ed.), pp. 246–83.

Kern, K. 2000. *Die Diffusion von Politikinformationen. Umweltpolitische Innovationen im Mehrebenensystem der USA*. Opladen: Leske und Budrich.

Kern, K., Jörgens, H., Jänicke, M. 2000. 'Die Diffusion umweltpolitischer Innovationen – Ein Beitrag zur Globalisierung von Umweltpolitik', *Zeitschrift für Umweltpolitik und Umweltrecht*, 4/2000, 507–46.

Kern, K. and Kissling-Näf, I. 2002. *Politikkonvergenz und Politikdiffusion durch Regierungs- und Nichtregierungsorganisationen, Ein internationaler Vergleich von Umweltzeichen*, Wissenschaftszentrum Berlin für Sozialforschung, Report FS II 02-302, Berlin.

Kerr, C. 1983. *The future of industrial societies: Convergence or continuing diversity?*, Cambridge: Harvard University Press.

KeTTHA 2010. *Malaysia to start feed-in tariff (FIT) for renewable energies by 2011*, KeTTHA press release. Kuala Lumpur: Ministry of Energy, Green Technology and Water.

Kjaer, C. 2004. 'Support mechanisms – Can one size fit all?', *Renewable Energy World*, March/April 2004, 48–59.

King, G., Keohane, R., Verba, S. 1994. *Designing social inquiry: Scientific inference in qualitative research*. Princeton: Princeton University Press.

Klein, A., Held, A., Ragwitz, M., Resch, G., Faber, T. 2006. Evaluation of different feed-in tariff design options – Best practise paper for the International Feed-in Cooperation. Available from http://www.worldfuturecouncil.org/fileadmin/user_upload/Miguel/best_practice_paper_final.pdf [accessed 30 September 2010].

Klein, A., Pfluger, B., Held, A., Ragwitz, M., Resch, G., Faber, T. 2008. Evaluation of different feed-in tariff design options – Best practise paper for the international Feed-in Cooperation, Second edition, October 2008. Available from http://www.feed-in-cooperation.org/images/files/best_practice_paper_2nd_edition_final.pdf [accessed 30 September 2010].

Klackenberg, D., Egenhofer, C., Kyriacos, G. 2004. *Rethinking the EU regulatory strategy for the internal energy market*, CEPS Task Force Report No. 52, December 2004.

Klessmann, C., Ensslin, C., Ragwitz, M., Resch, G. 2007. European renewable energy trade based on Guarantees of Origin (GOs) – concepts, critical issues, and recommendations for design, Berlin, Karlsruhe, Wien, Oktober 2007.

Klessmann, C., Nabe, C., Burges, C. 2008. 'Pros and cons for exposing renewables to electricity market risk – a comparison of market integration approaches in Germany, Spain, and the UK', *Energy Policy*, 36:10, 3646–61.

Kords, U. 1993. *Die Entwicklungsgeschichte des Stromeinspeisegesetzes vom 5.10. 1990. Ein Beispiel für die Mitwirkungsmöglichkeiten einzelner Abgeordneter an der Gesetzgebungsarbeit des Deutschen Bundestages.* Diplomarbeit (final year project), FFU, Freie Universität Berlin.

Knight, R.C./Montez, J.P./Knecht, F./Bouquet, T. (2005): Distributed Generation Connection Charging within the European Union – Review of Current Practices,Future Options and European Policy Recommendations. ELAP project (European Local Electricity Generation). Available from http://www.cogen.org/Downloadables/Publications/Grid_connection_charging_EU15_2005.pdf [accessed 20 October 2010].

Knill, C. 2001. *The Europeanisation of national administrations.* Cambridge: Cambridge University Press.

Knill, C. 2003. *Europäische Umweltpolitik. Steuerungsprobleme und Regulierungsmuster im Mehrebenensystem*, Opladen: VS Verlag für Sozialwissenschaften.

Knill, C. 2005. 'Introduction – Cross national policy convergence – Concepts, approaches and explanatory factors', *Journal of European Public Policy*, 12:5, 764–74.

Knill, C. and Lenschow, A. 1998. 'Coping with Europe: The implementation of EU environmental policy and administrative traditions in Britain and Germany', *Journal of European Public Policy*, 5:4, 595–614.

Knill, C. and Lenschow, A. 2005. 'Compliances, communication and competition: pattern of EU environmental policy making and their impact on policy convergence', *European Environment*, 15:2, 764–74.

Knill, C., Sommerer, T., Holzinger, K. 2008. 'Degree and direction of environmental policy convergence: analysis of aggregated data', in Holzinger, Knill and Arts (eds), pp. 98–143.

Koolwijk, J. 1974. 'Die Befragungsmethode', in Koolwijk and Wieken-Mayser, M. (eds), pp. 9–23.

Koolwijk, J. and Wieken-Mayser, M. (eds) 1974. *Technik der empirischen Sozialforschung, Volume 4, Erhebungsmethode: Die Befragung.* München, Wien: Oldenbourg.

Kords, U. 1993. *Die Entwicklungsgeschichte des Stromeinspeisegesetzes vom 5.10. 1990. Ein Beispiel für die Mitwirkungsmöglichkeiten einzelner Abgeordneter an der Gesetzgebungsarbeit des Deutschen Bundestages.* Diplomarbeit, Freie Universität Berlin.

Krägenow, T and Kade, C. 2008. Solarstrom spaltet Union, Financial Times Deutschland, 4 June 2008, p. 10.

Kratzat, M. and Langniß, O. 2007. *Proposal for a harmonised EU-wide feed-in tariff*, Centre for Solar Energy and Hydrogen Reseach Baden-Württemberg (ZSW), June 2007, Stuttgart.

Kromrey, H. 1998. *Empirische Sozialforschung*. 8th edition, Leske & Budrich: Opladen.

Laffan, B. and Shaw, C. 2005. Classifying and mapping of OMC in different policy areas, NEWGOV – New Modes of Governance, Reference number 02/ D09. Available from http://www.eu-newgov.org/database/DELIV/D02D09_ Classifying_and_Mapping_OMC.pdf [accessed 26 August 2010].

Lafferty, W.M. (ed.) 2004. *Governance for sustainable development. The challenge of adapting form to function*. Cheltenham: Edward Elgar.

Lamnek, S. 1995. *Qualitative Sozialforschung. Methoden und Techniken*. Third edition, Psychologie Verlags Union: Weinheim.

Lange, M. and Focken, U. 2005. *State-of-the-art in wind power prediction in Germany and international developments*, presentation at the second workshop of International Feed-In Cooperation, 16 December 2005, Berlin.

Lange, B., Cali, Ü., Ernst, B., Jursa, R., Moradi, J., Rohrig, K., Schlögl, F. 2006. *Wind Power Prediction in Germany – Recent Advances and Future Challenges*, Abstract for the European Wind Energy Conference (EWEC), Institut für Solare Energieversorgungstechnik e.V. (ISET), February 2006, Athens.

Langniss, O., Diekmann, J., Lehr, U. 2009. 'Advanced mechanisms for the promotion of renewable energy: Models for the future evolution of the German Renewable Energy Act', *Energy Policy*, 37:4, 1289–97.

Lauber, V. 2001. 'The different concepts of promoting RES-electricity and their political careers', in Biermann, Brohm and Dingwerth (eds), 2001, pp. 296–304.

Lauber, V. 2004. 'REFIT and RPS – options for a harmonized community framework', *Energy Policy*, 32:12, 1405–14.

Lauber, V. (ed.) 2005. *Switching to renewable power. A framework for the 21st century*. London: Earthscan.

Lauber, V. 2007. 'The politics of European Union policy on support schemes for electricity from renewable energy sources', in Mez (ed.), pp. 9–29.

Lauber, V. and Mez, L. 2004. 'Three decades of renewable electricity policies in Germany', *Energy & Environment (special issue)*, 15:4, 599–623.

Lauber, V. and Mez, L. 2007. 'Renewable electricity policy in Germany 1974–2005', in Mez (ed.), pp. 177–200.

Lazer, D. 2001. 'Regulatory interdependence and international governance', *Journal of European Public Policy*, 8:3, 474–92.

Lehman Brothers 2008. *Spain solar subsidy analysis, Equity research*, July 2008. New York: Lehman Brothers.

Lenschow, A., Liefferink, D., Veenman, S. 2005. 'When the birds sing. A framework for analysing domestic factors behind policy convergence', *Journal of European Public Policy*, 12:5, 797–816.

Lesieur, C. 2005. *The French feed-in tariff to support electricity from renewable energy sources*, Presentation at the second workshop of the International Feed-in Cooperation, December 2005, Berlin.

Lesser, J. and Su, X. 2008. 'Design of an economically efficient feed-in tariff structure for renewable energy deployment', *Energy Policy*, 36:3, 981–990.

Levi-Faur, D. (2002): *Herding towards a New Convention: On herds, shepherds and lost sheep in the liberalization of the telecommunications and electricity industries*, Politics Papers, Nuffield College, W6, February 2002. Available from http://www.nuffield.ox.ac.uk/Politics/papers/2002/w6/herding.pdf [accessed 20 October 2010].

Levi-Faur, D. 2005. 'The global diffusion of regulatory capitalism', *The Annals of the American Academy of Political and Social Science*, 598:1, 12–34.

Lewis, J. and Wiser, R. 2007. 'Fostering a Renewable Energy Technology Industry: An International Comparison of Wind Industry Policy Support Mechanisms', *Energy Policy*, 35:1, 1844–57.

Libertad Digital 2009. El Gobierno pone fin al déficit eléctrico y congela la tarifa a las rentas bajas, 30 April 2009. Available from http://www.libertaddigital.com/economia/el-gobierno-pone-fin-al-deficit-electrico-y-congela-la-factura-a-las-rentas-bajas-1276357914/ [accessed 20 October 2010].

Lipp, J. 2007. 'Lessons for effective renewable electricity policy from Denmark, Germany and the United Kingdom', *Energy Policy*, 35:11, 5481–95.

Long, N. van and Siebert, H. 1991. 'Institutional Competition Versus ex-ante Harmonisation. The Case of Environmental Policy'. *Journal of Institutional and Theoretical Economics*, 147, 296–311.

Lucas, H. 2007. *Viewpoint from the International Feed-In Cooperation*, presentation at the Futures-e Conference Harmonisation and Coordination of European RES-e support mechanisms, Instituto para la Diversifición y Ahorro de Energía, 20 June 2007, Brussels.

Lucas, H. 2009. *Background studies to the elaboration of the Renewable Energy Action Plan*, Presentation at the 7th workshop of the International Feed-in Cooperation on behalf of Spain, Instituto para la Diversifición y Ahorro de Energía, 28 & 29 May 2009, Ljubljana.

Luhmann, H.-J. 2009. FlexMechs abgeschrieben, Wuppertal Bulletin 12/2009, Wuppertal Institut für Klima, Umwelt, Energie GmbH, 23.

Lundqvist, L.J. 1974. 'Do political structures matter in environmental politics?, The case of air pollution control in Canada, Sweden, and the United States', *Canadian Public Administration*, 17:1, 119–41.

Mardones, I.G. 2000. *La rebaja de primas a las energías renovables estrangulará su futuro, según los productores*, El País, 4 January 2000, Madrid.

Markard, J. and Petersen, R. 2009. 'The offshore trend: Structural changes in the wind power sector', *Energy Policy*, 37:9, 3545–56.

Marsh, D. and Sharman, J. 2009. 'Policy diffusion and policy transfer', *Policy Studies*, 30:3, 269–88.

Martínez, S. 2009. *Energía geotérmica de alta entalpía, España tiene gran potencial pero falta legislación e inversion*, Expansion.com, 17 November 2009. Available from http://www.expansion.com/2009/11/17/empresas/energia/1258474915.html [accessed 3 November 2010].

Matlary, H. 1997. *Energy policy in the European Union.* Basingstoke (UK): Macmillan.

Maurer, P. 2006. *Umweltbeihilfen und Europarecht. Eine Untersuchung am Beispiel der Förderung der Stromerzeugung aus erneuerbaren Energien in Europa.* Berlin: Logos-Verlag.

May, H. 2010. 'Hermanns Schlacht', *Neue Energie*, 04/2010, 131–3.

Mayntz, R. 2002a. 'Zur Theoriefähigkeit makro-sozialer Analysen', in Mayntz (ed.), pp. 7–43.

Mayntz, R. (ed.) 2002b. *Akteure – Mechanismen – Modelle: Zur Theoriefähigkeit makro-sozialer Analysen.* Frankfurt am Main: Campus.

McKinsey 2010. *Transformation of Europe's power system until 2050*, September 2010. Düsseldorf: McKinsey.

MEEDDAT 2008a. *Développement des énergies renouvelables*, press release, 17 November 2008. Paris: Ministre de l'Ecologie, de l'Energie, du Développement durable et de l'Aménagement du Territoire.

MEEDDAT 2008b. Synthèse publique de l'étude des coûts de référence de la production électrique. Paris: Ministre de l'Ecologie, de l'Energie, du Développement durable et de l'Aménagement du Territoire.

MEEDDAT 2009a. *Programmation pluriannuelle des investissements de production d'électricité Période 2009 – 2020, Rapport au Parlement*, 3 July 2009. Paris: Ministre de l'Ecologie, de l'Energie, du Développement durable et de l'Aménagement du Territoire.

MEEDDAT 2009b. *Cahier des charges de l'appel d'offres portant sur des installations au sol de production d'électricité à partir de l'énergie solaire*, 20 April 2009. Paris: Ministre de l'Ecologie, de l'Energie, du Développement durable et de l'Aménagement du Territoire.

MEEDDAT 2010a. *Tarifs photovoltaïques*, press release, Cabinet du ministre d'État, 17 February 2010, Paris: Ministre de l'Ecologie, de l'Energie, du Développement durable et de l'Aménagement du Territoire.

MEEDDAT 2010b. *De nouveaux tarifs d'achat de l'électricité produite à partir de la biomasse, du solaire et de la géothermie*, Cabinet du ministre d'Etat, press release, 13 January 2010. Paris: Ministre de l'Ecologie, de l'Energie, du Développement durable et de l'Aménagement du Territoire.

MEEDDAT 2010c. *Adoption définitive du projet de loi portant engagement national pour l'environnement dit 'Grenelle 2'*, Cabinet du ministre d'Etat, press release, 29 June 2010. Paris: Ministre de l'Ecologie, de l'Energie, du Développement durable et de l'Aménagement du Territoire.

MEEDDAT 2010d. *National action plan for the promotion of renewable energies 2009–2020 in accordance with Article 4 of European Union Directive 2009/28/ EC*, Paris: Ministre de l'Ecologie, de l'Energie, du Développement durable et de l'Aménagement du Territoire.

Menanteau, P., Finon, D., Lamy, M. 2003. 'Price vesus quantities – choosing for promoting the development of renewable energy', Energy Policy, 31:8, 799–812.

Méndez, P. 2009. Las renovables baten sus marcas y generan ya el 30% de la electricidad, El País (edición imprensa), 9 March 2009. Available from http://www.elpais.com/articulo/sociedad/renovables/baten/marcas/generan/electricidad/elpepisoc/20090309elpepisoc_4/Tes [accessed 10 March 2009].

Mendonça, M. 2007. *Feed-in Tariffs – Accelerating the deployment of renewable energy*. Earthscan: London.

Mendonça, M. and Jacobs, D. 2009: 'Feed-in tariffs: the policy path to 100%', in: Droege (ed.), pp. 167–72.

Mendonça, M., Lacey, S., Hvelplund, F. 2009a. 'Stability, participation and transparency in renewable energy policy: Lessons from Denmark and the United States', *Policy and Society*, 27:4, 379–98.

Mendonça, M., Jacobs, D.; Sovacool, B. 2009b. *Powering the green economy – The feed-in tariff handbook*. Earthscan: London.

Menges, R. 1998. 'Staatlich garantierte Preise oder regenerativer Quotenhandel?', *Energiewirtschaftliche Tagesfragen*, 48:11, 687–93.

Mény, Y. Muller, P., Quermonne, J.-L. (eds) 1996. *Adjusting to Europe. The Impact of the European Union on National Institutions and Policies*. London: Routledge.

Meseguer, C. 2005. 'Policy Learning, Policy Diffusion, and the Making of a New Order', *The Annals of the American Academy of Political and Social Science*, 598:1, 67–82.

Meyer, N. 2003. 'European scheme for promoting renewables in liberalised markets', *Energy Policy*, 31:7, 665–76.

Mez, L. (ed.) 2007. Green Power Markets – Support schemes, case studies, and perspectives. Brentwood (UK): Multi-Science Publishing.

Mian, M. 2002. *Project economics and decision analysis, Volume I: Deterministic models*. Tulsa (OK): PennWell Books.

Mickwitz, P., Hyvättinen, H., Kivimaa, P. 2007. 'The role of policy instruments in the innovation and diffusion of environmentally friendlier technologies: popular claims versus case study experiences', *Journal of Cleaner Production*, 16:1, 162–70.

Midttun, A. and Gautesen, K. 2007. 'Feed in or certificates, competition or complementarity? Combining a static efficiency and a dynamic innovation perspective on the greening of the energy industry', *Energy Policy*, 35:3, 1419–22.

MINEFI 2002. *Programmation pluriannuelle des investissements de production électrique*, 29 Janvier 2002, Rapport au Parlement. Paris: Ministère de l'Économie, des Finances et de l'Industrie.

MINEFI 2006. *Rapport au Parlement Programmation pluriannuelle des investissements de production électrique, Période 2005–2015*, 6 June 2006. Paris: Ministère de l'Économie, des Finances et de l'industrie.

MITYC 2007a. *Propuesta de Real Decreto de retribución de la actividad de producción de energía eléctrica mediante tecnología solar fotovoltaica para instalaciones posteriores a la fecha límite de mantenimiento de la retribución del Real Decreto 661/2007, de 25 de mayo, para dicha tecnología, 27 de septiembre de 2007*. Madrid: Ministerio de Industria, Turismo y Comercio.

MITYC 2007b. *El Gobierno prima la rentabilidad y la estabilidad en el nuevo Real Decreto de energías renovables y cogeneración*, Ministerio de Industria, Turismo y Comercio, press release, 25 May 2007. Madrid: Ministerio de Industria, Turismo y Comercio.

MITYC 2008. *Planificación de los sectores de electicidad y gas, 2008–2016, Desarrollo de las redes de transporte*. May 2008. Madrid: Ministerio de Industria, Turismo y Comercio.

MITYC 2009a. *392 solicitudes inscritas en la primera convocatoria de la nueva regulación fotovoltaica*, press release, 19 February 2009. Madrid: Ministerio de Industria, Turismo y Comercio.

MITYC 2009b. *Industria incorpora 595 instalaciones fotovoltaicas al registro, press release*, 1 July 2009. Madrid: Ministerio de Industria, Turismo y Comercio.

MITYC 2010. Industria incorpora 848 nuevas plantas fotovoltaicas al listado de instalaciones con derecho a retribución, press realease, press release, 20 July 2010. Madrid: Ministerio de Industria, Turismo y Comercio.

MITYC and IDEA 2010. *Plan de acción nacional de energías renovables de España (PANER)*, 30 June 2010. Madrid: Ministerio de Industria, Turismo y Comercio.

Mitchell, C., Bauknecht, D., Connor, P.M., 2006. 'Effectiveness through risk reduction: a comparison of the renewable obligation in England and Wales and the feed-in system in Germany', *Energy Policy*, 34:3, 297–305.

Monteiro, C., Bessa, R., Miranda, V., Botterud, A., Wang, J., Conzelmann, G. 2009. *Wind Power Forecasting: State-of-the-Art 2009*, report ANL/DIS-10-1, Illinois: Argonne National Laboratory.

Montero, J.-P. 2002. 'Price versus quantities with incomplete enforcement', *Journal of Public Economics*, 85:3, 435–54.

Morris, C. 2009a. Grid parity, net metering and feed-in rates, PV Magazine, 03/2008, 16–19.

Morris, C. 2009b. The Spanish solar collapse, Online publication available from www.grist.org, [accessed 11 October 2009].

Morthorst P.E. 2000. 'The development of a green certificate market', *Energy Policy*, 28:15, 1085–94.

Morthorst, P. 2003. 'A green certificate market combined with a liberalised power market', *Energy Policy*, 31:13, 1393–402.

Morthorst, P. E, Jørgensen, B., Helby, P., Twidell, J. Hohmeyer, O., Mora, D., Auer, H., Resch, G., Huber, C., Schönbauer, C. 2005. *Support Schemes for Renewable Energy, A Comparative Analysis of Payment Mechanisms in the EU*, Brussels: European Wind Energy Association.

Morthorst, P.E., Wagemans, S., Purchala, K., Woyte, A., Ceña, A., Mora, Á., Cleijne, H., Kleinschmidt, C., Moldovan, N. (2007): *Detailed investigation of electricity market rules (D. 4.1.) – cases of France, Germany, Netherlands,*

Spain and Denmark, TradeWind project, edited by Poul Erik Morthorst, Risø National Laboratory, Technical University of Denmark, April 2007.

Müller, T. and Bitsch, C. 2007. 'Zur Vereinbarkeit einer europaweiten Einspeiseregelung mit dem europäischen Primärrecht – Ergebnisse der Begutachtung des BEE Modells „EU – FIT"', *Zeitschrift für Neues Energierecht*, 4/2007, 383–7.

Müller, T. and Bitsch, C. 2008. 'Grünstromzertifikate im EG-Binnenmarkt', *Zeitschrift für Neues Energierecht*, 3/2008, 197–202.

Müller-Langer, F., Witt, J., Thrän, D., Schneider, S., Baur, F., Koch, M., Fritsche, U., Wiegmann, K. 2007. *Monitoring zur Wirkung der Biomasse Verordnung, Endbericht (Kurzfassung), Forschungs- und Entwicklungsvorhaben (FKZ 20441133)*, March 2007, Leipzig.

Muñoz, M., Oschmann, V., Tábara, D. 2007. 'Harmonization of renewable electricity feed-in laws in the European Union', *Energy Policy*, 35:5, 3104–14.

Muro-Lara, J. 2006. *Investor Meeting the Netherlands and Germany*, presentation of Juan Muro-Lara, Group executive director, Acciona, 24 April 2006. Available from http://acciona.es/media/38985/pres_24abril2006.pdf [accessed 4 November 2010].

Natenco; WPD, Windstrom, Energy2Market, Nature Power Service & Management, in.power, Norderland, Green Wind Energy, mdp, e3, Pommer und Schwarz 2008. *Positionspapier für die Verordnungsermächtigung des Bundesministeriums für Umwelt, Naturschutz und Reaktorsicherheit zur Prämienregelung der Direktvermarktung (§64 des EEG)*. Position paper, 23 March 2009.

Nebreda Pérez, J. 2007. *Aspectos jurídicos de la producción eléctrica en régimen Especial – Puesta al día de la Ley 82/1980 al Real Decreto 661/2007 – Normativa estatal y autonómica comentada, a septiembre de 2007*. Navarra: Aranzadi.

Nedergaard, P. 2007. 'Maximizing policy learning in international committees: An analysis of the European Open Method of Coordination (OMC) committees', *Scandinavian Political Studies*, 30:4, 521–546.

Neue Energie 1998. 'Ja zum EU-Weißbuch, Nein zur Einspeise-Richtlinie', *Neue Energie*, 7/1998, 15.

Neue Energie 2000. 'EU-Kommissarin weist Richtlinien-Entwurf zurück', *Neue Energie*, 1/2000, 14–15.

Newbery, D. 2002. 'Regulatory challanges in European electricity liberalization', *Swedish Economic Policy Review*, 9:1, 9–43.

Nilsson, M., Nilsson, L., Ericsson, K. 2009. 'The rise and fall of GO trading in European renewable energy policy: The role of advocacy and policy framing', *Energy Policy*, 37:11, 4454–62.

Nitsch, J. 2008. Lead study 2008, Further development of the 'Strategy to increase the use of renewable energies' within the context of the current climate protection goals of Germany and Europe, Study commissioned by the German

Federal Ministry for the Environment, Nature Conservation and Nuclear Safety (BMU), October 2008. Stuttgart: Deutsches Zentrum für Luft- und Raumfahrt.

Oates, W.E. and Schwab, R.M. 1988. 'Economic Competition Among Jurisdictions. Efficiency Enhancing or Distortion Inducing?' *Journal of Public Economics*, 35:3, 333–54.

Timpe, C., Bergmann, H., Klann, U., Langniß, O., Nitsch, J., Cames, M., Voß, J.-P. 2001. *Umsetzungsaspekte eines Quotenmodells für Strom aus erneuerbaren Energien*, Abschlussbericht im Auftrag des Ministeriums für Wirtschaft und Verkehr Baden-Württemberg, Freiburg, Stuttgart, Heidelberg.

Oschmann, V. 2002. *Strom aus erneuerbaren Energien im Europarecht. Die Richtlinie 2001/77/EG des Europäischen Parlaments und des Rates zur Förderung der Stromerzeugung aus erneuerbarer Energiequellen im Elektrizitätsbinnenmarkt.* Baden-Baden: Nomos.

Oschmann, V. and Thorbecke, J. 2006. 'Erneuerbare Energien und die Förderung stromintensiver Unternehmen – Das Erste Gesetz zur Änderung des EEG', *Zeitschrift für Neues Energierecht*, 4/2006, 304–5.

Parisi, F., Schulz, N., Klick, J. 2001. *Two dimensions of regulatory competition*, Law and Economics Working Paper Series (03-2001), George Mason University.

Pause 2009. 'Die Reform der Förderung von Photovoltaikstrom in Spanien', *Energiewirtschaftliche Tagesfragen*, 59:8, 45–7.

Peters, B.G. 1998. *Comparative Politics: Theory and Methods.* Basingstoke: Macmillan.

Peterson, J. and Bomberg, E. 1999. Decision-making in the European Union, St. Martin's Press: New York.

Photon 2010. 'Jetzt verstehen wir, dass das ein Irrtum war', *Photon Germany*, March 2010, 37–9.

Pfeiffer, J. 2010. Statement from Dr Joachim Pfeiffer, Member of the German Parliament (CDU), at the conference '20 Jahre StrEG – 10 Jahre EEG – wie weiter?', BEE/Agentur für Erneuerbare Energien, 24 March 2010, Berlin.

Podewils, C. 2010a. 'Nährboden einer Industrie', *Photon Germany*, February 2010, 78–86.

Podewils, C. 2010b. 'Einspeisevergütung sinkt zum 1. Juli', *Photon Online*, 23 February 2010.

Podewils, C. 2010c. 'Solar schlägt Wind', *Photon Deutschland*, May 2010, 22–24.

Powell, W. and DiMaggio, P. (eds) 1991. *The new institutionalism in organizational analysis.* Chicago: Chicago University Press.

Pöyry 2010. Wind energy and electricity prices – Exploring the 'merit order effect', A literature review by Pöyry for the European Wind Energy Association, April 2010. Available from http://www.ewea.org/fileadmin/ewea_documents/documents/publications/reports/MeritOrder.pdf [accessed 4 November 2010].

Power in Europe 2007. 'Eurelectric warns of unfinished business, interview with Hans Ten Bergen, Secretary General, Eurelectric', *Power in Europe 510*, 4–5.

Presser, S. 2004. 'Methods for testing and evaluating survey questions', *Public Opinion Quarterly*, 68:1, 109–30.

Radaelli, C.M. 2005. 'Diffusion without convergence: how political context shapes the adoption of regulatory impact assessment', *Journal of European Public Policy*, 12:5, 924–43.

Ragwitz, M., Held, A., Resch, G., Faber, T., Huber, C., Haas, R. 2006. *Monitoring and evaluation of policy instruments to support renewable electricity in EU Member States*, Final Report, Federal Environmental Agency (Umweltbundesamt).

Ragwitz, M.; Held, A.; Resch, G.; Faber T., Haas, R.; Huber, C., Coenraads R., Voogt, M.; Reece, G.; Morthorst, P.E., Jensen, S.G., Konstantinaviciute, I., Heyder B. 2007a. *Assessment and optimisation of renewable energy support schemes in the European electricity market*, OPTRES final report, Karlsruhe, February 2007. Available online at http://www.optres.fhg.de/OPTRES_ FINAL_REPORT.pdf [accessed 5 March 2008].

Ragwitz, M., Resch, G., Schleich, J. 2007b. *Increased auctioning in the EU ETS and trade of guarantees of origin for renewables: A comparison of the impact on power sector producer rents*, Working paper Sustainability and Innovation, S 8/2007, Fraunhofer (ISI), Karlsruhe.

Ragwitz, M. and Sensfuss, F. 2008. *Stellungnahme des Fraunhofer Institut für System- und Innovationsforschung zum Entwurf eines Gesetzes zur Neuregelung des Rechts der Erneuerbaren Energien im Strombereich und zur Änderung damit zusammenhängender Vorschriften*, Kurzzusammenfassung, 30 April 2008, Karlsruhe: Fraunhofer ISE.

Ragwitz, M., del Río, P., Resch, G. 2009. 'Assessing the advantages and drawbacks of government trading of guarantees of origin for renewable electricity in Europe', *Energy Policy*, 37:1, 300–307.

REACT 2004a. *Installation of wind turbines trough the use of fixed ten year PPA and feed-in tariffs, case study #9, country: Greece*, Renewable Energy Action (Alterner 2002-157), 21 October 2004.

REACT 2004b. *Feed-in tariffs for wind energy, case study #25, country: Portugal*, Renewable Energy Action (Alterner 2002-157), 27 October 2004.

RECS International 2008. Legal dilemma in trade of renewable energy solved with transition period, press release, RECS International, 24 June 2008.

REE 2009. *El sistema eléctrico español en 2008*. Madrid: Red Eléctrica de España.

REE 2010. *El sistema eléctrico español en 2009*. Madrid: Red Eléctrica de España.

Reiche, D. 2004. *Rahmenbedingungen für erneuerbare Energien in Deutschland – Möglichkeiten und Grenzen einer Vorreiterpolitik*. Verlag Peter Lang: Frankfurt am Main.

Reiche, D. (ed.) 2005. *Handbook of Renewable Energies in the European Union – Case Studies of the EU15 States*. Frankfurt am Main: Peter Lang.

Reiche, D. and Bechberger, M. 2004. 'Policy differences in the promotion of renewable energies in the EU member states', *Energy Policy*, 32:7, 843–9.

Reiche, D. and Bechberger, M. (eds) 2006. Ökologische Transformation der Energiewirtschaft – Erfolgsbedingungen und Restriktionen. Berlin: Erich-Schmidt-Verlag.

Reiche, D. and Bechberger, M. 2006. 'Diffusion von Einspeisemodellen in der EU-25 als instrumenteller Beitrag zur Verbreitung erneuerbarer Energien', in Bechberger and Reiche (eds), pp. 199–205.

Reiche, D. and Jacobs, D. 2009. 'Erneuerbare Energien Politik in Norwegen – Eine kritische Bestandsaufnahme', *Energiewirtschaftliche Tagesfragen*, November 2009, 72–4.

REN21 2009. *Global Status Report – 2009 Update*. Paris: REN21.

REN21 2010. *Renewables 2010, Global Status Report*. Paris: REN21.

REN21 2011. *Renewables 2011, Global Status Report*. Paris: REN21.

Resch, G., Lopez-Polo, M.-A., Auer, H., Haas, R. 2005. *Electricity from renewable energy sources in EU-15 countries – A review of promotion strategies*, ReExpansion, March 2005. Available from http://www.ewea.org/fileadmin/ewea_documents/documents/projects/rexpansion/050531_Promotion_schemes.pdf [accessed 20 October 2010].

Resch, G., Faber, T., Haas, R., Huber, C., Ragwitz, M., Held, A., Morthorst, P.E., Jensen, S.G., Coenraads, R., Voogt, M., Reece, G., Konstantinaviciute, I., Heyder, B. 2007. *Recommendations for implementing effective & efficient renewable electricity policies,* Report of the IEE project OPTRES, Vienna, February 2007.

Resch, G., Faber, T., Ragwitz, M., Held, A., Panzer, C., Haas, R. 2008. *20% RES by 2020 – A balanced scenario to meet Europe's renewable energy target,* February 2008, Vienna (Austria). Available online at http://www.futures-e.org/ [accessed 30 October 2010].

Resch, G., Ragwitz, M., Rathmann, M., Panzer, C., Faber, T., Huber, C., Reece, G., Held, A., Haas, R., Morthorst, P., Grenna Jensen, S., Jaworski, L., Konstantinaviciute, I., Pasinetti, R., Vertin, K. (2009): *Action plan – Deriving a future European policy for renewable electricity*, Concise final report, Futures-e project, January 2009.

Rickerson, W. and Grace, R.C. 2007. *The Debate over Fixed Price Incentives for Renewable Electricity in Europe and the United States: Fallout and Future Directions*. March 2007, Washington DC: Heinrich Boll Foundation.

Riepe, M. 2004. *Die Verbreitung umweltpolitischer Maßnahmen unter besonderer Berücksichtigung der Einspeisegesetze für Strom aus erneuerbaren Energien in Deutschland und Frankreich*, Magisterarbeit, Environmental Policy Research Centre, Freie Universität Berlin.

Rigaud, C. 2006. 'Les nouveaux tarifs d'achat', *Systèmes Solaires*, 176, 16–17.

Ringel, M. 2006. 'Fostering the use of renewable energies in the European Union: the race between feed-in tariffs and green certificates', *Energy Policy*, 31:1, 1–17.

Ristau, O. 2008. Spaniens Solarbranche droht der Politik, Solarthemen 285, 3.

Rogers, E.M. 2003. *Diffusion and innovations*. Fifth edition, New York: Free Press.

Rooijen, S. and Wees, M. 2006. 'Green electricity policies in the Netherlands: an analysis of policy decisions', *Energy Policy*, 34:1, 60–71.

Ropenus, S. and Jensen, S.G. 2009. 'Support Schemes and Vertical Integration – Who Skims the Cream?', *Energy Policy*, 37:3, 1104–15.

Rose, R. 1991. 'What is lesson-drawing?', *Journal of Public Policy*, 11:1, 3–30.

Rostankowski, A.; Oschmann, V. 2009. 'Fit für die Zukunft? – Zur Neuordnung des EEG-Ausgleichsmechanismus und weiteren Reformansätzen', *Recht der Energiewirtschaft*, 12/2009, 361–8.

Rowlands, I. 2005. 'The European directive on renewable electricity – conflict and compromise', Energy Policy, 33:8, 965–74.

Ruete, M. 2008a. *Speech at the Renewable Energy Policy Workshop, Opening Session: The renewable energy framework Directive*, Sustainable Energy Week 2008, 30 January 2008, Brussels.

Rutschmann, I. 2007. 'Bei Vollgas ausgebremst', *Photon Germany*, 11/2007, 22–4.

RTE 2009a. *Bilan Prévisionel de l'équilibre offre-demande d'électricité en France, édition 2009*. Paris: Réseau de Transport d'Électricité.

RTE 2009b. *RTE met en service un nouveau dispositif de prévision de l'énergie éolienne et photovoltaïque*, press release, 30 November 2009. Paris: Réseau de Transport d'Électricité.

RTE 2010. *Le bilan électrique francais 2009*, 13 January 2010. Paris: Réseau de Transport d'Électricité.

Saint-Drenan, Y.-M., von Oehsen, A., Gerhardt, N., Sterner, M., Bofinger, S., Rohrig, K. 2009. *Dynamische Simulation der Stromversorgung in Deutschland nach dem Ausbauszenario der Erneuerbaren-Energien-Branche*, Abschlussbericht Fraunhofer Institut für Windenergie und Energiesystemtechnik (IWES), Kassel, FuE-Bereich Energiewirtschaft und Netzbetrieb, December 2009.

Sala-i-Martin, X. 1996. 'Regional cohesion: evidence and theories of regional growth and convergence', *European Economic Review*, 40:6, 1325–52.

Sandén, B. and Azar, C. 2005. 'Near-term technology policies for long-term climate targets – economy wide versus technology specific approaches', *Energy Policy*, 33:12, 1557–76.

Saravia, C., Rosell, D., Schug, A. 2010. 'Zunehmen trotz Diät', *Photon Deutschland*, January 2010, 17–19.

Sauter, R. 2006. 'Die Bedeutung exogenen Anpassungsdrucks für eine ökologische Transformation der Stromwirtschaft am Beispiel Frankreichs', in Reiche and Bechberger (eds), pp. 61–80.

Sawin, J. 2004. *National Policy Instruments – Policy Lessons for the Advancement and Diffusion of Renewable Energy Technologies around the World*, Thematic Beckground Paper, International Conference for Renewables, Bonn.

Scharpf, F. 1993. 'Positive und negative Koordination in Verhandlungssystemen', in: Héritier (ed.), pp. 57–83.

Schlandt, J. 2009. *Deutschlands strauchelnde Sonnenkönige*, Berliner Zeitung, 6 June 2009. Available from http://www.berlinonline.de/berliner-zeitung/ archiv/.bin/dump.fcgi/2009/0706/wirtschaft/0013/index.html [accessed 5 October 2010].

Schlögel, F. 2008. *ISET-Project: Distributed generation with high RES penetration*, Presentation at the 6th workshop of the International Feed-In Cooperation, 3–4 November 2008, Brussels.

Schmid, J., Rohrig, K., Braun, M., Gerhardt, N., Hochloff, P., Hoffstede, U., Lesch, K., Schlögl, F., Speckmann, M., Ritzau, M., Schrader, K., Rehfeldt, K., Diekmann, J., Nabe, C., Sensfuß, F., Erlich, I. 2009. *Wissenschaftliche Begleitung bei der fachlichen Ausarbeitung eines Kombikraftwerksbonus gemäß der Verordnungsermächtigung § 64 EEG 2009*, Zwischenbericht, Im Auftrag des Bundesministeriums für Umwelt, Naturschutz und Reaktorsicherheit, 20 April 2009.

Schmidt, F. 2008. *The Spanish Photovoltaic Market 2008/09 – New Framework Conditions, Updated Report Version in Accordance to the new Royal Decree 1578/2008*, EUDP Research, December 2008, Bonn.

Schreurs, M., Selin, H., VanDeveer, S. 2009a. *Transatlantic environment and energy politics: Comparative and international perspectives*. Farnham: Ashgate.

Schreurs, M., Selin, H., VanDeveer, S. 2009b. 'Expanding transatlantic relations: implications for environment and energy policies', in Schreurs et al. (eds). 2009, pp. 1–18.

Schreyer, M., Mez, L., Jacobs, D. 2008. *ERENE – European community for renewable energy*, Publication series on Europe – Volume 3, Heinrich Böll Foundation, June 2008.

Schubert, K. and Bandelow, N. (eds) 2003. Lehrbuch der Politikfeldanalyse. München: Oldenbourg.

Scott, N. 2007. *European practices with grid connection, reinforcement, constraint and charging of renewable energy projects*, 1 August 2007, Glasgow: Highlands and Islands Enterprise.

Scott, J. and Trubek, D. 2002. 'Mind the gap: law and new approaches to governing in the European Union', *European Law Journal*, 8:1, 1–18.

Sensfuß, F., Ragwitz, M., Kratzat, M., Langniß, O., Obersteiner, C., Müller, T., Merten, F., Fischedick, M. 2007. *Fortentwicklung des Erneuerbaren Energien Gesetzes (EEG) zur Marktdurchdringung erneuerbarer Energien im deutschen und europäischen Strommarkt*. Endbericht, Karlsruhe, September 2007.

Sensfuss, F., Ragwitz, M., Genoese, M. 2008. 'The Merit-order effect: A detailed analysis of the price effect of renewable electricity generation on spot market prices in Germany', *Energy Policy*, 36:8, 3076–84.

Sensfuss, F. and Ragwitz, M. 2009. *Entwicklung eines Fördersystems für die Vermarktung von erneuerbarer Stromerzeugung*, research paper 6, Internationale Energiewirtschaftstagung, TU Wien, 11–13 February 2009, Vienna.

SER 2010. *Energies renouvelables: Bilan fin 2009*, press dossier, 19 January 2010. Paris: Syndicat des Énergies Renouvelable.

SER-SOLER 2009. *Le développement du photovoltaïque en France*, March 2009, Commission Photovoltaïque. Paris: Syndicat des Énergies Renouvelables.

SER-SOLER 2010. *Etat du parc solaire photovoltaïque au 31 décembre 2009, Bilan de l'année 2009*, Commission Photovoltaïque. Paris: Syndicat des Énergies Renouvelables.

SER-SOLER 2011. *Etat du parc solaire photovoltaïque français, Bilan de l'année 2010*, Commission Photovoltaïque. Paris: Syndicat des Énergies Renouvelables.

Sijm, J. 2002. The *performance of feed-in tariffs to promote renewable electricity in European Countries*, The Energy Centre of the Netherlands (ECN), ECN-C-02-083, November 2002.

Simmons, B. and Elkins, Z. 2004. 'The globalisation of liberalisation: policy diffusion in the international policy economy', *American Political Science Review*, 98:1, 171–189.

Simmons, B., Dobbin, F., Garrett, G. (eds) 2008a. *The global diffusion of markets and democracy*. Cambridge: Cambridge University Press.

Simmons, B., Dobbin, F., Garrett, G. 2008b. 'Introduction: the diffusion of liberalization', in Simmons, Dobbin and Garrett (eds), pp. 1–63.

Sioshansi F. (ed.) 2008. *Comparative electricity markets – Design, Implementation, Performance.* Amsterdam: Elsevier.

Sohren, A. 2004. 'Es gibt kein naturgemäß überlegenes Fördermodell', *Erneuerbare Energien – das unabhängige Monatsmagazin für Zukunftsenergien*, 1/2004, 6–7.

Solano-Peralta, M, Moner-Girona, M., van Sark, G., Vallvè, X. 2009. '"Tropicalisation" of Feed-in Tariffs: A custom-made support scheme for hybrid PV/diesel systems in isolated regions', *Renewable and Sustainable Energy Review*, 13:9, 2279–94.

Solarthemen 2007. 'Feilschen um die EEG-Novelle', Solarthemen 267, 1.

Sötebier, J. 2003. 'Die Richtlinie zur Förderung der Stromerzeugung aus erneuerbaren Energiequellen im Elektrizitätsbinnenmarkt – eine rechtliche Analyse', *Zeitschrift für Umweltrecht*, 2/2003, 65–73.

Springmann, J. 2005. 'Die Förderung erneuerbarer Energien in der Stromerzeugung auf dem ordnungspolitischen Prüfstand', *Zeitschrift für Umweltpolitik und Umweltrecht*, 3/2006, 313–47.

SRU 2009. Weichenstellung für eine nachhaltige Stromversorgung, Thesenpapier, Mai 2009. Berlin: Sachverständigenrat für Umweltfragen.

Staiß, F. (ed.) 2003. Jahrbuch Erneuerbare Energien 2002/03, Radebeul: Bieberstein-Verlag.

Staiß, F. (ed.) 2000. Jahrbuch Erneuerbare Energien 2000, Radebeul: Bieberstein-Verlag.

Staiß, F., Schmidt, M., Musiol, F. 2007. *Forschungsbericht Vorbereitung und Begleitung der Erstellung des Erfahrungsberichtes 2007 gemäß Artikel 20*

EEG, November 2007. Available from http://www.erneuerbare-energien.de/files/pdfs/allgemein/application/pdf/eeg_forschungsbericht1_4.pdf [accessed 3 September 2010].

Stenzel, T.; Frenzel, A. 2008. 'Regulating technological change – The strategic reactions of utility companies towards subsidy policies in the German, Spanish and UK electricity markets', *Energy Policy*, 36:7, 2645–57.

Sterr-Kölln und Partner 2007. *Vergleich der Wirtschaftlichkeit eines beispielhaften Windparks nach deutschem und französischem Tarif*, Presentation at the Husum Wind, 20 September 2007, Husum (Germany).

Strom-News 2003. Wirtschaftsministerium: EEG-Härtefallklausel für energieintensive Betriebe kommt, 10 March 2003. available from http://www.strom-magazin.de/strommarkt/wirtschaftsministerium-eeg-haertefallklausel-fuer-energieintensive-betriebe-kommt_8789.html [accessed 12 April 2009].

Strang, D. and Meyer, J.W. 1993. 'Institutional conditions for diffusion', *Theory and Society*, 22:4, 487–511.

Suri, M., Huld T., Dunlop, E., Ossenbrink, H. 2007. 'Potential of solar electricity generation in the European Union member states and candidate countries', *Solar Energy 81*, 1295–305.

Systèmes Solaires 2009. 'Raccordements photovoltaïques: Bientôt la fin des bouchons?', *Le Journal des Énergies Renouvelables 191*, May/June 2009, 3–5.

Temblenque, L. 2007. *Spanish Renewable Regulation*, Presentation at the CEER conference 2007, Algiers (Algeria), 18 June 2007.

Tews, K. 2002a. 'Politiktranfer: Phänomen zwischen Policy-Lernen und Oktroi. Überlegungen zu unfreiwilligen Umweltpolitikimporten am Beispiel der EU-Osterweiterung', *Zeitschrift für Umweltpolitik und Umweltrecht*, 25:2, 173–201.

Tews, K. 2002b. Der Diffusionsansatz für die vergleichende Politikanalyse, Wurzeln und Potenziale eines Konzepts, Eine Literaturstudie, FFU-Report 2002-02, Freie Universität Berlin.

Tews, K. 2005a. 'Die Diffusion umweltpolitischer Innovationen im internationalen System: Eckpunkte eines Analysemodells'. in Tews and Jänicke (eds), pp. 25–54.

Tews, K. 2005b. 'The diffusion of environmental policy innovation. Cornerstones of an analytical framework', European Environment, 15:2, 63–79.

Tews, K. and Jänicke, M. (eds) 2005. *Die Diffusion umweltpolitischer Innovationen im internationalen System*. Wiesbaden: VS Verlag für Sozialwissenschaften.

Thomas, D. 2008. *Note on the main actual European feed-in tariff mechanisms and the resulted annual growth in MWp capacity*, 29 August 2008. Brussels: European Photovoltaic Industry Association.

Toke, D. 2005. 'Are green electricity certificates the way forward for renewable energy? An evaluation of the UK's Renewable Obligation in the context of international comparison', *Environment and Planning C: Government and Policy*, 23:3, 361–74.

Toke, D. 2008. 'The EU Renewables Directive – What is the fuss about trading?', *Energy Policy*, 36:8, 3001–8.

TradeWind 2009. Integration wind – Developing Europe's power market for the large-scale integration of wind power, February 2009, Brussels: European Wind Energy Association.

UK Government 2008a. *UK – proposed flexibility mechanism (leaked document)*, European Council, May 2008.

UK Government 2008b. *Explanatory note for UK flexibility proposal (leaked document)*, European Council, May 2008.

Van Sambeek, E. 2002. *The European Dimension of National Renewable Electricity Policy Making – An analysis of the Dutch experience*, ECN-RX-02-060, December 2002. Available from http://www.ecn.nl/docs/library/report/2002/rx02060.pdf [accessed 22 October 2010].

Vogel, D. 1995. *Trading Up: Consumer and Environmental Regulation in a Global Economy*. Cambridge: Harvard University Press.

Vogel, D. 1997. 'Trading up and governing across: Transnational governance and environmental protection', *Journal of European Public Policy*, 4:4, 556–71.

Voogt, M.H., Uyterlinde, M., de Noord, K., Skytte, L.H., Nielsen, M., Leonardi, M., Whiteley, M., Chapman 2001. *Renewable Energy Burden Sharing – REBUS – Effects of Burden Sharing and Certificate Trade on the Renewable Electricity Market in Europe*, ECN-C-01-030, Petten, The Netherlands.

Weidner, H. 2002. 'Capacity building for ecological modernization: lessons from cross-national research', *American Behavioural Scientist*, 45:9, 1340–68.

Weidner H. and Jänicke M. (eds). 2002. *National Environmental Policies: a Comparative Study of 17 Countries*. Springer: Berlin.

Weinhold, N. 2008. 'Runter vom Podest', *Neue Energie*, 02/2008, 52–6.

Weinberg, I. 1969. 'The problem of the convergence of industrial societies: A critical look at the state of the theory', *Comparative Studies in Society and History*, 11:1, 1–15.

Wenzel, B. and Nitsch, J. 2008. Ausbau erneuerbarer Energien im Strombereich, EEG-Vergütungen, Differenzkosten und –Umlagen sowie ausgewählte Nutzeneffekte bis 2030, Teltow/Stuttgart, December 2008.

Wenzel, B. and Nitsch, J. 2010. *Langfristszenarien und Strategien für den Ausbau der Erneuerbaren Energien in Deutschland bei Berücksichtigung der Entwicklung in Europa und global, Entwicklung der EEG-Vergütungen, EEG-Differenzkosten und der EEG–Umlage bis zum Jahr 2030 auf Basis des Leitszenario 2010*, DLR, Fraunhofer IWES, IFNE, Juni 2010.

Windpower Monthly 2004. 'Government Wields Heavy Hand in Spain – Tight Regulatory Control of New Windmarket Framework', *Windpower Monthly*, 04/2004, p. 29.

Wind Directions 2010. 'Spain flies high, Interview with the Spanish secretary of state for Energy, Pedro L. Marín Uribe', *Wind Directions*, April 2010, 46–8.

Winner, H. 2005. 'Has tax competition emerged in OECD countries? Evidence from panel data', *International Tax and Public Finance*, 12:5, 667–87.

Wiser, R., Barbose, G., Peterman, C. 2009. *Tracking the sun, The installed cost of photovoltaics in the US from 1998–2007*, Lawrence Berkeley National Laboratory, February 2009.

Wissen, R., Lienert, M., Gatzen, C. 2006. *Notwendigkeit und Ausgestaltung eines effizienten und nachhaltigen Fördersystems für erneuerbare Energien in Europa*, Endbericht (vorläufig), EWI: Köln.

Woyte, A. 2007. *Proposal for adaptation of market rules in order to remove power market barriers: Input for simulations (D7.4), WP7: Improved market mechanisms*, TradeWind project, October 2007.

WWEA 2010. *World wind energy report 2009*, March 2010, Bonn: World Wind Energy Association.

Zeitlin, J., Pochet, P., Magnusson, L. 2005. *The Open Method of Coordination in Action: The European Employment and Social Inclusion Strategies*. Brussels: Peter Lang.

Index

GLOBAL ENVIRONMENTAL GOVERNANCE SERIES

Full series list

Linking Trade, Environment, and
Social Cohesion
NAFTA Experiences, Global Challenges
*Edited by John J. Kirton and
Virginia W. Maclaren*

International Equity and Global
Environmental Politics
Power and Principles in US Foreign Policy
Paul G. Harris